NEUROSCIENCE OF PREJUDICE AND INTERGROUP RELATIONS

Psychological research on the origins and consequences of prejudice, discrimination, and stereotyping has moved into previously uncharted directions through the introduction of neuroscientific measures. Psychologists can now address issues that are difficult to examine with traditional methodologies and monitor motivational and emotional processes as they develop during ongoing intergroup interactions, thus enabling the empirical investigation of the fundamental biological bases of prejudice.

However, several very promising strands of research have largely developed independently of each other. By bringing together the work of leading prejudice researchers from across the world who have begun to study this field with different neuroscientific tools, this volume provides the first integrated view on the specific drawbacks and benefits of each type of measure, illuminates how standard paradigms in research on prejudice and intergroup relations can be adapted for the use of neuroscientific methods, and illustrates how different methodologies can complement each other and be combined to advance current insights into the nature of prejudice.

This cutting-edge volume will be of interest to advanced undergraduates, graduates, and researchers who study prejudice, intergroup relations, and social neuroscience.

Belle Derks is Assistant Professor in Social and Organizational Psychology at Leiden University, the Netherlands. She has published research articles and book chapters on the psychological, behavioral, and neural correlates of stigma and social identity threat. She has served as editor of the *Dutch Yearbook of Social Psychology*.

Daan Scheepers is Assistant Professor in Social and Organizational Psychology at Leiden University, the Netherlands. He has published several research articles on group processes and intergroup relations, with a focus on the physiological and psychological correlates of social identity threat. He is consulting editor of the *European Journal of Social Psychology* and has co-edited a special issue of *Group Processes & Intergroup Relations*.

Naomi Ellemers is Professor in Social and Organizational Psychology at Leiden University, the Netherlands. She has published extensively on prejudice and intergroup relations, and recently won a Spinoza award from the Dutch National Science Foundation (NWO) for her research on the use of neuroscientific methods in intergroup research. She has served as editor of the *Dutch Journal of Psychology* and the *Dutch Yearbook of Social Psychology*, was associate editor of the *British Journal of Social Psychology* and the *Journal of Personality and Social Psychology*, co-edited a special issue of the *Journal of Social Issues*, and is a member of the editorial board of the *Annual Review of Psychology*. She has co-edited books on stereotyping (1997), social identity (1999), and identity in organizations (2003).

NEUROSCIENCE OF PREJUDICE AND INTERGROUP RELATIONS

Edited by Belle Derks,
Daan Scheepers, and Naomi Ellemers

Leo Rodrigues

Psychology Press
Taylor & Francis Group
NEW YORK AND LONDON

First published 2013
by Psychology Press
711 Third Avenue, New York, NY 10017

Simultaneously published in the UK
by Psychology Press
27 Church Road, Hove, East Sussex BN3 2FA

Psychology Press is an imprint of the Taylor & Francis Group, an informa business

Library of Congress Cataloging-in-Publication Data
Neuroscience of prejudice and intergroup relations / edited by Belle Derks, Daan Scheepers, and Naomi Ellemers.
pages cm
1. Prejudices. 2. Intergroup relations. 3. Neuropsychology. I. Derks, Belle. II. Scheepers, Daan. III. Ellemers, Naomi.
BF575.P9N48 2013
155.9'2–dc23
2012048031

ISBN: 978-1-84872-998-8 (hbk)
ISBN: 978-1-84872-641-3 (pbk)
ISBN: 978-0-20312-463-5 (ebk)

Typeset in Bembo by Keystroke, Station Road, Codsall, Wolverhampton

SFI Certified Sourcing
www.sfiprogram.org
SFI-00453

Printed and bound in the United States of America
by Edwards Brothers, Inc.

CONTENTS

CONTRIBUTORS

Daniel L. Ames, Princeton University

David M. Amodio, New York University

Bruce D. Bartholow, University of Missouri, Columbia

Wendy Berry Mendes, University of California, San Francisco

Jim Blascovich, University of California, Santa Barbara

Jeanick Brisswalter, University of Nice Sophia Antipolis

Aïna Chalabaev, Paris West University

Mina Cikara, Carnegie Mellon University

Carsten K. W. De Dreu, University of Amsterdam

Belle Derks, Universiteit Leiden

Naomi Ellemers, Universiteit Leiden

Luke R. Enge, University of Texas at El Paso

Susan T. Fiske, Princeton University

Jennifer N. Gutsell, University of Toronto

Leor M. Hackel, New York University

Lasana T. Harris, Duke University

Michael Inzlicht, University of Toronto

Tiffany A. Ito, University of Colorado at Boulder

Jonathan W. Kunstman, Miami University, Ohio

Brenda Major, University of California, Santa Barbara

Rodolfo Mendoza-Denton, University of California, Berkeley

Elizabeth Page-Gould, University of Toronto

Jennifer A. Richeson, Northwestern University

Katie N. Rotella, Northwestern University

Philippe Sarrazin, Univ. Grenoble Alpes

Daan Scheepers, Universiteit Leiden

Jay J. Van Bavel, New York University

Félice Van Nunspeet, Universiteit Leiden

Y. Jenny Xiao, New York University

Michael A. Zárate, University of Texas at El Paso

1

THE "NATURE" OF PREJUDICE

What Neuroscience has to Offer to the Study of Intergroup Relations

Daan Scheepers, Naomi Ellemers, and Belle Derks

Introduction

Since the early 2000s, scholars with an interest in prejudice and intergroup relations have enthusiastically started to incorporate neuroscientific measures in their work (Amodio, 2008; Blascovich, Mendes, & Seery, 2002; Derks, Inzlicht, & Kang, 2008; Prentice & Eberhardt, 2008; Harris & Fiske, 2006; Lieberman, 2007; Van Bavel & Cunningham, 2011). What began as a small set of isolated efforts by a daring few has flourished into a vibrant area of research that has yielded exciting results, surprising conclusions, and intriguing possibilities for further examination. This development has raised curiosity among other scholars with an interest in adopting similar procedures to enhance and develop their own work. Unfortunately, however, it is not so easy to gain a clear picture of advances that have been made in this area of research during the last decade. Due to the rapid developments in this area, much of the empirical work is relatively recent and is published in a broad range of outlets, mostly as reports of separate studies addressing specific research questions. As a result, the broader picture painted by these separate findings and research lines is not always obvious, especially for those not (yet) familiar with the use of neuroscientific techniques.

The present volume aims to cater for those who seek an overview of what can be learned about prejudice and intergroup relations by the introduction of neuroscientific measures. At the same time, the work collected in this volume not only addresses the substantive advancements that have been made in this area of research, but also provides an introduction to the different methodologies that have been developed in this field, the types of questions each may help to address, and the types of challenges researchers face when building a bridge between social psychology and neuroscience.

This chapter first introduces the psychology of prejudice and aims to make a case for how neuroscience and psychophysiological measures might help to deal with some key current issues in this domain of research. We then present a general historic overview of the neuroscience of prejudice, followed by a more specific discussion of how we think this approach has led to new *measures*, new *questions*, new *answers*, but also new *challenges* for researchers trying to incorporate neuroscience methodology in their work. We conclude this chapter with an overview of the current volume.

Why Examine Prejudice—Still?

Prejudice denotes the tendency to evaluate or judge people negatively before we know them, merely because of their membership in a particular group or social category— based, for instance, on race, gender, or religion (Dovidio, Esses, Glick, & Hewstone, 2010; Nelson, 2009; Stangor, 2000). Two core characteristics of prejudice are its *pervasiveness* and its *complexity*. Prejudice is pervasive in the sense that it is of all times, is present in all cultures, and is directed toward all kinds of different groups in society. Prejudice is complex in that it involves explanatory factors at intrapersonal (e.g., biological), interpersonal, intergroup, and cultural levels. However, as we will argue below, it is precisely because of the pervasiveness and complexity of prejudice that more research is needed in this area, and why neuroscience and psychophysiological approaches seem to be particularly needed to shed light on some of the emerging core issues.

Initial analyses of prejudice sought to understand the phenomenon as stemming from a pathological personality (e.g., Adorno, Frenkel-Brunswik, Levinson, & Sanford, 1950), or as reflecting locally accepted lay explanations (e.g., "racial inferiority") for differences between specific groups in society (Pettigrew, 1958). In his landmark book, *The Nature of Prejudice*, Gordon Allport (1958) took a different approach; he explained prejudice as a logical by-product of the categorical thinking that people normally use to deal with complex information (see also Tajfel, 1978). Thus, Allport pointed out that prejudice essentially is a natural phenomenon, and went even as far as to argue that humans *are* "bundles of prejudice" (Allport, 1958, p. 4). Allport's early ideas resonate in our current understanding of social cognition, as category-based information is seen to play an important role in the perception and evaluation of specific individuals (Fiske & Neuberg, 1990; Macrae & Bodenhausen, 2001; Nelson, 2009).

A Google Scholar search with the terms "psychology" and "prejudice" yields 245,000 hits to scholarly publications on this topic. Surely this suggests that by now we should know all there is to be known about this topic? Indeed, since the late 1970s, theorists and researchers have made great progress in uncovering the psychological mechanisms that lie at the root of prejudice (Bargh, 1999; Bodenhausen & Macrae, 1998; Devine, 1989; Fiske, 2000), and have specified the prejudicial views that are held against different groups and how these can be measured (e.g., sexism, ageism, racism, nationalism; see Dovidio, Glick, & Rudman, 2005; Swim & Stangor, 1998). Other researchers have examined the motivational basis for prejudice (Abrams & Hogg, 1988; Fein & Spencer, 1997; Tajfel & Turner, 1979) and charted the different emotions involved in intergroup discrimination (Cottrell & Neuberg, 2005; Smith & Mackie, 2010). Additional efforts have focused on understanding how prejudice affects relations between individuals and groups in society (Brown, 2010; Levin & Van Laar, 2006; Shelton, Richeson, & Vorauer, 2006); and how the effects of prejudice can be reduced—for instance, through intergroup contact (Pettigrew & Tropp, 2008), or the creation of a common group identity (Gaertner & Dovidio, 2000).

All these efforts have led to the accumulation of evidence for the ubiquitous presence of prejudicial views across different societies and pertaining to different groups, which is in line with Allport's original conception of prejudice as a fact of life. This realization in turn has inspired social norms, policy guidelines, and formal legislation aiming to counteract the discriminatory implications of prejudicial thinking in terms of unequal outcomes. However, if we take seriously the characterization of prejudice as a natural human phenomenon, such external measures can hardly be expected to have an impact upon the emergence of prejudice itself. Indeed, one of the pernicious effects of societal shifts toward the ideal of equal treatment is that these did not eradicate prejudicial thinking; prejudice simply went underground instead (Dovidio & Gaertner, 1986). Admittedly, due to social norms and the desire to communicate in politically correct ways, "modern" expressions of sexism or racism are more masked and subtle (Dovidio & Gaertner, 1996). While this causes such communications to be less easily recognized as prejudicial, the endorsement of these views nevertheless co-varies with more blatant forms of discrimination (Dovidio & Gaertner, 1986). Perhaps more importantly, targets of prejudice tend to suffer just as much, if not more, from exposure to such modern prejudicial views, in part *because* of the more subtle and veiled nature. Instead of protesting against unequal treatment, they tend to lose self-confidence, start underperforming on important tasks, or slowly disengage from domains that might help them to improve their position in society (Ryan & Branscombe, in press).

Thus, research to date has yielded some highly intriguing (and to an extent disturbing) conclusions. Even if people are aware that they should not hold prejudicial views, or actively try to avoid thinking in these ways, they cannot escape the impact of categorical thinking as a basic cognitive shortcut in social evaluations (Bargh, 1999; Devine, 1989; Dovidio & Gaertner, 1996). Working on the assumption that prejudice cannot simply be evaded through deliberate decision making is the only way forward. However, it is precisely because of the changing (i.e., more subtle) appearance of prejudice— as well as an increased concern with appearing unprejudiced—that less reactive and more implicit measures of prejudice should be employed in order to move this field of research forward. As will become clear below, and in the chapters to follow, neuroscientific and psychophysiological methods form a valuable class of measures that can deal with the difficulty in measuring (subtle) prejudice. As such, these new methods may prove to be crucial in finding ways to reduce the expression of prejudice and its pernicious effects on those targeted by prejudice.

Apart from being pervasive, prejudice is also *complex*. Recent work in this area consistently points to the multi-layered nature of prejudice, involving the complex interplay between processes at the intrapersonal level (biology, affect [feeling], cognition), the interpersonal level (social judgment, dyadic interactions), the intergroup level (intergroup conflict, social norms), and the cultural level (political ideals, moral values). Of course, in a given situation, not all these processes at all these different levels may be equally important. Nevertheless, prejudice by definition always involves both (intra-)individual as well as social explanatory factors, if only because universal responses to situational contexts or task demands tend to be modified by social concerns having to do with self-presentation, self-protection, and social and cultural norms. Approaching each aspect one at a time—or opting to focus one's efforts on either one of the endpoints of a biology–culture continuum—impedes a full and complete understanding of these issues (Cacioppo, 2002).

The complexity of prejudice as a phenomenon that connects different levels of analysis also explains why we still do not know all there is to know about it. Recent developments in psychological research have opened up novel ways to study and understand the biological underpinnings of the processes involved at different levels (Amodio, 2008; Blascovich et al., 2002; Harris & Fiske, 2006; Van Bavel & Cunningham, 2011). Connecting these very basic individual factors to their social implications provides us with a new and exciting perspective that helps elucidate the role of prejudice as a factor in intergroup relations.

In summary, both the *pervasiveness* of prejudice, as well as its *complexity* explain why there is still a lot to discover about this important phenomenon. The pervasiveness of prejudice—and more, in particular, the fact that its appearance has changed into more subtle forms—motivates the need for more subtle and implicit measures (Dovidio & Gaertner, 1996). The complexity of prejudice illustrates the need to connect different levels of analysis (Cacioppo, 2002) and capitalize on the new insights that can be gained with the social neuroscience approach. The combination of these two features (connecting levels; and offering new, more subtle methodological tools) is precisely what the social neuroscience of prejudice is all about.

The Social Neuroscience of Prejudice

Recently, researchers have begun to address the "nature" of prejudice in a more literal way by exploring the biological and physiological basis of this important phenomenon (Prentice & Eberhardt, 2008). This was inspired at least in part by the realization that more traditional research tools—such as interviews, self-reported attitudes, and behavioral observations—only allow for the examination of more overt and explicit indicators of prejudice. Indeed, prejudicial expectations—more than other attitudes—are subject to (changing) social norms and desired self-views (e.g., appearing unbiased). This is why it was clear that additional measures were needed to gain a more complete understanding of the psychological processes underlying these overt responses. In particular, the interplay between self-presentational concerns on the one hand and the (in)ability to control or suppress prejudice on the other hand became a topic of interest in this field over the last decade. This has led to the development of various implicit cognitive measures, collaboration with cognitive scientists, and the adaptation of some of their standard measures for this purpose. In order to measure people's implicit intergroup bias, their motivation and ability to control this bias, and the cognitive depletion this causes, prejudice researchers are now using cognitive tasks like the implicit association task, the primed lexical decision task, the flankers task, and the Stroop task (e.g., Bartholow & Dickter, 2008; Dovidio, Kawakami, & Gaertner, 2002; Fazio, Sanbonmatsu, Powell, & Kardes, 1986; Greenwald, McGhee, & Schwartz, 1998).

As a relatively recent outcome of this broader development, researchers of prejudice and intergroup relations have also come to represent a substantial and promising strand in the emerging field of social neuroscience (Cacioppo, Visser, & Pickitt, 2005; Lieberman, 2007). Since the early 2000s, researchers with an interest in these topics have introduced methods like electroencephalography (EEG) and brain imaging techniques (fMRI) to address pre-conscious cognitive and emotional processes related to stereotyping and prejudice (Amodio et al., 2004; Ito & Urland, 2003), and to explain relations between different groups in society (Harris & Fiske, 2006). In addition, efforts have been made to establish how neuroendocrine measures (e.g., cortisol) can be employed to examine threat arising from the devaluation of one's group (Matheson & Cole, 2004; Mendes, Chapter 16 in this volume). Cardiovascular measures (electrocardiography, blood pressure, impedance cardiography) have become available to assess emotional and motivational states such as threat during inter-ethnic interactions (Mendes, Blascovich, Lickel, & Hunter, 2002) with indicators that also have important implications for psychological and physical well-being (Blascovich, 2008; Blascovich, Spencer, Quinn, & Steele, 2001). The added value of these indicators is clear, because they provide a truly novel perspective on the multi-layered nature of prejudice and its implications for social behavior. Indeed, the contribution that social neuroscience has to offer is of particular benefit for scientists with an interest in prejudice, as the questions they examine speak directly to the interplay between self-presentational concerns guiding overt responses on the one hand, and more automatically occurring or implicit biological processes on the other.

In view of the substantial investment in terms of time and resources that is needed to adapt these complex methodologies for the study of prejudice and intergroup relations, most researchers in this novel area of social neuroscience began their efforts by focusing on one specific type of methodology. The introduction of neuroscientific methods into this area of research is particularly challenging, because there is no natural match between the rich and complex social phenomena in the psychology of prejudice, and the reduced experimental conditions and controlled task settings required for the use of neuroscientific methodologies. It is therefore not surprising that initial efforts were primarily directed at examining whether such methods could even be used as *indicators* for relevant responses in this area of research (Blascovich et al., 2002; Guglielmi, 1999; Hart et al., 2000; Harris & Fiske, 2006; Ito & Urland, 2003; Phelps et al., 2000; Vrana & Rollock, 1998; Wheeler & Fiske, 2005). This caused researchers to seek out and discuss their efforts mostly with researchers who were familiar with this type of methodology (e.g., cognitive psychologists, biological psychologists) to help them develop suitable tools, rather than with those working on the same theoretical questions or societal problems. Even though the goals of this early work may seem limited in a sense in retrospect, at least compared to the modern social neuroscientific approach (focusing on processes and mediation; see Amodio, Chapter 9 in this volume, for a discussion of this shift), this pioneering work was crucial and groundbreaking. The pioneering demonstrations of concrete possibilities to assess responses to group-relevant information at a very direct physical level—without research participants even being aware that they displayed these responses—effectively opened up the neuroscientific toolbox for researchers of prejudice and intergroup relations.

The initial work to find and develop suitable measures was followed up by studies that considered physiological responses as potential *mediators* of more explicit responses, with the main ambition being directly to connect responses displayed at different levels of analysis (e.g., Richeson et al., 2003; see also chapters by Amodio [9], by Bartholow [10], by Ito [2], and by Van Bavel, Xiao, & Hackel [6], in this volume). Several researchers at this stage started to examine whether the physiological responses they assessed could be conceived as indicating the psychological process through which situational factors eventually impact upon overt behavior. Representative questions were, for instance, whether physiological stress responses to out-group representatives can explain hostile behavior toward them (Bijleveld, Scheepers, & Ellemers, 2012); or whether the observation of increased amygdala activity predicts displays of implicit or explicit bias against out-group members (Hart et al., 2000; Phelps et al., 2000). This type of work also revealed that it is often not so easy to establish direct empirical evidence for such relations—for discussions of this issue, see chapters by Major & Kunstman (17), by Mendes (16), by Scheepers (13), by Derks (3), and by Ito (2) in this volume. Although physiological and behavioral processes related to prejudice should theoretically be related to each other—and even if this actually is the case—they occur and need to be assessed at different levels of analysis, each of the processes requiring a different research approach. This is an issue we will return to below, under the heading "New Challenges."

Apart from demonstrating mediating processes for important intergroup phenomena, another development in the emerging neuroscience of prejudice concerns the demonstration that these processes themselves can also be moderated by the social context. That is, while initially it was mainly assumed that these more basic processes (e.g., amygdala activity as a marker of prejudice, or specific event-related brain potential [ERP] components as markers of categorization) were generic and stable, more recently it has become clear that these processes in themselves vary as a function of, for example, the salience of social identity, or the presence of threat (see chapters by Derks [3], by Van Bavel et al. [6], and by Gutsell & Inzlicht [15] in this volume).

During the last decade, work in this area evolved very rapidly, causing researchers to integrate and combine neuroscientific measures with behavioral, cognitive, and self-report measures, and to develop the more detailed knowledge about neuroscientific processes that also gives rise to new research questions. In the next sections, we provide a brief overview of these developments, to elucidate how the introduction of neuroscientific measures in research on prejudice and intergroup relations provides a different perspective that gives rise to *novel measures*, as well as *novel questions* that can yield *novel answers*. Before concluding the chapter with an overview of the current volume, we also point to some of the *novel challenges* that researchers face when integrating psychophysiological or neuroscience measures in their work.

Novel Measures

As indicated above, physiological measures can be used as additional and relatively unobtrusive *indicators* of psychological processes of interest, like prejudice (see Guglielmi, 1999, for an overview). When used this way, the main aim is to find ways to establish whether and how evidence for the operation of specific psychological processes can be obtained at a physiological level. For instance, although the hypothesis that stigma elicits threat responses during interactions between stigmatized and non-stigmatized individuals had already been proposed in the pioneering work by Goffman in the early 1960s (1963), it was not until the beginning of the 21st century that this idea was directly put to the test. That is, by applying the cardiovascular indices described by the biopsychosocial model (cardiac output, total peripheral resistance), Blascovich et al. (2002) revealed that interactions between stigmatized and non-stigmatized individuals elicit cardiovascular patterns indicative of threat motivational states in both parties involved. Thus, the introduction of cardiovascular measures indicating threat opened up new opportunities to assess directly, in an implicit way, psychological threat in relation to inter-ethnic interactions and coping with discrimination and devaluation of one's social identity—see the chapters by Blascovich (12), by Chalabaev, Sarrazin, & Brisswalter (8), by Major & Kunstman (17), by Mendes (16), by Mendoza-Denton & Page-Gould (14), and by Scheepers (13) in this volume.

Likewise, emergent knowledge of how functional magnetic resonance imaging (fMRI) indicates activation of specific brain areas raised the question of whether this might be used to examine the way we implicitly think of other groups and act on these ideas. Specifically, efforts were made to uncover whether different brain areas are involved in person perception, depending on whether people perceive an in-group or an out-group member. For instance, initial work examined whether the perception of African American faces would enhance amygdala activity indicated in fMRI, which would be consistent with the notion that the confrontation with out-group members raises negative emotions such as anxiety and hostility (Cunningham et al., 2004; Hart et al., 2000; Phelps et al., 2000; Wheeler & Fiske, 2005; see also Rotella & Richeson, Chapter 18 in this volume). Similarly, the research by Fiske and colleagues (see Fiske, Ames, Cikara, & Harris, Chapter 5 in this volume) has illustrated how activation in the medial prefrontal cortex (MPFC), a primary area involved in social cognition and person perception, varies as a function of the type of group to which a particular target belongs. Furthermore, in particular, members of groups who are perceived as scoring low on both *warmth* and *competence* (the homeless, drug addicts), are "dehumanized," meaning that they are denied unique human characteristics. The research by Fiske and colleagues has shown that this can also be seen at a brain level, in the sense that when one perceives members of such groups, this leads to lower activation in the MPFC compared to encounters with members of more positively stereotyped groups.

A third neuroscience method that has been used to examine prejudice concerns measures of electrical activity in the brain in response to specific events with electroencephalography (EEG)—so-called event-related brain potentials (ERPs). For example, EEG measures of the processing of in-group and out-group members have allowed for assessment of the degree to which faces of in-group and out-group targets are processed differently within the first hundreds of milliseconds of perception, thereby providing an online and pre-conscious indicator of social categorization (see Ito, Chapter 2 in this volume; Ito & Urland, 2003; Ito, Urland, Willadsen-Jensen, & Correll, 2005). The pioneering work by Ito and colleagues has revealed that people distinguish between faces of different gender and race even in the absence of explicit categorization goals, underlining the automatic nature of categorization (see Ito, Chapter 2 in this volume for an overview). Moreover, other researchers have used these ERP indicators to examine, for example, the role of implicit categorization in the control of racial bias (Amodio, 2010), and the effects of contextual manipulations such as social identity threat (Derks, Chapter 3 in this volume) and morality primes (Ellemers & Van Nunspeet, Chapter 11 in this volume) on implicit social categorization. These and other ERP components have been used to discern further among specific sub-processes involved in biased responses such as early social categorization, evaluative categorization, response monitoring, and control of biased responses, which are all relevant aspects of prejudice, but impossible to distill from just a reaction time—see chapters by Amodio (9), by Bartholow (10), by Derks (3), and by Ellemers and Van Nunspeet (11) in this volume. Another example of the application of EEG measurements to questions in relation to prejudice concerns the work on motor resonance ("mental mirroring" of the perceived movements of others) as an implicit indicator of perspective taking, depending on whether in-group or out-group targets are observed (Gutsell & Inzlicht, 2010; Chapter 15 in this volume).

From the discussion above, it also becomes clear that specific measures might be particularly well-suited to address specific questions within the neuroscience of prejudice. For example, whereas ERP measures may reveal *how* people attend to information about others (e.g., the extent to which they categorize them), fMRI indicates *which* specific brain areas are involved in person perception and prejudice (e.g., amygdala versus MPFC). Cardiovascular measures are particularly well-suited to measure threat (but also its motivational opposite, *challenge*; see below) during actual intergroup interactions, or to examine how stigmatized individuals cope with discrimination or identity threat. The current availability of a broad range of measures—which are all covered in diverse chapters of this book—offers increasing scope to triangulate specific response patterns or connect sequences of events occurring at different stages or different levels (Ellemers & Van Nunspeet, Chapter 11 in this volume).

It is also important to note at this point, however, that through the development of a neuroscience of prejudice, not only the measures, but also the manipulations within prejudice research have moved to another level. That is, apart from functioning as a dependent variable, neural and psychophysiological states have also been used as an independent variable (i.e., starting point) in research on prejudice. This work includes the administration of neuro-endocrine substances (e.g., oxytocin) to test whether such interventions may have a direct impact upon biological processes involved in intergroup discrimination (see De Dreu et al., 2010; De Dreu, Chapter 4 in this volume). The work by De Dreu and colleagues shows that administering oxytocin leads to strengthened in-group favoritism and, under threatening conditions, also out-group derogation.

Another example of research where brain processes form the independent rather than the dependent variable in the design is the work by Zárate on memory consolidation, sleep, and person perception in terms of individuation (i.e., unique personal characteristics) and stereotyping (see Zárate & Enge, Chapter 7 in this volume). Applying a neural model about the influence of sleep on memory consolidation, Zárate and colleagues have shown that more sleep leads to the integration of more individuating information in the impression one forms about a person. This research not only shows the creative (and low-tech/low-cost, compared to e.g., fMRI) adaptation of a neurological model and paradigm to study a social psychological question; it also illustrates how certain neural processes (induced by sleep) can form the independent variable in a research design on stereotyping.

The above examples all illustrate a prime methodological advantage of using neuroscience or psychophysiological measures in research on prejudice; namely, that such measures can more directly tap into crucial processes that are difficult or impossible to observe directly or to measure using self-reports. However, apart from forming unobtrusive and direct indicators of psychological states, there are additional advantages attached to the use of psychophysiological measures. For example, certain neuroscience measures, like event-related brain potentials (ERPs) can yield more *specific* information than traditional methodology, like reaction-time measures. As indicated above, ERPs can draw a more accurate picture of how prejudice unfolds, by allowing a distinction between different phases in information processing, like categorization, evaluation, and control (see chapters by Amodio [9], by Bartholow [10], by Derks [3], by Ellemers & Van Nunspeet [11], and by Ito [2] in this volume). In addition, many neuroscientific and psychophysiological measures can be measured in an online and continuous fashion. For example, cardiovascular indices of challenge and threat motivational states can be measured continuously, making it possible to examine how threat *develops* during inter-ethnic interactions (Blascovich, Chapter 12 in this volume), or how social identity threat can be turned into challenge (Scheepers, Chapter 13 in this volume).

Apart from being able to measure implicit processes more directly and more specifically, neuroscience measures can also provide invaluable information that helps to interpret multi-interpretable or counterintuitive results found with self-reports. One example of this is the work on *attributional ambiguity*, the phenomenon where members of stigmatized groups are often uncertain whether performance feedback they receive is based on their true performance or on their group membership—see chapters in this volume by Major and Kunstman (17), by Mendes (16), and by Mendoza-Denton and Page-Gould (14). As a result of attributional ambiguity, members of stigmatized groups do often show maintained levels of self-esteem after negative feedback, although a classic question is whether this is due to a defensive response (which suggests a state of threat); or due to successfully disengaging from the situation and discounting the negative feedback to external factors (which would be more in line with a state of challenge). Using cardiovascular measures of challenge and threat, Mendoza-Denton and Page-Gould (Chapter 14) were able to interpret the pattern of self-reports and found evidence for the latter discounting pattern by showing cardiovascular challenge responses after receiving negative status feedback. In addition, Major and Kunstman (Chapter 17), went one step further to show that it may actually be positive feedback in attributionally ambiguous situations that elicits the strongest threat response, particularly in group members expecting others to behave in a way that is "politically correct." This effect—at first sight somewhat counterintuitive—would have been much more difficult to show in the absence of physiological measures, while the interpretation in terms of attributional ambiguity would have been impossible in the absence of self-report measures (self-esteem, suspicion, emotions).

While the previous example illustrates how a physiological measure can help to interpret a self-report measure, self-report measures are also often used to interpret the effects of a physiological measure, especially in the case of a less specific physiological measure that relates to multiple psychological states (for a discussion, see Mendes, Chapter 16). For example, Chalabaev et al. (Chapter 8) describe research on stereotypes in the context of task performance. In the context of negative stereotypes about the out-group, Chalabaev and colleagues report reductions in heart rate variability, which is indicative of lowered parasympathetic arousal. However, by means of self-report measures, they were able further to interpret this effect in terms of more *effort* in the task, rather than in terms of the down-regulation of negative affect (see Chalabaev et al., Chapter 8 in this volume for details).

Raising Novel Questions that Yield Novel Answers

The availability of novel research paradigms and measures in turn elicited novel questions. This happened at a very basic level, when social psychologists—as novice users of neuroscientific methods—transformed and adapted existing research paradigms, due to their interest in different questions, and their search for different types of answers than the cognitive scientists who developed these tools. That is, whereas cognitive scientists had mainly focused on the role of different neural structures and processes in information processing, social neuroscientists were also interested in examining the malleability of these effects by social variables. Importantly, rather than demonstrating the lack of robustness in specific findings, charting contextual variations in people's neurophysiological response patterns informs us of relevant moderating factors.

For example, the likelihood that people engage in social categorization when presented with in-group and out-group stimuli is not as automatic as previously assumed, but depends on the social identity that is salient and the extent to which this identity is under threat (Derks, Chapter 3 in this volume). More specifically, the work by Derks showed that social categorization in in-group and out-group members, as measured by ERPs, increased when the in-group was under threat.

Another example is the work by Van Bavel et al. (Chapter 6 in this volume) who have shown that seemingly universal neural responses (such as activation of the amygdala when perceiving out-group members, or activation of the *fusiform face area* [FFA] when viewing faces) are moderated by the social identity that is salient at a given moment. More precisely, Van Bavel and colleagues showed that a relatively simple categorization in an artificial (i.e., minimal) group overrode the assumed automatic activation of the amygdala after perceiving racial in-group versus out-group members. A similar moderation by social identity was also shown for (increased) activation of the FFA after seeing racial in-group members. In this way, addressing a different part of the variance (due to the social context) in people's responses not only provides a more complete picture of the processes at work, but also offers scope for the development of contextual interventions that may target very basic psychological mechanisms relevant to the reduction of intergroup conflict or the achievement of social change.

Another example of the way in which neuroscience can add to building of social psychological theory concerns how new insights gained by measuring the antecedents and consequences of prejudice on a different level sometimes speak against current theories. This is the case, for instance, with theories on the regulation of implicit bias (see Amodio, Chapter 9; Bartholow, Chapter 10). It has traditionally been assumed that stereotypic knowledge is triggered automatically, and that only people who are subsequently motivated and able deliberately to control their biased expressions are able to override these responses (Devine, 1989; Monteith, 1993). However, research on the neural bases of prejudice control has challenged this perspective. Data show that non-deliberate processes are involved in detecting the need for control when stereotypes are triggered—just like the process of stereotype activation itself (Amodio, 2008; Bartholow & Dickter, 2008; see also Amodio, Chapter 9; and Bartholow, Chapter 10). Moreover, brain structures that are implicated in the regulation of bias such as structures in the frontal cortex show little direct connection with structures that are involved in automatic bias such as the amygdala. This speaks against the idea that the activation of automatic associations is overruled by the activation of brain structures that are responsible for control. Thus, the inclusion of a neural indicator of control led to the conclusion that prejudice regulation occurs non-consciously. Instead of deliberately down-regulating activated stereotypes, those who try to be unbiased detect the need to regulate their responses from the outset. Indeed, the recognition that prejudice regulation can function in this way also explains why some people (e.g., those that have a high intrinsic motivation to be unbiased) are better at overcoming implicit bias than others (Amodio, Devine, & Harmon-Jones, 2008). In summary, this work provides more specificity about the entire process through which stereotyping and prejudice unfold, a specificity that could not have been obtained by means of traditional methodology (e.g., a reaction time).

Another case in point relates to work showing that people who do not belong to our in-group seem less similar to us and receive less empathy. Although this finding has been repeatedly shown using affective and behavioral measures (e.g., Van der Schalk et al., 2011), recent social neuroscience work tells us more about *why* this is the case. One of the chapters in this book (Gutsell & Inzlicht, Chapter 15) presents a research program in which a neural indicator of mirror neuron activity when perceiving other individuals performing simple motor tasks is measured. Here it is shown that when we watch out-group members, we are less likely to mirror their actions than when we watch in-group members. Moreover, this tendency to mirror in-group but not out-group members is moderated by individual differences in racial bias, as well as instructions to take the perspective of out-group members. Thus, this research shows, once more, a more precise process that lies at the basis of prejudice and related intergroup phenomena, which could not have been identified without the aid of neuroscience methodology.

Another demonstration of how neuroscience methodology can shed light on the precise process though which intergroup behavior unfolds is formed by the work by Richeson on the ego-depleting character of intergroup contact (Richeson et al., 2003; see also Rotella & Richeson, Chapter 18 in this volume). In previous research, Richeson and colleagues had shown that after contact with a member of a racial minority group, members of the majority group (in particular, those who were relatively high in prejudice) show signs of ego depletion, as indicated by impaired performance on a Stroop task. Using fMRI, Richeson et al. (2003) were able to test two competing explanations for the effect (controlling prejudice during the interaction versus anxiety) against each other. By showing that the depleting effect of intergroup contact correlated with activation in the dorsolateral prefrontal cortex (which is implied in executive control), Richeson and her colleagues could champion the "controlling bias" explanation over the "negative emotions" explanation for ego depletion after intergroup contact.

Further development of new theory was inspired by the work that was initially motivated by the aim to measure threats to (social) identity in a more unambiguous and unobtrusive way using cardiovascular measures (Blascovich et al., 2002; see also chapters in this volume by Blascovich [12], by Major & Kunstman [17], by Mendes [16], by Mendoza-Denton & Page-Gould [14], and by Scheepers [13]). This work clearly revealed that threats to the value of one's group (*social identity threats*) are experienced as negative psychological threats that people are motivated to cope with. However, it also raised further questions: for instance, the question whether discrimination or negative feedback about one's group always leads to threat; or whether it can also, under particular circumstances, lead to challenge. As discussed by Scheepers (Chapter 13), it seems unlikely that all the different responses that members of low-status groups display in response to their thorny position (stronger commitment to the group, collective action, protest) are explained by threat. Indeed, the research by Scheepers showed that under particular circumstances (e.g., when group status was unstable), members of low-status groups (and in particular, those highly identified with the group) display a cardiovascular response pattern indicative of a challenge motivational state.

In addition, the realization that cardiovascular measures can be used to identify conditions that raise maladaptive stress responses opened up the possibility that there are very real health risks associated with the cumulative experience of psychological stress over time (e.g., due to biased treatment; see Blascovich et al., 2001). As a result, recent work in this area has led to emergent knowledge of endocrine and cardiovascular stress responses as risk factors for long-term physical health (Major, Mendes, & Dovidio, in press). To the extent that stigma and social exclusion have been documented as powerful causes of negative stress, this opens up a whole new area of interdisciplinary research and interventions, relating prejudice and intergroup relations to long-term health implications and early mortality due to cardiovascular diseases. Further developing knowledge of the different levels at which people may suffer from exposure to prejudice illustrates one of the most dramatic and far-reaching implications of social inequality and intergroup conflict.

New Challenges

The examples above show that the inclusion of neuroscience methodology substantially adds to the development and adjustment of theory concerning prejudice and intergroup relations. The insights gained from measuring how intergroup dynamics relate to psychophysiological processes, and the opportunity to connect social psychological theory to physiological concepts lead to important extensions of current theory. Nevertheless, the findings of interest sometimes are quite complex, while conducting neuroscience research also raises new challenges which are not always easily resolved.

One of these challenges is that empirical work aiming to connect responses at different levels directly (e.g., the neural or physiological level and the behavioral level) often reveals mixed results. Attempts to reconcile these issues have uncovered technical complexities that may impede researchers' abilities directly to establish evidence for such relations. For instance, the inclusion of multiple measures in a single study may raise problems in and of itself. In a very basic sense, for instance, requirements that optimize the reliability of physiological measures (e.g., fixating the body, preventing hand or eye movements), interfere with the measurement of other types of response (e.g., social interactions, task behavior). In a similar vein, tailoring experimental procedures to capture responses most effectively at one level may make them less suitable for the examination of the further implications of these responses. For instance, asking participants to categorize multiple faces into in-group versus out-group representatives precludes establishing a direct connection between categorization responses and observations of non-verbal behavior that require "richer" and more extended social interactions with the same target stimuli. Likewise, procedures to assess ERP responses reliably (e.g., by averaging responses over many trials) may extinguish task performance differences, which typically level off after a certain number of trials due to learning effects.

Even though these complications emerge at a methodological level, sometimes they tend to impede theoretical progress. For instance, the difficulties outlined above may make it difficult, if not impossible, to establish empirical evidence directly for the interrelation between different types of response (e.g., mediation of behavior through neural or physiological states)—even if such relations do exist. In the meantime, these efforts to understand how processes at different levels relate to each other have raised the more basic question as to whether it is even realistic to expect very complex and deliberate social behaviors (e.g., the tendency to engage in collective action) to show a one-on-one relation with a single specific physiological mediator (e.g., increased blood pressure as a result of social identity threat). While very basic physiological responses are likely to relate to broader motivations (such as the desire to change the status quo), in theory there is a range of potentially relevant courses of action that people might follow to enact this general motivation. Indeed, there are few grounds for assuming that basic biological stress responses should favor one specific behavioral strategy over another.

Despite these challenges, it should be noted that researchers have been extremely innovative and creative in developing and adapting paradigms to build the bridge between neuroscience and the psychology of prejudice. Examples of this include touching people while they are in an fMRI scanner as a way to manipulate intergroup contact, or even simulating more elaborative intergroup contact between two individuals lying in two scanners (see Rotella & Richeson, Chapter 18). Another example concerns the research described by Mendes (Chapter 16) in which real inter-ethnic interactions were studied by means of cardiovascular indices that were measured using ambulatory equipment. As a result of the basic electrical principles underlying cardiography, signal loss occurs when there is actual physical contact between individuals who are attached to this equipment. However, instead of treating this signal loss as an artifact, Mendes and her colleagues used it as an unobtrusive measure of physical contact (see Mendes, Chapter 16, for details). These, and several of the other paradigms that are described in the diverse chapters in this volume, show that the neuroscience of prejudice has not just yielded new theory, but also has led to new, clever research paradigms.

The Current Volume

The current volume celebrates the different ways in which neuroscience theory and methodology have enriched social psychological research on prejudice, but it also acknowledges the sometimes bumpy road leading to the bridge between these two different worlds. Combining neuroscientific methods with more "classic" self-reports or behavioral observations—instead of replacing them—offers a new way to examine how responses to social situations unfold at different levels. Rather than seeing neurocognitive measures as indicating people's "true" response or treating them primarily as process mediators of more overt behaviors, the combination of different measures reveals the parallel involvement of multiple systems, and points to the operation of recursive cycles in which people gradually become aware of their bodily state or automatic cognitive responses, or try to correct or suppress the implications of these immediate responses.

This book represents both the research and the personal stories behind this research by top researchers in the field of the neuroscience of prejudice and intergroup relations. By bringing together recent work in neuroscience in terms of its contribution to a common theoretical theme, this volume provides a unique overview of how different methods in social neuroscience can inform the development of novel theory on psychological processes related to prejudice. By asking the leading prejudice researchers from across the world who have carved out this field to tell their personal stories, this book provides a more integrated view on the specific benefits and limitations of each type of measure; illuminates how standard paradigms in research on prejudice and intergroup relations can be adapted for the use of neuroscientific methods; and helps to determine how different methodologies can complement each other and be combined to advance current insights into the nature of prejudice.

This book covers the entire process of prejudice, from social categorization and the development of in-group bias, to intergroup interactions and the impact of discrimination on its targets, using a variety of neuroscientific and psychophysiological measures. In the first section on *categorization and in-group favoritism*, Ito (Chapter 2) describes her pioneering work on ERPs as a way to measure social categorization. Derks (Chapter 3) then presents work on how these neural correlates of social categorization are influenced by social identity concerns like level of social identification and social identity threat. In Chapter 4, De Dreu focuses on the next step after categorization; namely, the development of in-group favoritism, as a function of oxytocin secretion.

The second section of the book handles *person perception and stereotyping*. In Chapter 5, Fiske et al. describe research on the stereotype content model, and present fMRI evidence that extremely negative out-groups (low in competence, low in warmth) are not even perceived as human; i.e., they are *dehumanized*. Van Bavel et al. (Chapter 6) then review a line of fMRI research demonstrating how social identity influences the neural correlates of group perception, even to such an extent that the salience of a novel and "minimal" social category can override the influence of racial in-group/out-group stimuli on amygdala and FFA activation. The research by Zárate and Enge (Chapter 7) shows the influence of memory consolidation through sleep on the use of individuating information in impression formation. Finally, in Chapter 8, Chalabaev et al. apply measures of heart rate variability to explain the performance-boosting influence that negative out-group stereotypes can have (stereotype lift).

After the research on the neural correlates of categorization, and the development of in-group bias and stereotyping, the third section addresses the ways in which people can *overcome implicit prejudice*. Amodio (Chapter 9) describes research showing how neuroscience methodology can be used to test different models about the activation and control of bias against each other. Then Bartholow (Chapter 10) presents ERP research on the different processes through which people (aim to) control biased responses, and the way alcohol consumption disrupts these processes. Finally, Ellemers and Van Nunspeet (Chapter 11) show how morality concerns (versus competence concerns) further moderate these preconscious control processes.

After learning about the development and expression of bias, in the fourth section we turn to the target's perspective and learn about the ways in which these targets *cope with prejudice and identity threat*. As a first contribution to this section, Blascovich (Chapter 12) presents the most recent description of the biopsychosocial model of challenge and threat, a model describing cardiovascular markers of these states which have repeatedly been used to study intergroup threat (this includes Blascovich's own research which is described in the chapter). Scheepers (Chapter 13) then reviews a line of research applying these cardiovascular measures to examine threats to social identity, and how these threats can be turned into challenges. Finally, in Chapter 14, Mendoza-Denton and Page-Gould apply cardiovascular indices of challenge and threat to study attributional ambiguity, in order to shed more light on the finding that members of stigmatized groups do often show unharmed self-esteem after receiving negative feedback.

In the final section, we turn to *intergroup interactions*. Gutsell and Inzlicht (Chapter 15) describe their research on motor resonance ("mental mirroring") to index perspective-taking in intergroup situations, as a first step in intergroup behavior, including intergroup interactions. Mendes (Chapter 16) then addresses diverse aspects of intergroup interactions (from the target, perceiver, and dyadic points of view), using neuro-endocrine and cardiovascular measures, and with a particular stress on meta-monitoring and over-correction behavior during such interactions. Major and Kunstman (Chapter 17) then present research on attributional ambiguity in targets of prejudice, and the role of suspicion in the (at first sight counterintuitive) finding that members of stigmatized groups show cardiovascular signs of threat after receiving *positive* feedback from members of non-stigmatized groups. Finally, the research by Rotella and Richeson (Chapter 18) nicely "closes the circle" of the section (and the book) by showing (using fMRI) how intergroup contact influences basic brain processes. In particular, Rotella and Richeson shed light on the neural basis of the ego-depleting effects of intergroup contact, and the development of (mis)trust as a function of perceived group victimhood.

Conclusion

The substantial developments reviewed above make it easy to forget that use of a neuroscientific perspective in research on prejudice and intergroup relations is still relatively new; and is, as yet, conducted by a relatively small group of active researchers. Indeed, whereas many are aware of this work at least at some level, and may speak about the progress that is being made with admiration, actually adopting neuroscientific methods into one's own research may pose a substantial hurdle. The researchers who have contributed to this volume have been quite candid in revealing some of the setbacks and difficulties they encountered when they set their first steps on this path, and fairly consider the drawbacks as well as the benefits of the methods they have used. This is not to discourage future researchers from choosing a similar path. On the contrary, we strongly believe that being aware beforehand of complications that may arise can only help other researchers make a more informed decision about which methodologies to adopt and why. The awareness that those who have achieved great successes with their innovative work also experienced disappointment and frustration, and sometimes had to invest many years of plain hard work to get to this stage, may provide solace to other researchers in need of patience and persistence until they get to the stage where their investment in the use of these complex methodologies starts to pay off. For those who ask themselves whether it is even worth trying, we think the answer is a resounding *yes*, because of the truly novel perspective that opens up when social psychological problems and research designs are combined with neuroscientific questions and measures. As such, we hope that this book inspires prejudice and intergroup relations researchers to incorporate neuroscience models and methods in their research.

References

Abrams, D., & Hogg, M. (1988). Comments on the motivational status of self-esteem in social identity and intergroup discrimination. *European Journal of Social Psychology, 18*, 317–334.

Adorno, T. W., Frenkel-Brunswik, E., Levinson, D. J., & Sanford, R. N. (1950). *The authoritatian personality*. New York: Harper.

Allport, G. (1958). *The nature of prejudice*. Garden City, NY: Doubleday Anchor.

Amodio, D. M. (2008). The social neuroscience of intergroup relations. *European Review of Social Psychology, 19*, 1–54.

Amodio, D. M. (2010). Coordinated roles of motivation and perception in the regulation of intergroup responses: Frontal cortical asymmetry effects on the P2 event-related potential and behavior. *Journal of Cognitive Neuroscience, 22*, 2609–2617.

Amodio, D. M., Devine, P. G., & Harmon-Jones, E. (2008). Individual differences in the regulation of intergroup bias: The role of conflict monitoring and neural signals for control. *Journal of Personality and Social Psychology, 94*, 60–74.

Amodio, D. M., Harmon-Jones, E., Devine, P. G., Curtin, J. J., Hartley, S. L., & Covert, A. E. (2004). Neural signals for the detection of unintentional race bias. *Psychological Science, 15*, 88–93.

Bargh, J. A. (1999). The cognitive monster: The case against the controllability of automatic stereotype effects. In S. Chaiken & Y. Trope (Eds.), *Dual process theories in social psychology* (pp. 361–382). New York: Guilford Press.

Bartholow, B. D., & Dickter, C. L. (2008). A response conflict account of the effects of stereotypes on racial categorization. *Social Cognition, 26*, 273–291.

Bijleveld, E., Scheepers, D., & Ellemers, N. (2012). The cortisol response to anticipated intergroup interactions predicts self-reported prejudice. *PLoS ONE*, 7, e33681.

Blascovich, J. (2008). Challenge, threat, and health. In J. Y. Shah & W. L. Gardner (Eds.), *Handbook of motivation science* (pp. 481–493). New York: Guilford Press.

Blascovich, J., Mendes, W. B., & Seery, M. D. (2002). Intergroup encounters and threat: A multi-method approach. In D. M. Mackie & E. R. Smith (Eds.), *From prejudice to intergroup emotions: Differentiated reactions to social groups* (pp. 89–109). New York: Psychology Press.

Blascovich, J., Spencer, S., Quinn, D., & Steele, C. (2001). African-Americans and high blood pressure: The role of stereotype threat. *Psychological Science, 12*, 225–229.

Bodenhausen, G. V., & Macrae, C. N. (1998). Stereotype activation and inhibition. In R. S. Wyer (Ed.), *Advances in social cognition* (Vol. 11, pp. 1–52). Hillsdale, NJ: Erlbaum.

Brown, R., (2010). *Prejudice, its social psychology*. Chichester: Wiley.

Cacioppo, J. T. (2002). Social neuroscience: Understanding the pieces fosters understanding the whole and vice versa. *American Psychologist, 57*, 819–830.

Cacioppo, J. T., Visser, P. S., & Pickitt, C. L. (Eds.). (2005). *Social neuroscience: People perceiving people*. Cambridge, MA: The MIT Press.

Cottrell, C. A., & Neuberg, S. L. (2005). Different emotional reactions to different groups: A sociofunctional threat-based approach to "prejudice." *Journal of Personality and Social Psychology, 88*, 770–789.

Cunningham, W. A., Johnson, M. K., Raye, C. L., Gatenby, J. C., Gore, J. C., & Banaji, M. R. (2004). Separable neural components in the processing of Black and White faces. *Psychological Science, 15*, 806–813.

De Dreu, C. K. W., Greer, L. L., Handgraaf, M. J. J., Shalvi, S., van Kleef, G. A., Baas, M., & Feith, S. W. W. (2010). The neuropeptide oxytocin regulates parochial altruism in intergroup conflict among humans. *Science, 328*, 1408–1411.

Derks, B., Inzlicht, M., & Kang, S. (2008). The neuroscience of stigma and stereotype threat. *Group Processes and Intergroup Relations, 11*, 163–181.

Devine, P. (1989). Stereotypes and prejudice: Their automatic and controlled components. *Journal of Personality and Social Psychology, 56*, 5–18.

Dovidio, J., Esses, V., Glick, P. S., & Hewstone, M. (Eds.). (2010). *Handbook of prejudice and discrimination*. London, UK: Sage.

Dovidio, J. F., & Gaertner, S. L. (Eds.). (1986). *Prejudice, discrimination and racism*. San Diego, CA: Academic Press.

Dovidio, J. F., & Gaertner, S. L. (1996). Affirmative action, unintentional racial biases, and intergroup relations. *Journal of Social Issues, 52*, 51–75.

Dovidio, J. F., Glick, P., & Rudman, L. A. (2005). *On the nature of prejudice: Fifty years after Allport*. Malden, MA: Blackwell Publishing.

Dovidio, J. F., Kawakami, K., & Gaertner, S. L. (2002). Implicit and explicit prejudice and interracial interaction. *Journal of Personality and Social Psychology, 82*, 62–68.

Fazio, R. H., Sanbonmatsu, D. M., Powell, M. C., & Kardes, F. R. (1986). On the automatic activation of attitudes. *Journal of Personality and Social Psychology, 50*, 229–238.

Fein, S., & Spencer, S. J. (1997). Prejudice as self-image maintenance: Affirming the self through derogating others. *Journal of Personality and Social Psychology, 73*, 31–44.

Fiske, S. T. (2000). Stereotyping, prejudice, and discrimination at the seam between the centuries: Evolution, culture, mind, and brain. *European Journal of Social Psychology, 30*, 299–322.

Fiske, S. T., & Neuberg, S. L. (1990). A continuum model of impression formation, from category-based to individuating processes: Influence of information and motivation on attention and interpretation. In M. P. Zanna (Ed.), *Advances in experimental social psychology* (Vol. 23, pp. 1–74). New York: Academic Press.

Gaertner, S. L., & Dovidio, J. F. (2000). *Reducing intergroup bias: The common ingroup identity model*. Philadelphia: Psychology Press.

Goffman, E. (1963). *Stigma: Notes on the management of spoiled identity*. Englewood Cliffs, NJ: Prentice-Hall.

Greenwald, A. G., McGhee, D. E., & Schwartz, J. K. L. (1998). Measuring individual differences in implicit cognition: The Implicit Association Test. *Journal of Personality and Social Psychology, 74*, 1464–1480.

Guglielmi, R. S. (1999). Psychophysiological assessment of prejudice: Past research, current status, and future directions. *Personality and Social Psychology Review, 3*, 123–157.

Gutsell, J. N., & Inzlicht, M. (2010). Empathy constrained: Prejudice predicts reduced mental simulation of actions during observation of outgroups. *Journal of Experimental Social Psychology, 46*, 841–845.

Harris, L. T., & Fiske, S. T. (2006). Dehumanizing the lowest of the low: Neuro-imaging responses to extreme outgroups. *Psychological Science, 17*, 847–853.

Hart, A. J., Whalen, P. J., Shin, L. M., McInerney, S. C., Fischer, H., & Rauch, S. L. (2000). Differential response in the human amygdala to racial outgroup vs ingroup face stimuli. *Neuroreport, 11*, 2351–2355.

Ito, T. A., & Urland, G. R. (2003). Race and gender on the brain: Electrocortical measures of attention to race and gender of multiply categorizable individuals. *Journal of Personality and Social Psychology, 85*, 616–626.

Ito, T. A., Urland, G. R., Willadsen-Jensen, E. C., & Correll, J. (2005). The social neuroscience of stereotyping and prejudice: Using event-related brain potentials to study social perception. In J. T. Cacioppo, P. S. Visser & C. L. Pickitt (Eds.), *Social neuroscience: People perceiving people* (pp. 189–208). Cambridge, MA: The MIT Press.

Levin, S., & Van Laar, C. (Eds.). (2006). *Stigma and group inequality: Social psychological perspectives*. Claremont, CA: Lawrence Erlbaum Associates Publishers.

Lieberman, M. D. (2007). Social cognitive neuroscience: A review of core processes. *Annual Review of Psychology, 58*, 259–289.

Macrae, C. N., & Bodenhausen, G. V. (2001). Social cognition: Categorical person perception. *British Journal of Psychology, 92*, 239–255.

Major, B., Mendes, W. B., & Dovidio, J. (in press). Intergroup processes that inform racial health disparities. *Health Psychology*.

Matheson, K., & Cole, B. (2004). Coping with a threatened group identity: Psychological and neuroendocrine responses. *Journal of Experimental Social Psychology, 40*, 777–786.

Mendes, W. B., Blascovich, J., Lickel, B., & Hunter, S. (2002). Challenge and threat during social interaction with white and black men. *Personality and Social Psychology Bulletin, 28*, 939–952.

Monteith, M. J. (1993). Self-regulation of prejudiced responses: Implications for progress in prejudice reduction efforts. *Journal of Personality and Social Psychology, 65*, 469–485.

Nelson, T. D. (2009). *Handbook of prejudice, stereotyping, and discrimination*. New York: Psychology Press.

Pettigrew, T. F. (1958). Personality and sociocultural factors in intergroup attitudes: A cross-national comparison. *Journal of Conflict Resolution, 2*, 29–42.

Pettigrew, T. F., & Tropp, L. R. (2008). How does intergroup contact reduce prejudice? Meta-analytic tests of three mediators. *European Journal of Social Psychology, 38*, 922–934.

Phelps, E. A., O'Connor, K. J., Cunningham, W. A., Funayama, E. S., Gatenby, J. C., Gore, J. C., & Banaji, M. R. (2000). Performance on indirect measures of race evaluation predicts amygdala activation. *Journal of Cognitive Neuroscience, 12*, 729–738.

Prentice, D. A., & Eberhardt, J. L. (2008). The neural underpinnings of group life. *Group Processes and Intergroup Relations, 11*, 139–142.

Richeson, J. A., Baird, A. A., Gordon, H. L., Heatherton, T. F., Wyland, C. L., Trawalter, S., & Shelton, J. N. (2003). An fMRI examination of the impact of interracial contact on executive function. *Nature Neuroscience, 6*, 1323–1328.

Ryan, M. K., & Branscombe, N. R. (Eds.). (in press). *The Sage handbook of gender and psychology*. London, UK: Sage.

Shelton, J. N., Richeson, J. A., & Vorauer, J. D. (2006). Threatened identities and interethnic interactions. *European Review of Social Psychology, 16*, 321–358.

Smith, E. R., & Mackie, D. M. (2010). Affective processes. In J. Dovidio, V. Esses, P. S. Glick & M. Hewstone (Eds.), *Handbook of prejudice and discrimination* (pp. 131–145). London, UK: Sage.

Stangor, C. (Ed.). (2000). *Stereotypes and prejudice: Essential readings*. Philadelphia: Psychology Press.

Swim, J. K., & Stangor, C. (Eds.). (1998). *Prejudice: The target's perspective*. San Diego, CA:

Academic Press.
Tajfel, H. (1978). *Differentiation between social groups: Studies in the social psychology of intergroup relations*. London, UK: Academic Press.
Tajfel, H., & Turner, J. C. (1979). An integrative theory of intergroup conflict. In W. G. Austin & S. Worchel (Eds.), *The social psychology of intergroup relations* (pp. 33–47). Monterey, CA: Brooks/Cole.
Van Bavel, J. J., & Cunningham, W. A. (2011). A social neuroscience approach to self and social categorisation: A new look at an old issue. *European Review of Social Psychology, 21*, 237–284.
Van der Schalk, J., Fischer, A. H., Doosje, B. J., Wigboldus, D., Hawk, S. T., Rotteveel, M., & Hess, U. (2011). Convergent and divergent responses to emotional displays of ingroup and outgroup. *Emotion, 11*, 286–298.
Vrana, S. R., & Rollock, D. (1998). Physiological response to a minimal social encounter: Effects of gender, ethnicity, and social context. *Psychophysiology, 35*, 462–469.
Wheeler, M. E., & Fiske, S. T. (2005). Controlling racial prejudice: Social-cognitive goals affect amygdala and stereotype activation. *Psychological Science, 16*, 56–63.

PART I

Categorization and In-group Favoritism

2

IMAGING THE PICTURES IN OUR HEADS

Using ERPs to Inform Our Understanding of Social Categorization

Tiffany A. Ito

Social behavior is complex and dynamic, which can create thorny measurement challenges for those interested in studying it (e.g., Cook & Selltiz, 1964). Asking individuals what they are feeling and thinking has many advantages, but we also know that social actors do not always have accurate introspective awareness of the causes of their behaviors (Nisbett & Wilson, 1977). Moreover, studying socially sensitive topics introduces concern that respondents are motivated to alter their true responses to appear more desirable (e.g., DeMaio, 1984; LaPiere, 1934). These issues have driven the field to important methodological innovations, such as the bogus pipeline (Jones & Sigall, 1971) and a variety of indirect and implicit measures (e.g., Fazio, Sanbonmatsu, Powell, & Kardes, 1986; Gaertner & McLaughlin, 1983; Gilbert & Hixon, 1991; Greenwald, McGhee, & Schwartz, 1998; Maass, Ceccarelli, & Rudin, 1996; Perdue & Gurtman, 1990).

These issues were particularly salient to me as a graduate student as I began initially to formulate research on the issues that interested me the most—stereotyping and prejudice. Today (2013), researchers might readily turn to reaction-time-based implicit measures, but the slow processing speed of computers in the early 1990s made implementing those measures a fairly substantial technical challenge. What I did have access to in the lab of my graduate advisor, Norman Miller, was the equipment for measuring facial electromyography (EMG). I thus began using EMG as a measure of incipient movement in the muscles associated with affective (emotional) expression as a way to index affective responses to intergroup situations.

Besides their availability, physiological measures attracted me for three additional reasons. First, because they can be obtained without participants' subjective awareness of the process of interest or awareness even of what exactly is being measured, they can be especially informative in situations where explicit and implicit responses diverge, or when participants might be motivated to modify their initial responses (Ito, 2010; Ito & Cacioppo, 2007). Second, they provided a way to combine my interest in social processes with my long-standing interest in biology and physiology. The latter topics had always been among my favorites in high school and college, so it was exciting to find that I could integrate them with my newfound interests in psychology (due to the vagaries of my high school curriculum, I received a solid foundation in biology, anatomy, and physiology, but took no classes in behavioral sciences). Many of the measures now widely included under the umbrella of social neuroscience also have exquisite temporal resolution and can be recorded on-line, as actual judgments and behaviors are occurring, thereby giving them direct access to the physiological processes that underlie judgments and decisions. As such, these measures have the potential to reveal directly the neural and physiological substrates of important social processes. Finally, I came to realize that explicitly integrative approaches like social neuroscience tend to be boundary expanding, potentially introducing not just new methods into our repertoire, but also additional theoretical insights (cf. Cacioppo & Berntson, 1992; see also Amodio, Chapter 9 in this volume). Personally, this translated into reading much more broadly than I probably would have done otherwise.

After initially using facial EMG and several autonomic nervous system measures such as heart rate and skin conductance to assess affective reactions to intergroup interactions in my graduate work (e.g., Vanman, Paul, Ito, & Miller, 1997), I started to explore the use of event-related brain potentials (ERPs) during a post-doctoral position with John Cacioppo. What appealed to me was the potential to switch from focusing on the affective outcomes of an intergroup situation to the mechanisms that drive such responses. Event-related brain potentials could assess a wider range of cognitive processes, and so promised to reveal information about a wider range of possible mechanisms (Ito & Cacioppo, 2007). As discussed in more detail below, there is reason to expect that many of the important mechanisms in stereotyping and prejudice will operate automatically, which also made ERPs appealing because of their high temporal resolution (in the order of milliseconds) and their ability to access mental operations occurring very early in processing. Finally, like many other physiological measures, ERPs can be recorded without informing participants of what is being assessed or requiring them to report their reactions accurately or honestly, making these measures quite useful in assessing both explicit and implicit processes (Ito, 2010; Ito & Cacioppo, 2007).

ERPs: A Methods Primer

ERPs are changes in brain electrical activity that occur in response to discrete events such as a stimulus or onset of a response. They are recorded non-invasively from the surface of the scalp and are thought to reflect summated postsynaptic potentials from large sets of synchronously firing neurons in the cerebral cortex

(Fabiani, Gratton, & Coles, 2000). Time-locked positive and negative deflections in the ERP waveform are referred to as *components*,[1] which are assumed to reflect distinct information-processing operations (Gehring, Gratton, Coles, & Donchin, 1992). Quantification of component latency and amplitude, therefore, provides information about the timing and degree of various psychological operations. The naming of a component often indicates both its polarity and mean latency. For example, the P200 is a positive-going component occurring at roughly 200 milliseconds (ms) whereas the N400 is a negative-going component occurring around 400 ms.

Event-related brain potentials are derived by extracting relatively short (e.g., <1,000 ms) time-locked intervals from the electroencephalogram (EEG), which measures continuous voltage changes in brain activity over time. The EEG is recorded with electrodes resting on the surface of the scalp, typically placed in accordance with published guidelines to standardize recording locations across participants and labs (American Encephalographic Society, 1994). Several companies currently make stretch-nylon caps into which electrodes are sewn in the standardized locations, greatly facilitating the quick placement of many electrodes (e.g., 64–264) with measurement of just a few scalp landmarks. Signals are then passed to powerful bioamplifiers which magnify the relatively small signals present at the surface of the scalp 10,000–50,000 times. They are then digitally sampled at a high frequency and stored on a computer hard drive in conjunction with markers signaling important events (e.g., presentation of a stimulus).

Off-line, the data go through a number of post-collection processing stages to extract the event-related activity from the background EEG. Because the ERP signals are quite small (a few μVs), these processing stages also typically include steps to reduce sources of noise such as background EEG and movement. One such source of noise is movement of the eyes, which creates changes in the voltage gradient across the scalp that can be detected by scalp electrodes, thereby obscuring the ERP signal. Reduction of this source of noise is typically accomplished by explicitly recording horizontal (e.g., due to saccades) and vertical (e.g., due to blinks) eye movements from electrodes placed near the eyes, and then either manually eliminating trials with large eye-movement artifacts or applying an algorithm to reduce the impact of the eye movements on scalp recordings (e.g., Gratton, Coles, & Donchin, 1983; for a more detailed discussion, see Luck, 2005). The application of digital filters can also reduce noise, assuming the noise is in a frequency range different from the ERP activity (see Luck, 2005, for a discussion of the potentially distorting effects of filters). The ERP activity of interest in social and cognitive neuroscience typically occurs in the 0.01–30 Hz range (Luck, 2005; Marshall-Goodell, Tassinary, & Cacioppo, 1990), so filters are often applied to attenuate activity outside those ranges. Next, data from the same psychological condition are averaged together for each participant. To the degree that non-ERP activity varies randomly from trial to trial, it will average to 0, whereas the activity time-locked to the event of interest will remain. Finally, the amplitude and latency of portions of the ERP waveform are quantified from the individual condition averages for each subject (for a discussion of quantification issues, see Fabiani et al., 2000; Luck, 2005).

The Automaticity of Category-based Judgments

While my graduate work was focused primarily in the substantive area of stereotyping and prejudice, my initial work using ERPs during my post-doctoral training focused on issues of evaluative categorizations. Here we used ERPs to obtain unique evidence of the degree to which people automatically classify stimuli in terms of their evaluative tone and respond more strongly to negative than positive input (e.g., Ito, Larsen, Smith, & Cacioppo, 1998; Ito & Cacioppo, 2000). When I had developed this greater facility with ERPs, I was eager to adapt the ERP paradigms I had been working with to focus once again on questions of stereotyping and prejudice. Dominant at this time were models of person perception differentiating between relatively more individuated versus category-based judgments (e.g., Bodenhausen & Macrae, 1998; Brewer, 1988; Bruner, 1957; Fiske & Neuberg, 1990; Macrae & Bodenhausen, 2000; Stangor, Lynch, Duan, & Glass, 1992). A critical assumption in many of these models was the automatic nature of category-based judgments, with mere exposure to category exemplars assumed to activate stereotypical associations (Bargh, 1984, 1999; Devine, 1989; Dovidio, Evans, & Tyler, 1986). At the time, though, this assumption had not been rigorously tested. In fact, with the methods available at the time, it was difficult to assess the automaticity of categorization independent from the effects of stereotyping and prejudice (cf., Macrae, Bodenhausen, & Milne, 1995). The sensitivity of ERPs to implicit judgments, and the ability to obtain such measures without requiring participants to make explicit stereotype or prejudice-related judgments made them perfect for testing this assumption of automatic social categorization.

We initially approached this question with a fairly straightforward paradigm of simply showing participants pictures of individuals who varied in social category while recording ERPs. We focused initially on differences in race and gender, showing pictures of Black and White college-aged men and women. To ensure that the neural responses we were recording reflected social categorization, we gave participants an explicit categorization task; half explicitly categorized the faces in terms of race, while the others categorized in terms of gender (Ito & Urland, 2003). Most participants in these studies identified their ethnicity as White.

We focused on those ERP components occurring within the first few hundred milliseconds of perception which had been shown to vary as a function of covert orienting to task-relevant and/or salient features (Crowley & Colrain, 2004; Czigler & Geczy, 1996; Eimer, 1997; Kenemans, Kok, & Smulders, 1993; Naatanen & Gaillard, 1983; Wijers, G. Mulder, Okita, L. Mulder, & Scheffers, 1989). Consistently across numerous studies, we find that racial and gender categorization modulate neural responses very early in perception (Correll, Urland, & Ito, 2006;

Ito & Tomelleri, 2013; Ito & Urland, 2003, 2005; Kubota & Ito, 2007; Willadsen-Jensen & Ito, 2006, 2008; see also the chapters by Amodio [9], Bartholow [10], and Derks [3] in this volume). As can be seen in Panel A of Figure 2.1, race effects are seen as early as 170 ms in the P200 component where our White subjects show larger responses to out-group Black than to White faces. This effect continues into the next component, the N200, which typically has a mean peak latency around 260 ms, but the direction of the effect changes, with larger N200s to White faces than to Black faces. Similar effects are seen when White participants view faces of Asians and Whites: P200s are larger to out-group Asian faces; and N200s are larger to in-group White faces. Target gender also modulates neural responses, with larger P200s to male than female faces, but larger N200s to female than male faces. This can be seen in Panel B of Figure 2.1. Gender effects are not moderated by participant gender.

Of particular relevance to the question of whether social category information is processed automatically, race and gender modulate neural activity regardless of what task participants are performing. We concluded this first by showing that ERP responses are sensitive to race and gender regardless of whether participants are explicitly categorizing faces in terms of race or gender (Ito & Urland, 2003). It is the case, though, that both tasks orient participants' attention to a categorical distinction. It is possible that highlighting one dimension (e.g., race) sensitizes perceivers to other categorical distinctions (e.g., gender). To test the automaticity of social category encoding further, we therefore thought it important to measure sensitivity to categorical distinctions when attention was more explicitly directed away from this information. This was achieved in some studies by focusing on a more person-based judgment (e.g., asking participants to make snap judgments about personality) (Ito & Urland, 2005). Others studies adopted the strategy of de-emphasizing the social nature of the stimuli altogether by having participants locate non-social physical features on the stimuli (e.g., a white dot) (Ito & Tomelleri, 2013; Ito & Urland, 2005; cf. Macrae, Bodenhausen, Milne, Thorn, & Castelli, 1997). In both cases, the same ERP modulations as a function of race and gender are observed, suggesting that perception of social category information is fairly ubiquitous, occurring even when perceivers do not have the goal of consciously attending to it.

In addition to supporting the automaticity of social categorization, we looked for ways to demonstrate its conceptual independence from stereotyping. One way we pursued this was to simultaneously assess the effects of task manipulations on categorization and stereotyping. Independence is implied if a manipulation affects one, but not the other. We did this in a study in which ERPs were recorded during an implicit stereotyping task (Ito & Tomelleri, 2013). A sequential priming task was used in which participants had to categorize pictures of handguns and insects as such shortly after the presentation of a Black or White prime (Judd, Blair, & Chapleau, 2004). This was done under two different face-processing conditions: Participants either categorized the face primes in terms of race or they performed

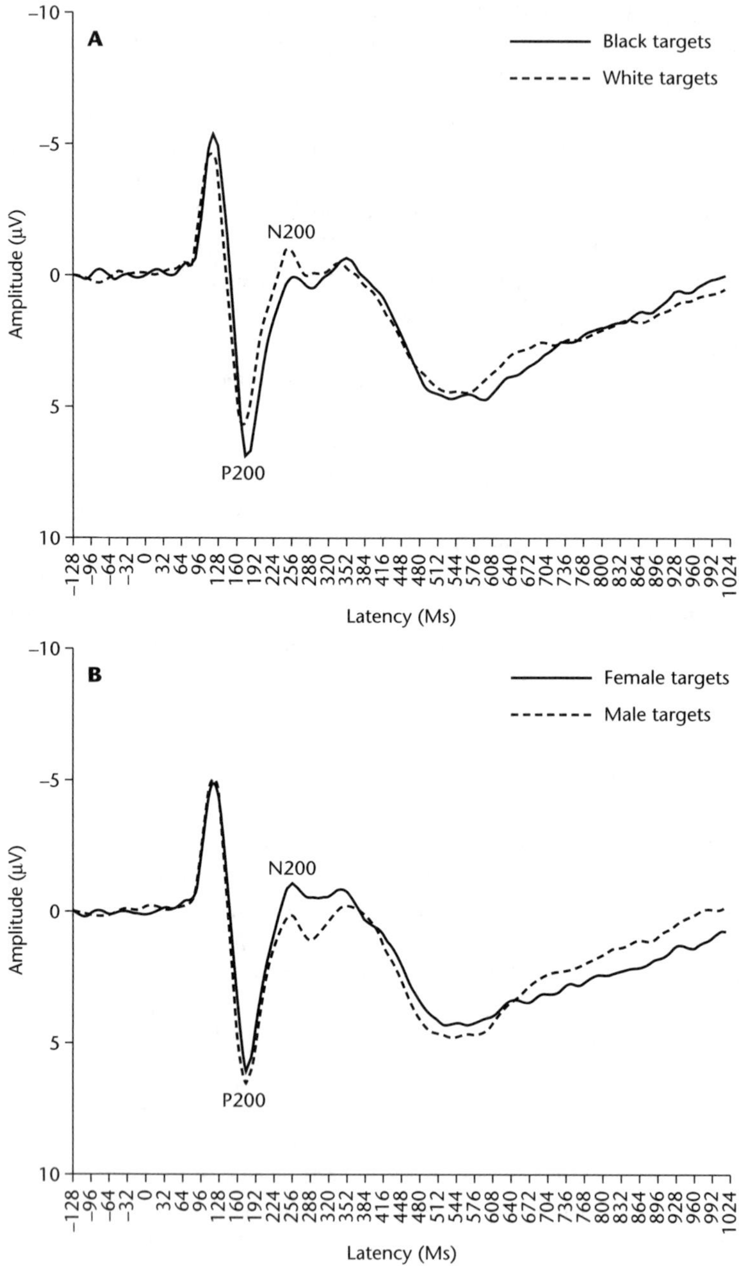

FIGURE 2.1 P200 and N200 Responses to Faces of Different Races and Genders

Note: Panel A shows responses as a function of target race, and Panel B shows responses as a function of target gender. Waveforms are from the central scalp area (electrode Cz).

Reprinted with permission from Ito & Urland (2003). Copyright American Psychological Association.

a feature detection task of looking for white dots on the stimulus. Whereas P200 and N200 responses to the primes were identical regardless of whether participants were mentally classifying them in terms of race or visually searching for a white dot, implicit stereotyping in the form of faster and more accurate responses to guns following the Black primes occurred only when participants were encoding the primes in terms of race. Said differently, social category information was initially encoded to the same degree regardless of task, but implicit stereotyping only occurred when racial information was explicitly attended.

Social Category Effects or Physical Feature Effects?

Concluding that social category information is processed across a range of tasks implies that perception is influenced by the properties of the individual being perceived more than by the goals of the perceiver, but it is important to ask what aspects of the target are driving these effects. Our assumption has been that perceivers are reacting to the social meaning of the people they are seeing, but it is also possible that the ERP responses simply reflect low-level visual differences that happen to co-vary with the categorical distinction. Take the case of race. Black and White faces probably differ on average in luminance, and these physical properties could be driving the ERP effects. It is not uninteresting to know what physical dimensions differentiate members of social groups, but responding simply to physical differences is meaningfully different from being sensitive to the social nature of the categorical distinction.

Our initial assumption was that the automatic attentional differentiation we were observing reflects the meaning perceivers ascribe to these social categories, but we knew it was important to verify this empirically. Our very first studies in this area provided some initial evidence. In Ito and Urland (2003), we conducted two studies that used the same Black and White male and female faces as stimuli, but varied whether the images were in color or grayscale. For the latter, mean luminance was equated across social categories. Race and gender were still readily discriminable among the grayscale images, but an important source of potential difference among the groups (luminance) was removed. Nevertheless, the same pattern of ERP responses was seen for the color and grayscale images.

In Ito and Urland (2003), we used the strategy of preserving the social meaning of the stimuli (e.g., participants could still readily distinguish between Blacks and Whites) while equating low-level visual properties. In a subsequent study, we employed the opposite strategy—we obscured the social meaning of the stimuli by rendering them unidentifiable as faces, yet maintained the physical differences. This was done by inverting and blurring color photos of Black and White men (Kubota & Ito, 2007). The same procedure was applied to an equal number of pictures of dog and cat faces, which served as control stimuli. Participants then performed a spatial location task, judging whether the picture was to the right or left of fixation. The morphology of the ERP waveforms recorded during this task was completely different from those recorded when stimuli are clearly identifiable as faces. We also failed to obtain any difference in responses to the Black and White faces (nor were there any differences in response between the human and animal faces).

As a third strategy, we have manipulated perceived group membership, and then looked for corresponding changes in ERP responses. We did this by creating a set of racially ambiguous Black–White faces which were then presented with either the label "Black" or "White." Behavioral responses to the faces differed as a function of the label, with more negative implicit evaluations activated by racially ambiguous faces labeled "Black" as compared to "White." The ERP responses were similarly responsive to the labels. Racially ambiguous faces labeled as Black elicited ERP responses similar to those elicited by racially prototypical Black faces. When these same faces were labeled as "White," they elicited ERP responses similar to those elicited by racially prototypical White faces (Willadsen-Jensen & Ito, 2013). That ERP responses differed as a function of the social meaning of the faces, even though they possessed the same physical features in both labeling conditions, is clear evidence that these ERP responses reflect social category effects and not just low-level featural effects.

Finally, another piece of relevant information comes from a recent study in which we examined the lateral occipital P1 ERP response to Black and White faces. The P1 is sensitive to visual properties such as luminance (e.g., Handy & Khoe, 2005; Rousselet, Pernet, Bennett, & Sekuler, 2008), so if image intensity was influencing our other ERP effects, we would expect P1 differences to Black and White faces (cf., Brebner, Krigolson, Handy, Quadflieg, & Turk, 2011). There were, instead, no P1 race effects (Senholzi & Ito, 2013).

The studies in this section present a clear and consistent picture. Neural modulations sensitive to race and gender are not simply a reflection of low-level visual properties that happen to differ among different social groups. By showing images in grayscale instead of color, we are able to retain the social meaning of the stimuli in the sense that they are readily identifiable in terms of race and gender, but we reduce low-level physical differences among the groups. In this case, ERP differences as a function of race and gender are still obtained. By blurring and inverting the faces, we rendered them unidentifiable in terms of race while retaining the low-level physical differences. In this case, no target race differences in P200 or N200 responses were observed. Finally, when we change the social meaning of a face, via changes in the categorization of racially ambiguous faces, we get a corresponding change in the ERP responses. These results together indicate to us that perceivers are responding automatically to the social meaning of the categories—for example, that this is a person who belongs to the group "Black," and not just that this is a stimulus that has less luminance than another.

In-group/Out-group or Target Group Effects

Another interesting question raised by these results is the degree to which they reflect the in-group/out-group relationship between the perceiver and target or a target group effect. In the former case, neural responses should differ depending on whether the target shares social category membership with the participant. In the latter, the effects should be invariant across participant race and gender, instead being driven by some property of the target group. The studies reviewed in the previous section demonstrate that low-level visual features of the target groups are unlikely to drive the ERP responses—suggesting that the effects are driven by the in-group/out-group relation—but it is possible that ERP responses are sensitive to other properties of the target group (e.g., the content of their stereotype), making viable the possibility of a target group effect.

We have been somewhat limited in our ability to disentangle in-group/out-group from target group effects with respect to race because of relatively little racial diversity in our sample of research participants, but results from a sample of Asian participants support an in-group/out-group interpretation (Willadsen-Jensen & Ito, 2008). In this study, Asian participants showed larger P200s to out-group Whites as compared to in-group Asian faces, reversing the typical pattern shown by White participants (cf. Willadsen-Jensen & Ito, 2006). The N200 results showed an interesting sensitivity to racial context. When the Asian and White faces were viewed embedded in a background of mostly Asian faces, N200s were larger to in-group Asian faces. That is, the pattern of both P200 and N200 responses was opposite to what White participants show, demonstrating a clear in-group/out-group effect. However, when the faces were viewed within a context of mostly White faces, the Asian participants showed the same N200 effect as White participants: larger N200s to *White* faces. A similar contextual manipulation has been used with White participants, yet yields no effect on P200 and N200 responses (Ito & Urland, 2003, 2005; Willadsen-Jensen & Ito, 2006). This greater sensitivity to context on the part of Asian participants is intriguing, perhaps indicating that greater experience operating both within predominantly White and predominantly Asian contexts confers the ability to adopt different racial perspectives depending on the situation. In particular, a predominantly White context may signal the need for greater relative attention to Whites.

Returning to the question of whether racial modulation of ERP responses reflects differential sensitivity to in-group and out-group members or a target group effect, the results support the former. Rather than showing the same pattern of responses that White participants do, Asian participants reverse their responses, with the twist that context affects their N200 responses. Another study with Black participants shows similarly larger P200s to out-group Whites and larger N200s to in-group Blacks (Dickter & Bartholow, 2007).[2]

But while race effects differ as a function of shared group membership with the target, gender effects have shown no such in-group/out-group effects in our research. In all studies, both male and female participants show larger P200s to males, but larger N200s to females (Ito & Urland, 2003, 2005; but see Dickter & Bartholow, 2007; van Hooff, Crawford, & van Vugt, 2011). It is not clear why gender shows a target group effect, but race shows sensitivity to the in-group/out-group relation. While race and gender have similar effects in some situations (e.g., both dimensions are processed automatically), there are also meaningful differences between them (e.g., intergroup contact is typically much higher for gender than race, and sex has inherent biological significance that race does not; cf. A. Smedley & B. Smedley, 2005). Such differences may contribute to the diverging effects of race and gender on attention.

Mapping ERPs to Behavior

As can be seen, numerous studies converge to demonstrate that perceivers distinguish among members of different social groups quite quickly and automatically. Part of the theoretical import of this finding is that the activation of group-based associations is presumed to depend on such categorization. A direct relation between the encoding of social category information and category-based responses such as stereotyping is, therefore, implied. Examining this relation has formed the second component of our research program.

In this research, we have focused primarily on how the race differences seen in the N200 predict behavioral manifestations of category-based responding. This was based on the particular pattern of N200 results we have obtained in our studies varying race, with larger N200s to racial in-group members. What piqued our interest was the overlap between this effect and the very robust cross-race effect in perception in which out-group faces are processed in a less configural, less individuated, and more race-specifying way than in-group faces (Hugenberg, Miller, & Claypool, 2007; Levin, 1996, 2000; Michel, Corneille, & Rossion, 2007). We wondered whether the greater attention we see to in-group faces in the N200 plays a role in this behavioral outcome. If so, we would expect the degree to which N200 responses are larger to racial in-group than out-group members to relate to the degree to which in-group members are responded to in a more individuated way, whereas out-group members elicit category-based responses.

This has been tested in three studies that all use the general strategy of simultaneously recording ERP and behavioral responses in a task where the behavioral measure reflects an aspect of category-based responding. In the first, we examined the activation of stereotypical associations in a first-person shooter simulation (Correll, Park, Judd, & Wittenbrink, 2002), measuring the speed with which participants make decisions to shoot armed and unarmed Blacks and Whites. Participants in this paradigm are typically faster to make the shoot response if the target is Black as compared to White, but faster to make the non-shoot response if the target is White as compared to Black, presumably reflecting racial stereotypes about danger and violence. A second study focused on race-based evaluative associations using a sequential priming paradigm (Fazio, Jackson, Dunton, & Williams, 1995). Participants were primed with faces of Black and White men, and then made evaluative categorizations of positive and negative words. Black primes typically facilitate responses to negative as compared to positive words. In the final study, we focused on a robust tendency for perceivers to be faster at determining the racial identity of out-group than in-group members (Levin, 1996; Stroessner, 1996), an effect thought to reflect greater salience of race-specifying information for out-group members at the expense of more individuating information. In all studies, the majority of participants were White, so White primes or targets were the racial in-group for nearly all participants.

At the mean level in all three studies, N200s were larger to in-group White than out-group Black faces, but it was the variability in this effect that interested us. Across studies, the size of these race differences predicted the degree of category-based responding. That is, larger relative N200s to White than to Black faces were associated with:

1. faster shooting of armed Blacks than Whites, but faster non-shoot responses to unarmed Whites than Blacks (Correll et al., 2006);
2. greater facilitation of negative than positive responses following Black primes (Willadsen-Jensen & Ito, 2013); and
3. faster racial categorization of Blacks versus Whites (Kubota & Ito, 2007).

This pattern of results is consistent with the possibility that differential N200 responses to in-group and out-group members reflect differences in the degree to which they are processed at a more individuated versus superficial level. Up until this point in time, though, these data have been correlational, focusing on the relation between spontaneously displayed differences in N200 responses and spontaneously displayed differences in categorical responding. The studies just reviewed also involved situations in which participants made fairly simple speeded decisions (i.e., with a time deadline and asked to be as fast as possible); in some cases, the decisions were not even about the individuals (e.g., judgments were made about words in Willadsen-Jensen & Ito, 2013). We wished to strengthen our conclusion further with an experimental manipulation implemented in a context in which participants were able to give more thoughtful consideration of individuating factors.

In a subsequent study, we first encouraged some White participants to think about an out-group Black man in a more individuated manner. This was done by having some participants first write about one Black and one White male using the first person, imagining looking at the world through the eyes of the targets and experiencing it in their shoes (Kubota, 2010). Other participants adopted a third-person perspective. We then recorded ERPs as all participants made a series of behavioral and attitudinal judgments about the two targets (e.g., "spends an hour in a coffee shop with friends," "likes to question authority"). We reasoned that prior experience looking at the world through the eyes of the targets would lead participants in the first-person perspective condition to engage in more thoughtful, individuated reasoning when making judgments about both the in-group and out-group targets. For those in the third-person condition, we expected a more typical tendency to use superficial category-based information when responding to the out-group members. Importantly, if the N200 is sensitive to these differences in processing, we would expect increased N200s to the out-group Black target only in the first-person perspective condition. This is what we found. The N200s were larger to in-group Whites than out-group Blacks among those who had previously adopted a third-person perspective; but among those who had previously taken the perspective of these individuals, N200s were equally large to the in-group and out-group members.

These results are interesting in light of the studies reviewed earlier suggesting that racial category information is encoded automatically, regardless of task. Kubota (2010) revealed a condition under which differential attention to in-group and out-group members can be eliminated. However, this elimination of racial differences was confined to the N200. In Kubota (2010), there was still a racial difference in the P200, with larger P200s to out-group Blacks, even among those who had taken a first-person perspective. Such results speak to a dissociation in the processes reflected in the P200 and N200, with the P200 potentially reflecting a more obligatory differentiation between in-group and out-group members; while the N200 reflects a difference that occurs in many situations, but that can be overcome with strong enough motivation. This conclusion must necessarily be tentative at this point as it awaits confirmation through further research.

Outstanding Questions

The preceding section demonstrates relatively recent efforts to map ERP and behavioral responses directly, implying an inverse relation between N200 amplitude and category-based responses. We have much less evidence about the meaning of our P200 responses. As noted, P200 and N200 amplitude responded differently to a perspective-taking manipulation in Kubota (2010), suggesting they are sensitive to different psychological processes. Generally speaking, P200 amplitude has been associated with attentional orientation (Crowley & Colrain, 2004). In the context of social perception, P200 amplitude has also been associated with perceptions of threat and negativity (Schutter, de Haan, & van Honk, 2004). Our larger P200s to racial out-group members might, therefore, reflect greater attention orientation to novel individuals, individuals with more negative associations, and/or individuals perceived as more threatening. This might parsimoniously be considered to reflect a vigilance response.

Another outstanding question is the relation between our social categorization effects on the N200 and other manipulations that affect the N200. In particular,

N200 amplitude has also been shown to vary as a function of response conflict, with larger N200s on high-conflict trials (e.g., larger N200s in a flanker task on trials in which the central target and flankers are incompatible; see Kopp, Rist, & Mattler, 1996; van Veen & Carter, 2002). It is also sensitive to manipulations of local stimulus probability, with larger N200s to less frequent stimuli (Bartholow et al., 2005; Nieuwenhuis, Yeung, Van Den Wildenberg, & Ridderinkhof, 2003). Dickter and Bartholow (2010) recently simultaneously manipulated all three factors—race, compatibility, and probability. Race effects still emerged, with larger N200s to White than Black faces, but the effects were also moderated by compatibility and probability. The authors interpreted the results as demonstrating that N200 amplitude is jointly affected by multiple psychological processes, but more research is needed along these lines to understand more fully how multiple psychological states can impact upon the same ERP component.

Challenges of ERP Research

The results reviewed here point to the ability of neuroscience measures to expand our understanding of social processes. I have already noted considerable benefits to using ERPs to study social categorization, including their ability to:

1. index implicit processes;
2. measure responses without participants' conscious awareness of what is being measured;
3. reflect information-processing operations on-line as they are occurring; and
4. to do so with high temporal resolution.

There are also, as might be expected, some limitations. Most obviously, specialized equipment and expertise are required. The reliance on signal averaging to reduce noise requires the presentation of many trials in each experimental condition (typically 30+). This is easily achieved in our studies of social categorization through the presentation of many target individuals, but it can be challenging to gather responses to more complicated or idiosyncratic situations that might plausibly be encountered only a few times in a single experimental context (e.g., encountering someone who behaves in a particularly stereotype-disconfirming manner). Many factors such as the expense of having multiple recording systems and the desire to minimize movement artifacts usually result in the collection of data from participants in isolation. This has made it challenging to acquire ERPs during actual social interactions, but it is certainly possible to structure the experimental context to involve social cognitive processes. For example, paradigms have been designed that invoke feelings of social evaluation, exclusion, competition, and comparison (Amodio, Kubota, Harmon-Jones, & Devine, 2006; Boksem, Kostermans, Milivojevic, & De Cremer, 2012; Crowley et al., 2009; Rigoni, Polezzi, Rumiati, Guarino, & Sartori, 2010; White et al., 2012).

It is also worth noting that measuring the brain does not necessarily confer greater inferential power. Social neuroscience is promising because it can offer new insights into the mechanisms that underlie behavior, but having access to that information does not provide fundamentally more important data than the behavior itself. Implicating a brain area or neural process in a psychological phenomenon is just the beginning. We ultimately need to know what functional role each area plays. Answering this question will only be done with close integration with more traditional behavioral methods and with a clear grounding in existing behavioral (as well as neuroscience) theory.

Unique Contributions of ERP to Understanding Social Categorization

The goal of the research reviewed here has been to leverage the features of ERP measures to inform our understanding of key aspects of impression formation. One critical assumption strongly supported by this research is the automatic encoding of racial and gender distinctions. This social category information affects attentional processes very quickly in perception, as evidenced by target race and gender modulation of the P200 component, which tends to peak within 200 ms. It is also encoded relatively automatically, as reflected by significant race and gender effects both when participants are explicitly attending to social category distinctions, as well as when they are focused on more individuating information or on non-social information. Finally, our results support the distinction between categorization and subsequent processes like stereotype activation and application. Manipulating how participants encoded Black and White face primes had no effect on the ERP responses to the faces—neural responses differentiated between racial in-group and out-group members both when participants encoded the faces in terms of race, and when they simply searched the pictures for a non-social stimulus. By contrast, the racial information activated stereotypical associations only in the race-encoding condition, as seen in a facilitation of responses to handguns following Black primes (Ito & Tomelleri, 2013). These results show that differentially encoding in-group and out-group members is not isomorphic with activating out-group stereotypes.

In addition to providing evidence about the identity and ordering of information-processing operations that contribute to person perception, this research also reveals specific information about the timing at which critical operations occur. Recall that in Kubota (2010), a first-person perspective-taking exercise produced the typically larger P200s to racial out-group members, but eliminated the larger N200s to in-group members. Such a result may indicate the way in which personalization affects perception—leaving intact initial encoding of category information, but then affecting a subsequent stage of processing.

One way to look at these results is as showing that category information "gets in" under most circumstances. That is, across most task manipulations, we find that participants differentially encode members of different racial and gender groups. This may speak to the difficulty of overcoming category-based responding, as the information on which this is based is probably typically available, but it also has implications for how we think about reducing intergroup bias. For some time, stereotyping and prejudice have been viewed as directly resulting from categorization (e.g., Tajfel, 1969, 1970). That even very arbitrary and trivial group classifications—so-called minimal group manipulations—can foster in-group bias (Brewer, 1979; Diehl, 1990; Wilder, 1986) implicates the categorization process as a fundamental component of between-group antipathies. This, in turn, drove a focus on the reduction of category salience as critical for the reduction of stereotyping and prejudice (Brewer & Miller, 1984, 1988; Gaertner, Mann, Murrell, & Dovidio, 1989). It is only more recently that researchers have returned to the possibility of relatively positive intergroup responses even in the face of strong categorical boundaries (Wolsko, Park, Judd, & Wittenbrink, 2000). Our results provide unique insights relevant to this latter view, suggesting that a complete lack of attention to social categories, at least in the case of highly practiced distinctions like race, may be very difficult to achieve. This implies a focus on intergroup bias reduction that centers on issues such as how stereotypes become activated and applied, rather than on the elimination of categorization.

Notes

1 The term "components" has various meanings in this literature, with morphological features of the waveform providing only one means of identifying components. The interested reader might consult Fabiani et al. (2000) for more details.
2 There was no contextual manipulation in this study (Dickter & Bartholow, 2007), so we cannot determine whether Black participants show the same influence of racial context on perception as the Asian participants in Willadsen-Jensen and Ito (2008).

Suggestions for Further Reading

Bartholow, B. D., & Amodio, D. M. (2009). Using event-related brain potentials in social psychological research. In E. Harmon-Jones & J. S. Beer (Eds.), *Methods in social neuroscience* (pp. 198–232). New York: Guilford Press.
Highly accessible, expanded discussion of many technical issues in ERP recording.

Fiske, S. T. (1998). Stereotyping, prejudice, and discrimination. In D. T. Gilbert, S. T. Fiske & G. Lindzey (Eds.), *The handbook of social psychology* (pp. 357–411). New York: McGraw-Hill.
Comprehensive review of the major theoretical models of stereotyping and prejudice.

Ito, T. A., & Cacioppo, J. T. (2007). Attitudes as a mental and neural state of readiness: Using physiological measures to study implicit attitudes. In B. Wittenbrink & N. Schwartz (Eds.), *Implicit measures of attitudes* (pp. 125–158). New York: Guilford Press.
Broad overview of the use of physiological measures to study implicit and explicit processes in social psychology.

Macrae, C. N., & Quadflieg, S. (2010). Perceiving people. In S. T. Fiske, D. T. Gilbert & G. Lindzey (Eds.), *Handbook of social psychology* (5th ed., pp. 428–463). New York: Wiley.
Provides an overview of the person perception literature.

References

American Encephalographic Society (1994). Guidelines for standard electrode position nomenclature. *Journal of Clinical Neurophysiology, 11*, 111–113.

Amodio, D. M., Kubota, J. T., Harmon-Jones, E., & Devine, P. G. (2006). Alternative mechanisms for regulating racial responses according to internal vs. external cues. *Social Cognitive and Affective Neuroscience, 1*, 26–36.

Bargh, J. A. (1984). Automatic and conscious processing of social information. In R. S. Wyer, Jr. & T. K. Srull (Eds.), *Handbook of social cognition* (Vol. 3, pp. 1–43). Hillsdale, NJ: Erlbaum.

Bargh, J. A. (1999). The cognitive monster. The case against the controllability of automatic stereotype effects. In S. Chaiken & Y. Trope (Eds.), *Dual-process theories in social psychology* (pp. 361–382). New York: Guilford Press.

Bartholow, B. D., Pearson, M. A., Dickter, C., Sher, K. J., Fabiani, M., & Gratton, G. (2005). Strategic control and medial frontal negativity: Beyond errors and response conflict. *Psychophysiology, 42*, 33–42.

Bodenhausen, G. V., & Macrae, C. N. (1998). Stereotype activation and inhibition. In R. S. Wyer, Jr. (Ed.), *Stereotype activation and inhibition* (pp. 1–52). Mahwah, NJ: Lawrence Erlbaum.

Boksem, M. A. S., Kostermans, E., Milivojevic, B., & De Cremer, D. (2012). Social status determines how we monitor and evaluate our performance. *Social, Cognitive and Affective Neuroscience*, 7, 304–313.

Brebner, J. L., Krigolson, O., Handy, T. C., Quadflieg, S., & Turk, D. J. (2011). The importance of skin color and facial structure in perceiving and remembering others: An electrophysiological study. *Brain Research, 1388*, 123–133.

Brewer, M. B. (1979). In-group bias in the minimal intergroup situation: A cognitive-motivational analysis. *Psychological Bulletin, 86*, 307–334.

Brewer, M. B., & Miller, N. (1984). Beyond the contact hypothesis: Theoretical perspectives on desegregation. In N. Miller & M. B. Brewer (Eds.), *Groups in contact: The psychology of desegregation* (pp. 281–302). Orlando, FL: Academic Press.

Brewer, M. B., & Miller, N. (1988). Contact and cooperation: When do they work? In P. A. Katz & D. A. Taylor (Eds.), *Eliminating racism: Profiles in controversy* (pp. 315–328). New York: Plenum Press.

Brewer, M. B. (1988). A dual process model of impression formation. In R. Wyer and T. Scrull (Eds.), *Advances in social cognition* (Vol. 1, pp. 1–36). Hillsdale, NJ: Erlbaum.

Bruner, J. S. (1957). On perceptual readiness. *Psychological Review, 64*, 123–151.

Cacioppo, J. T., & Berntson, G. G. (1992). Social psychological contributions to the decade of the brain: Doctrine of multilevel analysis. *American Psychologist, 47*, 1019–1028.

Cook, S. W., & Selltiz, C. (1964). A multiple-indicator approach to attitude measurement. *Psychological Bulletin, 62*, 36–55.

Correll, J., Park, B., Judd, C. M., & Wittenbrink, B. (2002). The police officer's dilemma: Using ethnicity to disambiguate potentially threatening individuals. *Journal of Personality and Social Psychology, 83*, 1314–1329.

Correll, J., Urland, G. R., & Ito, T. A. (2006). Event-related potentials and the decision to shoot: The role of threat perception and cognitive control. *Journal of Experimental Social Psychology, 42*, 120–128.

Crowley, K. E., & Colrain, I. M. (2004). A review of the evidence for P2 being an independent component process: Age, sleep and modality. *Clinical Neurophysiology, 115*, 732–744.

Crowley, M. J., Wu, J., McCarty, E. R., David, D. H., Bailey, C. A., & Mayes, L. C. (2009). Exclusion and micro-rejection: Event-related potential response predicts mitigated distress. *Neuroreport*, *20*, 1518–1522.

Czigler, I., & Geczy, I. (1996). Event-related potential correlates of color selection and lexical decision: Hierarchical processing or late selection? *International Journal of Psychophysiology*, *22*, 67–84.

DeMaio, T. J. (1984). Social desirability and survey measurement: A review. In C. F. Turner & E. Martin (Eds.), *Surveying subjective phenomena* (Vol. 2, pp. 257–282). New York: Russell Sage Foundation.

Devine, P. G. (1989). Stereotypes and prejudice: Their automatic and controlled components. *Journal of Personality and Social Psychology*, *56*, 5–18.

Dickter, C. L., & Bartholow, B. D. (2007). Event-related brain potential evidence of ingroup and outgroup attention biases. *Social, Cognitive, and Affective Neuroscience*, *2*, 189–198.

Dickter, C. L., & Bartholow, B. D. (2010). Ingroup categorization and response conflict: Interactive effects of target race, flanker compatibility and infrequency on N2 amplitude. *Psychophysiology*, 47, 596–601.

Diehl, M. (1990). The minimal group paradigm: Theoretical explanations and empirical findings. In W. Stroebe & M. Hewstone (Eds.), *European review of social psychology* (pp. 263–292). Chichester, UK: Wiley.

Dovidio, J. F., Evans, N., & Tyler, R. B. (1986). Racial stereotypes: The contents of their cognitive representations. *Journal of Experimental Social Psychology*, *22*, 22–37.

Eimer, M. (1997). An event-related potential (ERP) study of transient and sustained visual attention to color and form. *Biological Psychology*, *44*, 143–160.

Fabiani, M., Gratton, G., & Coles, M. G. H. (2000). Event-related brain potentials. In J. T. Cacioppo, L. G. Tassinary & G. G. Berntson (Eds.), *Handbook of psychophysiology* (2nd ed., pp. 53–84). Cambridge, UK: Cambridge University Press.

Fazio, R. H., Jackson, J. R., Dunton, B. C., & Williams, C. J. (1995). Variability in automatic activation as an unobtrusive measure of racial attitudes: A bona fide pipeline? *Journal of Personality and Social Psychology*, *69*, 1013-1027.

Fazio, R. H., Sanbonmatsu, D. M., Powell, M. C., & Kardes, F. C. (1986). On the automatic activation of attitudes. *Journal of Personality and Social Psychology*, 50, 229–238.

Fiske, S. T., & Neuberg, S. L. (1990). A continuum of impression formation, from category-based to individuating processes: Influences of information and motivation on attention and interpretation. *Advances in Experimental Social Psychology*, *23*, 1–73.

Gaertner, S. L., & McLaughlin, J. P. (1983). Racial stereotypes: Associations and ascriptions of positive and negative characteristics. *Social Psychology Quarterly*, *46*, 23–30.

Gaertner, S. L., Mann, J., Murrell, A., & Dovidio, J. F. (1989). Reducing intergroup bias: The benefits of recategorization. *Journal of Personality and Social Psychology*, *57*, 239–249.

Gehring, W. J., Gratton, G., Coles, M. G. H., & Donchin, E. (1992). Probability effects on stimulus evaluation and response processes. *Journal of Experimental Psychology: Human Perception and Performance*, *18*, 198–216.

Gilbert, D. T., & Hixon, J. G. (1991). The trouble of thinking: Activation and application of stereotypic beliefs. *Journal of Personality and Social Psychology*, 60, 509–517.

Gratton, G., Coles, M. G. H., & Donchin, E. (1983). A new method for the off-line removal of ocular artifacts. *Electroencephalography and Clinical Neurophysiology*, *55*, 468–484.

Greenwald, A. G., McGhee, D. E., & Schwartz, J. L. K. (1998). Measuring individual differences in implicit cognition: The Implicit Association Test. *Journal of Personality and Social Psychology*, *74*, 1464–1480.

Handy, T. C., & Khoe, W. (2005). Attention and sensory gain control: A peripheral visual process? *Journal of Cognitive Neuroscience, 17*, 1936–1949.

Hugenberg, K., Miller, J., & Claypool, H. (2007). Categorization and individuation in the CR recognition deficit: Toward a solution to an insidious problem. *Journal of Experimental Social Psychology, 43*, 334–340.

Ito, T. A. (2010). Implicit social cognition: Insights from social neuroscience. In B. Gawronski & K. Payne (Eds.), *Handbook of implicit social cognition: Measurement, theory, and applications* (pp. 80–92). New York: Guilford Press.

Ito, T. A., & Cacioppo, J. T. (2000). Electrophysiological evidence of implicit and explicit categorization processes. *Journal of Experimental Social Psychology, 36*, 660–676.

Ito, T. A., & Cacioppo, J. T. (2007). Attitudes as a mental and neural state of readiness: Using physiological measures to study implicit attitudes. In B. Wittenbrink & N. Schwartz (Eds.), *Implicit measures of attitudes* (pp. 125–158). New York: Guilford Press.

Ito, T. A., Larsen, J. T., Smith, N. K., & Cacioppo, J. T. (1998). Negative information weighs more heavily on the brain: The negativity bias in evaluative categorizations. *Journal of Personality and Social Psychology, 75*, 887–900.

Ito, T. A., & Tomelleri, S. (2013). Seeing is not stereotyping: Categorization and stereotype activation are functionally independent. Manuscript under review.

Ito, T. A., & Urland, G. R. (2003). Race and gender on the brain: Electrocortical measures of attention to race and gender of multiply categorizable individuals. *Journal of Personality and Social Psychology, 85*, 616–626.

Ito, T. A., & Urland, G. R. (2005). The influence of processing objectives on the perception of faces: An ERP study of race and gender perception. *Cognitive, Affective, and Behavioral Neuroscience, 5*, 21–36.

Jones, E., & Sigall, H. (1971). The bogus pipeline: A new paradigm for measuring affect and attitude. *Psychological Bulletin, 76*, 349–364.

Judd, C. M., Blair, I. V., & Chapleau, K. M. (2004). Automatic stereotypes versus automatic prejudice: Sorting out the possibilities in the Payne (2001) weapon paradigm. *Journal of Experimental Social Psychology, 40*, 75–81.

Kenemans, J. L., Kok, A., & Smulders, F. T. Y. (1993). Event-related potentials to conjunctions of spatial frequency and orientation as a function of stimulus parameters and response requirements. *Electroencephalography and Clinical Neurophysiology, 88*, 51–63.

Kopp, B., Rist, F., & Mattler, U. (1996). N200 in the flanker task as a neurobehavioral tool for investigating executive control. *Psychophysiology, 33*, 282–294.

Kubota, J. T. (2010). *The neural correlates of categorical and individuated impressions*. Doctoral dissertation, University of Colorado. Retrieved from Dissertations and Theses database. (UMI No. 3433360, at http://gradworks.umi.com/3433360.pdf).

Kubota, J. T., & Ito, T. A. (2007). Multiple cues in social perception: The time course of processing race and facial expression. *Journal of Experimental Social Psychology, 43*, 738–752.

LaPiere, R. T. (1934). Attitudes vs. actions. *Social Forces, 14*, 230–237.

Levin, D. T. (1996). Classifying faces by race: The structure of face categories. *Journal of Experimental Psychology: Learning, Memory, and Cognition, 22*, 1364–1382.

Levin, D. T. (2000). Race as a visual feature: Using visual search and perceptual discrimination tasks to understand face categories and the cross-race recognition deficit. *Journal of Experimental Psychology: General, 129*, 559–574.

Luck, S. J. (2005). *An introduction to the event-related potential technique*. Cambridge, MA: The MIT Press.

Maass, A., Ceccarelli, R., & Rudin, S. (1996). Linguistic intergroup bias: Evidence for in-group protective motivation. *Journal of Personality and Social Psychology, 71*, 512–526.

Macrae, C. N., & Bodenhausen, G. V. (2000). Social cognition: Thinking categorically about others. *Annual Review of Psychology, 51*, 93–120.

Macrae, C. N., Bodenhausen, G. V., & Milne, A. B. (1995). The dissection of selection in person perception: Inhibitory processes in social stereotyping. *Journal of Personality and Social Psychology, 69*, 397–407.

Macrae, C. N., Bodenhausen, G. V., Milne, A. B., Thorn, T. M. J., & Castelli, L. (1997). On the activation of social stereotypes: The moderating role of processing objectives. *Journal of Experimental Social Psychology, 33*, 471–489.

Marshall-Goodell, B. S., Tassinary, L. G., & Cacioppo, J. T. (1990). Principles of bioelectrical measurement. In J. T. Cacioppo and L. G. Tassinary (Eds.), *Principles of psychophysiology: Physical, social, and inferential elements* (pp. 113–148). New York: Cambridge University Press.

Michel, C., Corneille, O., & Rossion, B. (2007). Race categorization modulates holistic face encoding. *Cognitive Science, 31*, 911–924.

Naatanen, R., & Gaillard, A. W. K. (1983). The orientating reflex and the N2 deflection of the event-related potential (ERP). In A. W. K. Gaillard & W. Ritter (Eds.), *Tutorials in ERP research: Endogenous components* (pp. 119–141). New York: North-Holland Publishing Company.

Nieuwenhuis, S., Yeung, N., Van Den Wildenberg, W., & Ridderinkhof, K. R. (2003). Electrophysiological correlates of anterior cingulate function in a go/no-go task: Effects of response conflict and trial type frequency. *Cognitive, Affective, and Behavioral Neuroscience, 3*, 17–26.

Nisbett, R. E., & Wilson, T. D. (1977). Telling more than we can know: Verbal reports on mental processes. *Psychological Review, 84*, 231–259.

Perdue, C. W., & Gurtman, M. B. (1990). Evidence for the automaticity of ageism. *Journal of Experimental Social Psychology, 26*, 199–216.

Rigoni, D., Polezzi, D., Rumiati, R., Guarino, R., & Sartori, G. (2010). When people matter more than money: An ERPs study. *Brain Research Bulletin, 81*, 445–452.

Rousselet, G. A., Pernet, C. R., Bennett, P. J., & Sekuler, A. B. (2008). Parametric study of EEG sensitivity to phase noise during face processing. *BMC Neuroscience, 9*, 98–120.

Schutter, D. J. L. G., de Haan, E. H. F., & van Honk, J. (2004). Functionally dissociated aspects in anterior and posterior electrocortical processing of facial threat. *International Journal of Psychophysiology, 53*, 29–36.

Senholzi, K. B., & Ito, T. A. (2013). Structural face encoding: How task affects the N170's sensitivity to race. Manuscript under review.

Smedley, A., & Smedley, B. D. (2005). Race as biology is fiction, racism as a social problem is real: Anthropological and historical perspectives on the social construction of race. *American Psychologist, 60*, 16–26.

Stangor, C., Lynch, L., Duan, C., & Glass, B. (1992). Categorization of individuals on the basis of multiple social features. *Journal of Personality and Social Psychology, 62*, 207–218.

Stroessner, S. J. (1996). Social categorization by race or sex: Effects of perceived non-normalcy on response times. *Social Cognition, 14*, 247–276.

Tajfel, H. (1969). Cognitive aspects of prejudice. *Journal of Social Issues, 25*, 79–98.

Tajfel, H. (1970). Experiments in intergroup discrimination. *Scientific American, 223*, 96–102.

van Hooff, J. C., Crawford, H., & van Vugt, M. (2011). The wandering mind of men: ERP evidence for gender differences in attention bias towards attractive opposite sex faces. *Social Cognitive and Affective Neuroscience, 6*, 477–485.

Vanman, E. J., Paul, B.Y., Ito, T. A., & Miller, N. (1997). The modern face of prejudice and structural features that moderate the effect of cooperation on affect. *Journal of Personality and Social Psychology, 73*, 941–959.

van Veen, V., & Carter, C. S. (2002). The timing of action-monitoring processes in the anterior cingulate cortex. *Journal of Cognitive Neuroscience, 14*, 593–602.

White, L. O., Wu, J., Borelli, J. L., Rutherford, H. J. V., David, D. H., Kim-Cohen, J. . . . Crowley, M. J. (2012). Attachment dismissal predicts frontal slow-wave ERPs during rejection by unfamiliar peers. *Emotion, 12*, 690–700.

Wijers, A., Mulder, G., Okita, T., Mulder, L. J. M., & Scheffers, M. (1989). Attention to color: An analysis of selection, controlled search, and motor activation, using event-related potentials. *Psychophysiology, 26*, 89–109.

Wilder, D. A. (1986). Social categorization: Implications for creation and reduction of intergroup bias. In L. Berkowitz (Ed.), *Advances in experimental social psychology*. Orlando, FL: Academic Press.

Willadsen-Jensen, E. C., & Ito, T. A. (2006). Ambiguity and the timecourse of racial perception. *Social Cognition, 24*, 580–606.

Willadsen-Jensen, E. C., & Ito, T. A. (2008). A foot in both worlds: Asian Americans' perceptions of Asian, White, and racially ambiguous faces. *Group Processes and Interpersonal Relations, 11*, 182–200.

Willadsen-Jensen, E. C., & Ito, T. A. (2013). The effect of context on responses to racially-ambiguous faces: Changes in perception and evaluation. Manuscript under review.

Wolsko, C., Park, B., Judd, C. M., & Wittenbrink, B. (2000). The influence of ideological and normative contexts on the interethnic perceptions of White American youth. *Journal of Personality and Social Psychology, 78*, 635–654.

3

THE IMPLICIT EFFECTS OF SOCIAL IDENTITY

Measuring Early Social Categorization with Event-related Brain Potentials

Belle Derks

One of the most influential theories in the domain of intergroup behavior is Social Identity Theory (Tajfel & Turner, 1979, 1986; Turner, Hogg, Oakes, Reicher, & Wetherell, 1987). The main idea underlying this perspective is that because people base part of their identity on the groups to which they belong, they are motivated to see this social identity as having positive value. To this end, people:

1. categorize others as belonging to the in-group or out-group;
2. compare their in-group to relevant out-groups; and
3. strive to achieve positive in-group distinctiveness.

Since the early 1970s, the social identity perspective has inspired a large body of research, uncovering, for example, that social identification leads group members to favor their in-group over other groups, that status differences between groups motivate low-status groups to improve their standing and high-status groups to protect their outcomes, and illuminating how individuals cope with stigma and social identity threat (Crocker & Major, 1989; Derks, Scheepers, Van Laar, & Ellemers, 2011; Ellemers, Spears, & Doosje, 2002; Scheepers, 2009).

Inspired by the social identity perspective, in recent years, I have studied how members of groups with low status are affected by stigma and negative stereotypes, and how they cope with this social identity threat (Derks, Ellemers, Van Laar, & de Groot, 2011; Derks, Scheepers et al., 2011; Derks, Van Laar, & Ellemers, 2006, 2007, 2009; Derks, Van Laar, Ellemers, & de Groot, 2011; Van Laar, Derks, Ellemers, & Bleeker, 2010). What I came to realize over time, however, is that some of the key concepts on which social identity theory is based, like social categorization and social identity threat, have remained relatively implicit and understudied in research on social identity. I observed that social identity researchers often assume that their results (e.g., stronger in-group bias, reduced self-esteem and motivation) are a consequence of stronger social categorization or a response to the experience of social identity threat. However, rarely did I come across studies that actually addressed these processes in their own right (see also Scheepers, Chapter 13 in this volume).

In this chapter I present a new research line that was motivated by this realization. In this project, I measure social categorization—i.e., the process by which people distinguish between in-group- and out-group members—more directly by using event-related brain potentials (ERPs). This work reveals that social identity not only affects our visible behavior and the attitudes we report, but also has an impact on more 'upstream' cognitive processes that occur within the first hundreds of milliseconds in which we encounter others. This has important implications; for example, for prejudice control and the improvement of intergroup relations.

Using Psychophysiological Measures to Study Key Social Identity Concepts

I am interested in what happens with people when a group to which they belong does not contribute positively to their social identity. Interestingly, people are quite resourceful in coping with these threats to their identity. One of our research lines, for example, examined women who had been successful in male-dominated work environments such as the police force (Derks, Ellemers et al., 2011; Derks, Van Laar et al., 2011). Here we found that some women cope with this environment that threatens their social identity by showing "queen bee" responses; distancing themselves from other women and communicating highly stereotypic perceptions of their female subordinates as a way to secure their unique individual position. In another research line, we tested the effectiveness of different strategies to cope with social identity threat (i.e., self-affirmation versus group-affirmation) in preserving performance motivation and group-serving behavior (Derks et al., 2006, 2007, 2009). Although both research lines yielded new insights into how group members respond when a part of their identity is under threat, the key motivational force that we hypothesized to underlie these effects remained relatively implicit. That is, in our research endeavors, we assumed that coping responses were motivated by social identity threat, and we sometimes found evidence from self-report measures of self-esteem and affect (emotion) that participants indeed suffered from induced threats to their identity. However, sometimes we found exactly the opposite of what we predicted, suggesting that respondents might be under-reporting their experience as a way of coping with threat—see also Major and Kunstman (Chapter 17), and Mendoza-Denton and Page-Gould (Chapter 14) in this volume. Moreover, we wondered how the experience of social identity threat was affected by our explicit probing: Would people have realized their disadvantage, would they have specified their affective response if we had not asked them to reflect on this? Might this have steered their coping responses in a different direction? What we were looking for were more implicit measures of social identity threat, measures that did not rely on participants' willingness and ability to report their thoughts and feelings.

I was lucky that at this time one of my colleagues at Leiden University, Daan Scheepers, had set up a new research lab that focused on exactly this problem: assessing social identity threat more implicitly with cardiovascular measures. Daan had started measuring the cardiovascular effects of status hierarchies that threatened social identity (e.g., stable low status, unstable high status) and showed that threats to social identity can be assessed at a physiological level (inducing high blood pressure and cardiovascular threat responses; see Scheepers, 2009; Scheepers & Ellemers, 2005). When Daan and I combined cardiovascular measurements with my previous work on self-affirmation versus group-affirmation as ways of coping with social identity threat (Derks et al., 2009), we indeed found the more direct evidence for which I was looking: These coping strategies reduced social identity threat, leading to a cardiovascular response indicative of challenge (Derks, Scheepers et al., 2011). Moreover, we had been able to show these effects in an implicit way, not having to ask participants to reflect upon their situation and how it affected them, nor relying on their ability and willingness to report this. I was intrigued by the opportunities offered by psychophysiological measures, and was eager to learn more.

The opportunity to do this soon came when I was able to spend a couple of months in Michael Inzlicht's lab at the University of Toronto. Michael had just installed an electroencephalography (EEG) lab and invited me to make myself familiar with this new method and conduct a study with him. Although that first study unfortunately did not yield the results we hoped for, I learned a great deal and was "hooked." I spent time reading up on all the possibilities that EEG measures offered, and started seeing how this method could help me study a second, and even more basic, social identity concept that had remained relatively elusive for lack of direct measurements: social categorization.

Social Categorization and Intergroup Behavior

As indicated above, social identity theory proposes social categorization as the foundation on which creation of social identity and the resulting group behavior are based. In order for individuals to show group behavior, they have to self-categorize as group members rather than as individuals, and act upon their social, rather than their personal, identity (Turner et al., 1987). The first step in determining one's social identity is social categorization; establishing who is part of the in-group and who is not. As such, social categorization is often seen as "the root of all evil," leading individuals to show in-group favoritism and discriminate against members of the out-group (Bodenhausen & Macrae, 1998; but see Oakes, 2004). Not surprisingly then, many interventions hoping to reduce stereotyping or improve intergroup relations try to diminish social categorization. This is attempted, for example, by inviting people to individualize targets rather than perceiving them as interchangeable group members; by altering self-categorizations; by emphasizing people's individual identities; or by creating superordinate identities as a way to reduce social categorization in terms of subgroups (Gaertner, Mann, Murrell, & Dovidio, 2001). The general idea underlying these interventions is that by reducing social categorization, we can ameliorate relationships between groups and reduce intergroup conflict and discrimination.

Given the central role of social categorization in the social identity perspective, one would expect many studies to be directed at testing exactly how and when people categorize others. Quite surprisingly, however, most research on social categorization has examined the *consequences* of social categorization rather than the process itself. In many studies, social categorization is manipulated as an independent variable to determine how it affects subsequent processes such as intergroup discrimination (Tajfel, Billig, Bundy, & Flament, 1971), face memory (Hugenberg & Sacco, 2008), perceived intragroup interdependence (Platow, Grace, & Smithson, 2012), and intergroup trust and conflict (Dovidio & Gaertner, 2010).

Studies that are presented as focusing on social categorization as a dependent variable actually often rely on downstream processes that are seen to accompany social categorization, such as in-group favoritism, stereotype activation, or the activation of associated knowledge (Blair, Judd, & Fallman, 2004; Bodenhausen & Macrae, 1998; Fiske & Neuberg, 1990; Kawakami & Dovidio, 2001; Macrae, Bodenhausen, Milne, Thorn, & Castelli, 1997). The studies in which social categorization is examined more directly have done so with a range of measures varying from the quite explicit, such as the allocation of faces to categories (Dotsch, Wigboldus, & van Knippenberg, 2011); to more indirect and implicit measures, like the "who said what paradigm" (Taylor, Fiske, Etcoff, & Ruderman, 1978); or measures of response-time advantages in judging own-race compared to other-race faces (e.g., Ma, Yang, & Han, 2011).

As with the measures used to assess social identity threat, I felt that these measures were either too indirect or too explicit. I wondered what we would find if we could measure social categorization directly by assessing the degree to which people differentiate between targets of different social categories, without having to ask them to perform this operation for us and without them being aware that this was what we were trying to establish. This time the solution came in the form of ERPs.

Measuring Social Categorization with Event-related Brain Potentials

In my search to find ways of assessing key social identity concepts in more direct ways, I came across a line of work which measured how people process pictures of faces belonging to different social categories (e.g. Black/White; male/female) using ERPs. These are brain responses to specific events, be they stimuli that are presented to a participant (stimulus-locked ERPs) or the responses participants provide themselves (response-locked ERPs). Event-related brain potentials are distilled from continuous EEG measures of brain activity during task performance. In our case, by presenting a certain facial category many times (30–40) and averaging the time window surrounding this event over all trials, it was possible to extract the brain's response to this focal stimulus, while cancelling out random noise due to concurrent brain responses and environmental influences that are unrelated to target processing (Bartholow & Amodio, 2009; Luck, 2005).

Electroencephalography (EEG) is measured with electrodes placed over (standardized) positions on the participant's scalp that pick up the electric activity in the outer layers of the brain. These are easy to apply, and relatively non-invasive and inexpensive (compared to, for example, fMRI measurements). Important for my purposes, unlike functional magnetic resonance imaging (fMRI), EEG has a very high temporal resolution (milliseconds [ms] rather than seconds), which allows for the measurement of different stages in information processing rather than simply yielding a single end-product (e.g., activation of a specific area in the brain or a focal behavioral response). Additionally, measuring ERPs enables the assessment of implicit information processing, because it is possible to detect differential neural processing of different types of stimuli without having to ask participants to evaluate or even to respond to these stimuli. The ERP waveform that is observed after averaging a time window surrounding the event of interest (e.g., –300–1,000 ms), depicts average brain activity in microvolts and is made up of positive- and negative-going peaks (or components) that are thought to reflect different information-processing operations (e.g., categorization, evaluation, conflict detection, performance monitoring). These peaks are often named after their polarity—P (positive) or N (negative)—and the approximate latency of their occurrence (e.g., P200/P2 for the positive peak occurring around 200 ms after the event of interest).

In recent years, ERP research on the perception of in-group and out-group faces has shown evidence of social categorization already occurring in the first 300 ms of encountering these faces. This pattern of early social categorization has been found particularly in three components: the N1 (a negative component peaking around 100 ms); the P2 (a positive component peaking between 150 and 250 ms); and the N2 (a negative component peaking between 200 and 400 ms) (see Ito & Bartholow, 2009, for a review). These peaks have been associated with selective attention, with higher amplitudes indicating greater attention (Luck, 2005).

When people process faces that differ according to some social category (e.g., race, ethnicity), typically they first focus on out-group faces before turning their attention to in-group faces: The two early components N1 and P2 usually show higher amplitudes to out-group targets than to in-group targets; whereas the later N2 component is commonly larger for in-group targets than for out-group targets (Dickter & Bartholow, 2007; Ito & Bartholow, 2009; Ito & Urland, 2003, 2005;

see also Ito, Chapter 2 in this volume). In previous work, the N1 and P2 have been associated with vigilance for threatening and distinctive cues (e.g., negative facial expressions; see Bar-Haim, Lamy, & Glickman, 2005). In the context of social categorization, they have been interpreted as reflecting vigilance for out-group targets. The N2, which peaks slightly later, has been interpreted as indicating deeper attention to familiar faces like in-group faces (Ito & Bartholow, 2009). Importantly, differential amplitudes to faces of different social groups in N1, P2 and N2 components have been shown to depend on the perceiver's group membership: Whereas Whites show greater P2 amplitudes to Black rather than White faces and greater N2 amplitudes to White rather than Black faces, Black participants show exactly the opposite pattern (Dickter & Bartholow, 2007). These findings suggest that differentiation in these early components reflects broader social categorization in terms of in-group and out-group, rather than more narrow distinctions between Black and White faces (e.g., on the basis of visual figure-ground contrasts). Finally, social categorization in these ERP components occurs regardless of processing objective or task, making it a valuable measure when one wants to study social categorization without giving participants the processing objective of attending to group membership or explicitly asking them to categorize faces.

Toward a Dynamic Model of Social Identity and Social Categorization

Possibly due to the inability to assess social categorization directly and unobtrusively, up until now, social identity researchers have generally conceived of social categorization as a fixed starting point after which other social identity processes (like in-group evaluation, experiences of social identity threat, and coping responses) unfold sequentially (Tajfel & Turner, 1979). Implicitly, the assumption has been that social categorization is a "cold" cognitive perceptual process that takes place automatically once people self-categorize at a group level rather than an individual level (Turner, Oakes, Haslam, & McGarty, 1994). The psychological meaning of the group to the individual or the content of group membership within a specific context was thought to follow from this initial process and determine the *consequences* of social categorization (e.g., identity threat, collective protest), rather than indicating social categorization itself (Ellemers et al., 2002; Tajfel & Turner, 1986). However, the opportunity to assess social categorization with ERPs allows us to test a more dynamic model of the role of social categorization in social identity, in which the starting point (i.e., social categorization) is influenced by processes that can only be observed at a later stage (e.g., identity threat) as much as the other way around.

In order to test this idea, one needs to examine whether implicit social categorization is indeed a stable process that occurs automatically (which has been assumed in previous work on early social categorization in ERPs; see, for example,

Ito & Urland, 2003, 2005; Ito, Chapter 2 in this volume); or whether variables that change the meaning of social identity affect the degree to which people categorize in-group and out-group members in early social perception. In order to test this hypothesis, we performed several studies in which we tested the effects of two separate social identity variables on early social categorization; namely, social identification (an individual-level variable) and social identity threat (a contextual variable).

Does Social Identification Affect Early Social Categorization?

One of the key concepts in the social identity framework is social identification, the degree to which people feel connected and committed to their group (Ellemers, Spears, & Doosje, 1997). In addition to contextual cues that induce people to self-categorize as individuals or group members, group members tend to be more protective of their group and behave more in line with group goals the more they identify with it (Ellemers et al., 2002). Moreover, in contrast to low identifiers, on explicit measures, high identifiers are motivated to present their group as separate and different from other groups in order to establish a clearly defined social identity. Research has already demonstrated that, at an explicit level, high identifiers are more likely to perceive similarity between themselves and their in-group, to see their group as homogeneous, and to stress differences between their own group and other groups. Low identifiers, on the other hand, tend to stress the heterogeneity of the in-group and emphasize differences between themselves and other group members (Derks, Ellemers et al., 2011; Derks, Van Laar et al., 2011; Ellemers et al., 1997; Ellemers et al., 2002). High identifiers are also more protective of their group and show in-group over-exclusion: On overt measures, they take more time to determine whether someone is an in-group member and categorize more targets as out-group members than do low identifiers (Castano, Yzerbyt, Bourguignon, & Seron, 2002). In a first study, we therefore tested whether this increased motivation to differentiate the in-group from the out-group found on explicit measures also appears in the early perceptual categorization of in-group and out-group members displayed by high identifiers.

To test the prediction that stronger social identification with one's ethnic group is related to stronger implicit social categorization of members of this ethnic in-group relative to an ethnic out-group, we invited native Dutch students to our lab and asked them to indicate their ethnic identification on a self-report scale (Derks, Stedehouder, & Ito, 2013). Then, after applying the EEG equipment, we asked them to perform a sequential priming task (Fazio, Jackson, Dunton, & Williams, 1995) in which positive and negative words were primed with pictures of Dutch women (in-group targets) or Muslim women wearing a headscarf (out-group targets). Participants were asked to respond to the words (but not the pictures), enabling us to measure spontaneous rather than forced social categorization with the ERP responses to these pictures. Finally, we also measured self-reported in-group favoritism by asking participants to rate all in-group and out-group targets on likeability.

While our results revealed no differentiation on self-reported in-group favoritism, we did find the predicted effects on the ERP measures of social categorization: Extending similar findings from an American Black/White racial context to a European context with different target groups, N2 amplitude was larger in response to Dutch faces than to Muslim faces. Most importantly, however, this effect was stronger for participants with relatively high levels of ethnic identification compared to those who had indicated they were not highly identified with the Dutch. In fact, participants who reported ethnic identification at one standard deviation below the mean did not show any evidence of social categorization.

From this first study, we could thus conclude that the relationship between social categorization and social identification is more dynamic than was previously assumed: Social identification not only affects how we behave toward in-group and out-group members once this categorization has taken place (Ellemers et al., 2002), but the extent to which a specific group is important to our identity also affects the tendency to engage in social categorization itself. The more strongly we identify with our group, the more we are inclined to differentiate the people we encounter in terms of this categorization, even when we do not have the explicit goal to do so. Importantly, we were unable to detect similar differentiation patterns on the more overt evaluation measures, indicating that the effects of social identification are not always captured by self-report measures.

Does Social Identity Threat Enhance Implicit Social Categorization?

Apart from showing that individual differences in social identification affect the inclination to engage in social categorization, in a second line of work, we examined whether implicit social categorization is also affected by contextual variables that change the meaning of one's social identity. In earlier work, we had examined how group members are affected by contextual threats to their social identity like negative stereotypes or low status relative to other groups. There we focused on the negative impact of social identity threat on affective, behavioral, and cardiovascular responses (Derks, Scheepers et al., 2011; Derks et al., 2009; Derks, Van Laar et al., 2011). We observed that low and high identifiers experience threats to identity differently and, as a result, deal with these threats in different ways. The next question I wanted to examine was how early in information processing social identity threat would have its effect.

A sequential model of social identity would predict that threats to identity affect responses *after* social categorization has taken place. In this model, when social rather than personal identity is primed, group members categorize automatically, regardless of whether social identity is threatened or not. This would mean that social identity threat does not affect 'upstream' social categorization directly, but only impacts upon 'downstream' coping responses. On the other hand, a dynamic model of social identity would predict that threats to social identity shape our social perceptions, inducing group members to categorize even more when under threat. In this scenario, social categorization would not simply be a relatively involuntary operation, but would also be the outcome of motivational concerns triggered by identity threat.

To test these alternative hypotheses, we performed two ERP experiments in which we manipulated two different types of social identity threat—group-value threat (Derks, Stedehouder, & Ito, 2013) and distinctiveness threat (Derks, Domen, & Ito, 2013). In the first experiment, we invited Muslim students to our lab, made their Muslim identity salient and placed them in one of two conditions. In the high-threat condition, participants were asked to recall a situation in which other people had been negative about their Muslim identity, and to describe their subsequent thoughts and feelings. In the control condition, we tried to prevent any threat to group value by asking participants to describe a situation in which other people had been positive about their Muslim identity. Afterwards, participants again performed the sequential priming task with pictures of Dutch and Muslim women while their EEG was measured, and rated individual in-group and out-group targets in terms of likeability.

The results revealed that, even though social identity was made salient in both conditions, social categorization was indeed affected by the presence versus absence of group-value threat. This was evident in the N1 and N2 components. For Muslim participants in the control condition, the N1 was higher for Muslim (in-group) targets than for Dutch (out-group) targets. However, when Muslims were under threat, this pattern reversed completely, resulting in a higher N1 for Dutch (out-group) targets than for Muslim (in-group) targets. Additionally, social identity threat affected social categorization in the N2 component, with participants in the threat condition showing stronger differentiation in amplitude for in-group compared to out-group targets than participants in the control condition. Interestingly, in contrast to the Dutch participants in Study 1, for the Muslim participants in this study, we did find a significant effect of group-value threat on self-reported in-group favoritism: Manipulating social identity threat resulted in higher in-group favoritism compared to the control condition. Possibly, Muslim participants thought it was less socially undesirable to favor the in-group (due to its disadvantaged position in society) than did Dutch participants. However, self-reported in-group favoritism and social categorization in the ERPs were unrelated. This suggests that although both implicit and explicit forms of intergroup differentiation were enhanced following social identity threat, expressions of in-group favoritism were determined by other factors than social categorization per se.

In a final experiment (Derks, Domen, & Ito, 2013), we examined the interactive effect of social identification and social identity threat, this time manipulating 'distinctiveness threat' (Branscombe, Ellemers, Spears, & Doosje, 1999). Apart from threat to group value, group members can be threatened by informing them that their group is insufficiently distinct from other groups. In response, group members tend to emphasize differences between the in-group and out-group even more strongly ("the reactive distinctiveness hypothesis"; Jetten, Spears, & Postmes, 2004). This particular kind of social identity threat is experienced mostly by high identifiers, and has been shown to lead to attitudinal and behavioral forms of intergroup differentiation (Jetten, Spears, & Manstead, 1999; Wilson & Hugenberg, 2010). But would such a lack of group distinctiveness also lead highly identified group members to categorize more strongly even in the first hundreds of milliseconds upon perceiving in-group and out-group members? In Study 3, we tested the effects of distinctiveness threat and social identification on Dutch participants' social categorization of Dutch and Moroccan male faces.

After measuring their ethnic identification, we manipulated distinctiveness threat by presenting participants with information about the alleged (lack of) distinctiveness in opinions between native and Moroccan Dutch on several cultural issues. In the control condition, we told participants that native and Moroccan Dutch have substantially different opinions on these issues. In the distinctiveness threat condition, however, participants were told that the opinions of Moroccan Dutch about societal issues are quite similar to those endorsed by native Dutch. Following this manipulation, participants again went through the sequential priming paradigm with Dutch and Moroccan male faces, while EEG was recorded.

Results extended findings from previous behavioral studies. On the N1 component, social categorization was found, but only in one condition: Participants who had indicated themselves to be relatively highly identified with the Dutch responded to the distinctiveness threat with increased social categorization (higher amplitudes to Moroccan faces than to Dutch faces). However, the ERP findings enabled us to see that in the control condition, where group distinctiveness was affirmed, high identifiers showed no evidence of categorization. They responded in a similar way to participants who identified relatively little with their ethnic group; these did not show categorization regardless of the condition they were in. Thus, informing members of the ethnic majority that an ethnic minority group is not so different from them after all, does not alleviate social categorization, but can actually intensify it. This finding is not only of theoretical interest, but also has important practical implications.

Together, the results from these last two studies show that threats to social identity have their effect very early in information processing: Only when social identity is chronically important or when it is under threat, do group members respond by categorizing in-group and out-group members. Early social categorization, as measured with ERPs, is therefore not a universal process that is fully driven by target characteristics such as race or the presence or absence of a headscarf (Ito & Urland, 2003, 2005). Instead, the degree to which people spontaneously distinguish between in-group and out-group members is shaped by the strength of their social identity and their motivation to protect it from threats to the group's value and distinctiveness. This implies that social identity processes affect people much earlier in information processing than we were previously able to show. Moreover, social identity is more dynamic than was previously thought: Increased social categorization does not only affect downstream processes like amplified perceptions of and responses to social identity threat (as was found in previous studies). Instead, downstream perceptions and motivations also feed back into unconscious forms of social categorization to affect the degree to which people perceive their social world through the lens of their social identity.

What Have We Learned?

When I embarked on this project, my aim was to examine whether people's social identity affects their early social perception. Stimulated by previous neuroscience research which had revealed early social categorization processes in social perception with ERPs, but had characterized this process as relatively uniform, I wondered how universal early social categorization, in fact, is. To my mind, measuring social categorization with ERPs has been worthwhile because of the results' implications for social identity theory and intergroup relations more broadly, as well as their implications for cognitive neuroscience.

Implications for Social Psychology

First, this research has provided new insights into how social identity affects social perception, by revealing the dynamic nature of social identity processes. One of the great benefits of using ERPs is that no previous study has been able to measure social categorization as directly, implicitly, and early in information processing as we did in our current studies. The findings show how ingrained our social identity is in our self-definition and information processing: When a group is important to our identity, we are more likely unconsciously to perceive the world in terms of "us" versus "them." Moreover, when this social identity is threatened, either because the value or because the distinctiveness of our group is called into question, we react by perceiving this distinction even more intensely. Importantly, this happens spontaneously, outside our awareness, without making the conscious decision to respond in this way. These results offer a new perspective on social identity processes, revealing that social categorization is not simply a starting point that drives intergroup perceptions and behavior. Instead, social categorization seems to be affected by the quality of intergroup relations within a given context just as much as the other way around.

The results of our studies are informative in that they shed light on affective and behavioral results found previously in social identity research. For example, it has been found that low and highly identified group members differ in the ways in which their social identity can become threatened and how they cope with this threat (Branscombe et al., 1999; Derks et al., 2009). High identifiers respond strongly to threats to the value and distinctiveness of their group (i.e., negative group stereotypes, forced assimilation into superordinate categories) and are more likely to show collective responses (e.g., protest, helping in-group members, sticking together). Low identifiers, in contrast, seem more worried about being categorized into a group they do not want to belong to ("categorization threat"), and respond by distancing themselves from other group members and pursuing individual mobility. These results reveal how cognitive processes contribute to these effects. More specifically, our current studies show that high identifiers are more likely spontaneously to perceive the world in terms of us and them, and are even more likely to do so when they experience group-value or distinctiveness threats. Their collective responses to these forms of threat therefore may not only reflect the conscious concern they feel for their group, but possibly they also stem from high identifiers' implicit propensity to focus on the situation in intergroup rather than individual terms. Low identifiers, on the other hand, are less likely implicitly to categorize people in this way, even when their group is under threat. When low identifiers are confronted with others who categorize them as group members, this may be a source of threat. Not only because low identifiers consciously prefer to be seen as individuals, but possibly also because other people's views conflict with how they intuitively perceive the situation themselves.

In a similar vein, the results presented here help to explain why interventions aiming to ameliorate intergroup relations by focusing on individual or superordinate identities rather than subgroup identities may have counterproductive effects (Dovidio, Gaertner, & Saguy, 2007; Hornsey & Hogg, 2000). Given that high identifiers categorize in-group and out-group members intuitively, outside their awareness, and are even more likely to do so when the distinctiveness of their group is threatened, it is unlikely that interventions that invite them to focus on superordinate categorizations and downplay their group's distinctiveness will ameliorate this. Our results may help to explain why it is often more effective to focus members of different groups on superordinate identities, while at the same time acknowledging their subgroup identity (Hewstone & Brown, 1986). Such a "dual categorization" intervention would not threaten in-group distinctiveness; nor does it conflict with the implicit social categorization that group members perceive and is therefore less likely to intensify the implicit social categorization that highly identified group members tend to derive from their subgroup identity. In future research, we aim to test this hypothesis directly by examining the effects of different types of intervention (individuation, common identity, dual identity) on implicit social categorization as measured with ERPs.

Implications for Cognitive Neuroscience

In addition to yielding novel insights for social psychology, our research adds to the field of cognitive neuroscience by increasing insight into the psychological, top-down effects of social identity factors on early cognition (see also Van Bavel, Xiao, & Hackel, Chapter 6 in this volume). Most cognitive neuroscience research focuses on how people process the world, how information processing is affected by bottom-up variables (e.g., stimulus or task characteristics), and what function different kinds of brain structures and ERP components have in information processing. It is up to (social) psychologists to reveal how these processes in turn are affected by the psychological state or context that people are in. Regarding the ERP components I studied, until very recently the assumption was that early social categorization in the N1, P2, and N2 components was universal and driven by characteristics of the target (race, gender) and task (e.g., determining race versus the presence of a dot on the face; see Ito & Urland, 2003, 2005). The research done by social psychologists who started to examine these same components and relate them to contextual manipulations and individual difference variables has revealed that early social categorization is not so compulsory after all. Instead, the degree to which people categorize this early in perception is affected by individual differences such as prejudice levels (Correll, Urland, & Ito, 2006; Ito & Urland, 2005) and social identification as well as contextual factors that affect people's motivation, such as focusing them on their morality (Ellemers & Van Nunspeet, Chapter 11 in this volume) and inducing social identity threat as has been discussed in this chapter. This growing body of work adds to the conclusion that even very early cognitive processes are affected by motivation and the social context in which they occur (see also Amodio, Chapter 9 in this volume).

When Should You Include Social Neuroscience Measures in Your Research?

For researchers, learning about new research methods and including them in their toolbox can be an exciting and rewarding endeavor in itself. However, you have to keep in mind that there are also costs associated with including neuroscience methods in your research. For one, in my case, it has taken me several years to find my way in the vast field of cognitive neuroscience literature; build an EEG lab; adjust my research paradigms to procedures that would allow for measurements of the ERPs of interest; and to learn how to measure, analyze, and interpret ERP results. In this process, I was fortunate to get help in the form of research grants and advice from experts in the field. However, before embarking on such a journey, it is crucial to consider what measures are most conducive to your research program, and whether you actually need expensive neuroscience tools to answer your questions in the first place.

An increasing number of psychologists look to neuroscience for more implicit operationalizations of the hypothesized mediators of their behavioral effects, as if it is somehow more meaningful and more valuable if we can explain people's behavior with processes that occur in their body or outside their awareness. However, we have to keep in mind that neuroscience measures are not necessarily superior, but that they simply measure processes on a different level. When considering the use of neuroscience measures, one has to have a clear conception of the research question that needs to be answered and which measure is most suitable to answer this question.

One important thing to consider is that, just like other measures, neuroscience measures only tap into a small part of the cognitive and affective processes that lead to complex behavior. This is an advantage when looking for specificity, but can be a drawback if your aim is to explain complex behavior. In my case, I started measuring social categorization with ERPs because it seemed the most direct measure of social categorization that I could find, and it allowed me to assess social categorization without having to draw participants' conscious attention to social categories. Moreover, there was already quite a lot of work on social categorization in ERPs on which I could base my study designs.

However, for other research questions, it makes less immediate sense to include neuroscience measures. Behavioral responses are often determined by many different processes (e.g., how I respond to in-group members is determined by whether I categorize them as such, my evaluations of in-group members, whether the context allows me to express this bias, etc.). Neuroscience measures often only tap into one specific aspect of the information-processing stream (e.g., social categorization, target evaluation). Trying to explain behavioral results with specific neural processes is therefore not always feasible. In theory, I might want to examine why women who turned into "queen bees" on their way to the top show more social categorization in their explicit job evaluations of female and male subordinates. Although it could be interesting in itself to test whether queen bees are more likely spontaneously to categorize others according to gender, I would not predict this very specific and implicit upstream process to be the key explanation for why queen bees show gender bias in their overt self-reports. Rather, self-report measures of more downstream self-presentational concerns (e.g., the threat of being seen as a typical woman) are more likely candidates for significant mediation of the queen bee effect. In this case, ERP indicators could improve insight into the phenomenon on a different level, but should not be thought of as the primary mediator or the focal cause of the process of interest.

Sometimes social psychologists turn to neuroscience measures because they see these measures as inherently more sound and reliable. This may be true in so far as it is difficult for participants to influence their implicit or automatic physiological responses. However, when it comes to interpreting results, neuroscience results are sometimes just as open to interpretation and debate as are behavioral and self-report results. For example, although ERP components are thought to underlie separate cognitive operations, for many components, the specific interpretation is still under debate. Concerning my own research, it is clear from previous work that the N1, P2, and N2 components show evidence of social categorization in terms of differential attention to in-group and out-group targets. What is less clear, however, is how we should interpret each of these individual components (N1, P2, N2). What does it mean that we found an effect of social identification only on the N2 component, but not on the earlier components? Why did the effects of social identity threat appear somewhat earlier on the N1 component, but not in the later peaks? Although over time, as the body of research on these components grows, more will become clear about the information-processing operations that underlie each component (see also Ito & Bartholow, 2009), currently whether the different components can be interpreted beyond a simple "differential attention" interpretation is still under investigation. Some ERP components have been studied more than others (e.g., the error-related negativity, the lateralized readiness potential), so when thinking about measuring ERPs, it is useful to consider focusing on ERPs that are relatively well-studied and therefore well-understood.

Conclusion

The research presented in this chapter illustrates the potential of combining theoretical insights and research methods from social psychology and cognitive neuroscience. On the one hand, this approach benefits social psychology by expanding our ability to examine the role of more "upstream" implicit cognitive processes in social behavior. On the other hand, social neuroscience introduces a top-down perspective on how psychology affects cognition in a research field that has mainly conceptualized these processes as bottom-up. As such, social neuroscience holds great promise for our understanding of human behavior as it offers the best of both worlds.

Suggestions for Further Reading

Bartholow, B. D., & Amodio, D. M. (2009). Using event-related brain potentials in social psychological research: A brief review and tutorial. In E. Harmon-Jones & J. S. Beer (Eds.), *Methods in social neuroscience* (pp. 198–232). New York: Guilford Press.

This chapter offers an excellent and highly comprehensible introduction to the measurement of event-related potentials.

Ito, T. A., & Urland, G. R. (2003). Race and gender on the brain: Electrocortical measures of attention to the race and gender of multiply categorizable individuals. *Journal of Personality and Social Psychology*, *85*(4), 616–626.

This research paper was the first to show social categorization based on gender and race in ERPs. In two experiments, the authors reveal social categorization in the N100, P200, and N200 (N1, P2, and N2) components and show that both types of categorization occur regardless of whether participants are instructed to attend to gender or to race.

Ito, T. A., & Bartholow, B. D. (2009). The neural correlates of race. *Trends in Cognitive Sciences*, *13*, 524–531.

This article gives an overview of the neural correlates of racial categorizations, among them ERPs, and offers an interpretation of the different ERP components based on research so far.

Jetten, J., Spears, R., & Postmes, T. (2004). Intergroup distinctiveness and differentiation: A meta-analytic integration. *Journal of Personality and Social Psychology*, *86*, 862–879.

This empirical article offers a brief background into the social identity perspective and social categorization and presents a meta-analysis on the effects of threats to group distinctiveness.

References

Bar-Haim, Y., Lamy, D., & Glickman, S. (2005). Attentional bias in anxiety: A behavioral and ERP study. *Brain and Cognition, 59*, 11–22.

Bartholow, B. D., & Amodio, D. M. (2009). Using event-related brain potentials in social psychological research: A brief review and tutorial. In E. Harmon-Jones & J. S. Beer (Eds.), *Methods in social neuroscience* (pp. 198–232). New York: Guilford Press.

Blair, I. V., Judd, C. M., & Fallman, J. L. (2004). The automaticity of race and Afrocentric facial features in social judgments. *Journal of Personality and Social Psychology, 87*, 763–778.

Bodenhausen, G. V., & Macrae, C. N. (1998). Stereotype activation and inhibition. In R. S. Wyer, Jr. (Ed.), *Stereotype activation and inhibition* (pp. 1–52). Mahwah, NJ: Lawrence Erlbaum Associates Publishers.

Branscombe, N. R., Ellemers, N., Spears, R., & Doosje, B. (1999). The context and content of social identity threat. In N. Ellemers, R. Spears & B. Doosje (Eds.), *Social identity: Context, commitment, content* (pp. 35–58). Oxford, UK: Basil Blackwell.

Castano, E., Yzerbyt, V., Bourguignon, D., & Seron, E. O. (2002). Who may enter? The impact of in-group identification on in-group/out-group categorization. *Journal of Experimental Social Psychology, 38*, 315–322.

Correll, J., Urland, G. R., & Ito, T. A. (2006). Event-related potentials and the decision to shoot: The role of threat perception and cognitive control. *Journal of Experimental Social Psychology, 42*, 120–128.

Crocker, J., & Major, B. (1989). Social stigma and self-esteem: The self-protective properties of stigma. *Psychological Review, 96*, 608–630.

Derks, B., Domen, I., & Ito, T. A. (2013). We are not like them: ERP evidence for increased ingroup/outgroup categorization following group distinctiveness threat. Manuscript in preparation.

Derks, B., Ellemers, N., Van Laar, C., & de Groot, K. (2011). Do sexist organizational cultures create the queen bee? *British Journal of Social Psychology, 50*, 519–535.

Derks, B., Scheepers, D., Van Laar, C., & Ellemers, N. (2011). The threat vs. challenge of car parking for women: How self- and group affirmation affect cardiovascular responses. *Journal of Experimental Social Psychology, 47*, 178–183.

Derks, B., Stedehouder, J., & Ito, T. A. (2013). How social identity processes modify face perception: An ERP study of social categorization. Manuscript in preparation.

Derks, B., Van Laar, C., & Ellemers, N. (2006). Striving for success in outgroup settings: Effects of contextually emphasizing ingroup dimensions on stigmatized group members' social identity and performance styles. *Personality and Social Psychology Bulletin, 32*, 576–588.

Derks, B., Van Laar, C., & Ellemers, N. (2007). Social creativity strikes back: Improving motivated performance of low status group members by valuing ingroup dimensions. *European Journal of Social Psychology, 37*, 490–493.

Derks, B., Van Laar, C., & Ellemers, N. (2009). Working for the self or working for the group: How self- versus group affirmation affects collective behavior in low-status groups. *Journal of Personality and Social Psychology, 96*, 183–202.

Derks, B., Van Laar, C., Ellemers, N., & de Groot, K. (2011). Gender-bias primes elicit queen-bee responses among senior policewomen. *Psychological Science, 22*, 1243–1249.

Dickter, C. L., & Bartholow, B. D. (2007). Racial ingroup and outgroup attention biases revealed by event-related brain potentials. *Social Cognitive and Affective Neuroscience, 2*, 189–198.

Dotsch, R., Wigboldus, D. H. J., & van Knippenberg, A. (2011). Biased allocation of faces to social categories. *Journal of Personality and Social Psychology, 100*, 999–1014.

Dovidio, J. F., & Gaertner, S. L. (2010). Intergroup bias. In S. T. Fiske, D. T. Gilbert & G. Lindzey (Eds.), *The handbook of social psychology, Vol. 2* (5th ed., pp. 1084–1121). Hoboken, NJ: John Wiley & Sons Inc.

Dovidio, J. F., Gaertner, S. L., & Saguy, T. (2007). Another view of 'we': Majority and minority group perspectives on a common ingroup identity. *European Review of Social Psychology, 18*, 296–330.

Ellemers, N., Spears, R., & Doosje, B. (1997). Sticking together or falling apart: In-group identification as a psychological determinant of group commitment versus individual mobility. *Journal of Personality and Social Psychology, 72*, 617–626.

Ellemers, N., Spears, R., & Doosje, B. (2002). Self and social identity. *Annual Review of Psychology, 53*, 161–186.

Fazio, R. H., Jackson, J. R., Dunton, B. C., & Williams, C. J. (1995). Variability in automatic activation as an unobtrusive measure of racial attitudes: A bona fide pipeline? *Journal of Personality and Social Psychology, 69* , 1013–1027.

Fiske, S. T., & Neuberg, S. L. (1990). A continuum of impression formation, from category-based to individuating processes: Influences of information and motivation on attention and interpretation. In M. P. Zanna (Ed.), *Advances in experimental social psychology* (Vol. 23, pp. 1–74). New York: Academic Press.

Gaertner, S. L., Mann, J., Murrell, A., & Dovidio, J. F. (2001). Reducing intergroup bias: The benefits of recategorization. In M. A. Hogg & D. Abrams (Eds.), *Intergroup relations: Essential readings* (pp. 356–369). New York: Psychology Press.

Hewstone, M., & Brown, R. (1986). Contact is not enough: An intergroup perspective on the "contact hypothesis." In M. Hewstone & R. Brown (Eds.), *Contact and conflict in intergroup encounters* (pp. 1–44). Cambridge, MA: Basil Blackwell.

Hornsey, M. J., & Hogg, M. A. (2000). Subgroup relations: A comparison of mutual intergroup differentiation and common ingroup identity models of prejudice reduction. *Personality and Social Psychology Bulletin, 26*, 242–256.

Hugenberg, K., & Sacco, D. F. (2008). Social categorization and stereotyping: How social categorization biases person perception and face memory. *Social and Personality Psychology Compass, 2*, 1052–1072.

Ito, T. A., & Bartholow, B. D. (2009). The neural correlates of race. *Trends in Cognitive Sciences, 13*, 524–531.

Ito, T. A., & Urland, G. R. (2003). Race and gender on the brain: Electrocortical measures of attention to the race and gender of multiply categorizable individuals. *Journal of Personality and Social Psychology, 85*, 616–626.

Ito, T. A., & Urland, G. R. (2005). The influence of processing objectives on the perception of faces: An ERP study of race and gender perception. *Cognitive, Affective and Behavioral Neuroscience, 5*, 21–36.

Jetten, J., Spears, R., & Manstead, A. S. R. (1999). Group distinctiveness and intergroup discrimination. In N. Ellemers, R. Spears & B. Doosje (Eds.), *Social identity: Context, commitment, content* (pp. 107–126). Oxford, UK: Basil Blackwell.

Jetten, J., Spears, R., & Postmes, T. (2004). Intergroup distinctiveness and differentiation: A meta-analytic integration. *Journal of Personality and Social Psychology, 86*, 862–879.

Kawakami, K., & Dovidio, J. F. (2001). The reliability of implicit stereotyping. *Personality and Social Psychology Bulletin, 27*, 212–225.

Luck, S. J. (2005). *An introduction to the event-related potential technique.* Cambridge, MA: The MIT Press.

Ma, Y., Yang, S., & Han, S. (2011). Attitudes influence implicit racial face categorization in a perceptual task. *Group Processes and Intergroup Relations, 14*, 887–899.

Macrae, C. N., Bodenhausen, G. V., Milne, A. B., Thorn, T. M. J., & Castelli, L. (1997). On the activation of social stereotypes: The moderating role of processing objectives. *Journal of Experimental Social Psychology, 33*, 471–489.

Oakes, P. (2004). The root of all evil in intergroup relations? Unearthing the categorization process. In M. B. Brewer & M. Hewstone (Eds.), *Social cognition* (pp. 102–119). Malden, UK: Blackwell Publishing.

Platow, M. J., Grace, D. M., & Smithson, M. J. (2012). Examining the preconditions for psychological group membership: Perceived social interdependence as the outcome of self-categorization. *Social Psychological and Personality Science, 3*, 5–13.

Scheepers, D. (2009). Turning social identity threat into challenge: Status stability and cardiovascular reactivity during inter-group competition. *Journal of Experimental Social Psychology, 45*, 228–233.

Scheepers, D., & Ellemers, N. (2005). When the pressure is up: The assessment of social identity threat in low and high status groups. *Journal of Experimental Social Psychology, 41*, 192–200.

Tajfel, H., Billig, M. G., Bundy, R. P., & Flament, C. (1971). Social categorization and intergroup behaviour. *European Journal of Social Psychology, 1*, 149–178.

Tajfel, H., & Turner, J. C. (1979). An integrative theory of intergroup conflict. In W. G. Austin & S. Worchel (Eds.), *The social psychology of intergroup relations* (pp. 33–47). Monterey, CA: Brooks/Cole Publishing Company.

Tajfel, H., & Turner, J. C. (1986). The social identity theory of intergroup behavior. In S. Worchel & W. G. Austin (Eds.), *The psychology of intergroup relations* (pp. 7–24). Chicago: Nelson-Hall.

Taylor, S. E., Fiske, S. T., Etcoff, N. L., & Ruderman, A. J. (1978). Categorical and contextual bases of person memory and stereotyping. *Journal of Personality and Social Psychology, 36*, 778–793.

Turner, J. C., Hogg, M. A., Oakes, P. J., Reicher, S. D., & Wetherell, M. S. (1987). *Rediscovering the social group: A self-categorization theory*. New York: Basil Blackwell.

Turner, J. C., Oakes, P. J., Haslam, S. A., & McGarty, C. (1994). Self and collective: Cognition and social context. *Personality and Social Psychology Bulletin, 20*, 454–463.

Van Laar, C., Derks, B., Ellemers, N., & Bleeker, D. (2010). Valuing social identity: Consequences for motivation and performance in low status groups. *Journal of Social Issues, 66*, 602–617.

Wilson, J. P., & Hugenberg, K. (2010). When under threat, we all look the same: Distinctiveness threat induces ingroup homogeneity in face memory. *Journal of Experimental Social Psychology, 46*, 1004–1010.

4

OXYTOCINERGIC CIRCUITRY MOTIVATES INTRAGROUP COOPERATION AND INTERGROUP COMPETITION

Carsten K. W. De Dreu

Humans form long-term attachments to others with whom they are close (Mikulincer & Shaver, 2007), empathize with others (Batson, 1998), and sacrifice their immediate self-interests to promote the overarching interests of the groups and communities to which they belong (Dawes, 1980; Komorita & Parks, 1995; Ostrom, 1998; Penner, Dovidio, Piliavin, & Schroeder, 2005). Their extraordinary capacity to create and maintain social life may take rather extreme forms—to promote a common cause, or to protect their group against enemy forces, humans may sacrifice their lives; for example, as soldiers or suicide bombers (Kruglanski, Chen, Deschesne, Fishman, & Orehek, 2009; Stouffer et al., 1949; Tobena, 2009). These and related observations fit the idea, however, that throughout evolution, pro-social behavior promotes the functioning of the individual's in-group, which in turn provides for levels of security and prosperity well beyond what individuals could possibly achieve alone (Bowles & Gintis, 2011; Henrich, N. & Henrich, J., 2007).

Here I explore an intriguing implication of these observations, namely that:

1. If self-sacrifice and cooperation has survival functionality, the human brain must have evolved to sustain and motivate such pro-social tendencies.
2. We should be able to identify neurobiological systems and circuitries that modulate cooperative decision making.

This possibility occurred to me in 2008, when I was preparing a review on social conflict that, among other things, integrated work on social dilemmas and the emerging literatures on the neurobiological drivers and correlates of trust and human pro-sociality (De Dreu, 2010). With interest I read a pioneering paper by

Kosfeld, Heinrichs, Zak, Fischbacher and Fehr (2005), showing that males displayed higher levels of trust toward their protagonist when they were given a nasal dose of oxytocin, a nine-amino-acid neuropeptide that functions as both hormone and neurotransmitter and is known primarily for its reproductive functions (e.g., Carter, Grippo, Pournajafi-Nazarloo, Ruscio, & Porges, 2008). To me, this finding was particularly striking for two reasons. First, I had no idea that such complex human behavior like trust in others could have such a straightforward relationship with a neuropeptide that is produced in the hypothalamus and projected into a broad spectrum of brain areas, like the amygdala and the hippocampus, and interacts with reward-processing circuitries like the inferior frontal gyrus, the caudate nucleus, and the nucleus accumbens (Donaldson & Young, 2008; Ludwig & Leng, 2006; Skuse & Gallagher, 2008). Second, I found this finding hard to reconcile with my own research, and that of many others. In fact, throughout the years, I had come to understand human trust and cooperation as far from indiscriminate, but rather as behavioral tendencies that are quite fragile, and highly contingent upon features of one's protagonist; contextual factors like incentive structures; and individual differences like identification with one's group, one's chronic need for affiliation, and one's social value orientation (for reviews, see De Dreu, 2010; De Dreu, Weingart, & Kwon, 2000; Ellemers, De Gilder, & Haslam, 2004).

The "solution" to this apparent conundrum came when I read an article by Halevy, Bornstein, and Sagiv (2008) in which they developed their so-called Intergroup Prisoner's Dilemma–Maximizing Differences Game (IPD–MD). In this game, participants are split into two three-person groups and each individual is given a personal endowment (e.g., US$10) that they are allowed to keep. They are allowed also to invest in a within-group pool, so that each dollar invested generates 50 cents to each in-group member, the investor included. The within-group pool has the structure of a social dilemma—each in-group member is best off when others invest and he or she does not, yet all are worse off when no one invests compared to when all in-group members invest. Investments in the within-group pool reflect trust and self-costly cooperation (cf., in-group love; in-group favoritism; Brewer, 1999). In addition to the within-group pool, however, participants could also invest in a between-group pool. Each dollar invested now would yield a return of 50 cents to the in-group (as with the within-group pool), yet would also create a cost of 50 cents imposed on each out-group member. Contributing to the between-group pool could thus be motivated only by spite—the desire to hurt the out-group (cf., out-group hate, out-group derogation; Brewer, 1999). Using this game, Halevy and colleagues showed, among other things, that humans contributed much more to in-group love than to out-group hate, yet did contribute more than nothing to out-group hate. Human cooperation served the in-group, not the overarching collective of in- and out-group together (which would have required no investment in out-group hate whatsoever; Halevy et al., 2008).

The question could now be refined as follows—could it be that oxytocin, that ancient and evolutionary highly preserved molecule, served to promote the functioning of one's in-group in particular; and if so, what type of response would it generate toward more or less rivaling out-groups? Could it be that oxytocin's functions include something that is called parochial altruism—the tendency to help and promote in-group members through within-group cooperation and, if needed, out-group aggression? And if so, would this show up not only in behavioral tendencies such as investments in within-group and between-group pools in experimental games, but also in in-group favoritism toward natural in-groups and out-group derogation toward natural out-groups? Could it be that the oxytocinergic circuitry is the primary neurobiological basis of a suite of socio-cognitive and behavioral tendencies in intergroup relations so well documented in social psychological research? It was these questions that became the core focus of the O-team,[1] and the results of the ensuing research program will be reviewed here.

As an outline of the things to come, I will begin with a brief primer of oxytocin—what is it really, and how can it be measured or manipulated? Subsequently, I follow the structure of a recent review published elsewhere (De Dreu, 2012a). Specifically, the second section of this chapter examines how hypothalamic release of oxytocin modulates two critical functions underlying parochial cooperation; namely, in-group trust, and concern for (members of) the in-group. The third section reviews the research evidence for these propositions and additionally reveals that oxytocin motivates non-cooperation toward rivaling out-groups especially when out-groups threaten one's in-group. The final section summarizes the main conclusions and identifies avenues for new research into the link between oxytocin and the regulation of intra- and intergroup relations.

Primer on (Measuring and Manipulating) Oxytocin

As noted above, oxytocin is a neuropeptide that is produced in the hypothalamus and projected into a broad spectrum of brain areas, like the amygdala and the hippocampus; it interacts with reward-processing circuitries like the inferior frontal gyrus, the caudate nucleus, and the nucleus accumbens (Donaldson & Young, 2008; Ludwig & Leng, 2006; Skuse & Gallagher, 2008). To assess its effects on brain and behavior, three methods are commonly applied, and these will be reviewed briefly here.

The first method is to correlate targeted brain or behavior responses with genotyping of oxytocin receptor polymorphisms. Oxytocin receptor (OXTR) genes are localized in a single copy to chromosome 3 of the human genome. Genetic variation of the OXTR genes rs53576 and rs1042778 have been associated with variations in social behavior (e.g., Roderigues, Saslow, Garcia, John, & Keltner, 2009; also see Tost et al., 2010; Israel et al., 2009), although other studies failed to observe consistent relations between OXTR polymorphism and social decision making (e.g., Apicella et al., 2010). More work on OXTR thus is needed before we can draw conclusions about its effects on human social behavior in either interpersonal or intergroup settings.

The second method is to correlate targeted brain or behavior responses with levels of oxytocin in blood plasma. For example, Morhenn, Park, Piper, and Zak (2008) drew 20 ml of blood from healthy males both prior to and 40 minutes after they received a 15-minute massage by a professional massage therapist. Results showed that both pre- and post-massage, blood plasma oxytocin averaged around 190 pg/ml. When, in addition to the massage, participants also received a nice gesture from an unknown protagonist (in a so-called trust game, see below), post-test plasma oxytocin significantly increased to approximately 220 pg/ml. In naturalistic settings, Feldman, Weller, Zagoory-Sharon, and Levine (2007) showed that blood plasma oxytocin during pregnancy and post-partum predicted maternal bonding behaviors (e.g., gaze, vocalizations, positive affect), affectionate touch, and attachment-related thoughts. Gordon, Zagoory-Sharon, Leckman, and Feldman (2010) added to this that paternal oxytocin levels correlated with the degree of stimulatory parenting behaviors (e.g., proprioceptive contact, tactile stimulation, and object presentation); and Feldman, Gordon, Schneiderman, Weisman, and Zagoory-Sharon (2010) showed that parental and infant oxytocin levels were correlated and these related, in turn, to greater affect synchrony and infant social engagement. In short, there is mounting evidence that warm physical touch and positive social interaction with close others correlates with elevated levels of oxytocin in blood plasma.

Correlational methods have a number of drawbacks, including the fact that they prohibit conclusions about cause and effect. Furthermore, in this particular case, there is reason to assume that blood plasma oxytocin levels are a poor indicator of the level of central oxytocin present in the human brain (Campbell, 2010; Heinrichs, von Dawans, & Domes, 2009). Finally, the method is rather invasive, requiring medical personal drawing blood at specific intervals in time. It is hard to tell what the effects of such invasive methodology are on human social cognitive and behavioral functioning. For these and related reasons, it is thus fortunate that more recently it has become possible to detect oxytocin in saliva samples, and to correlate variations in saliva oxytocin with brain and/or behavior responses on experimental tasks (Weisman, Zagoory-Sharon, & Feldman, 2012). I expect this novel technique to be of great help in establishing links between fluctuating oxytocin and behavioral tendencies in both experimental and natural settings.

Correlational methods can be complemented by experimental manipulations of the level of oxytocin in the brain through nasal spray, and to assess possible activation effects on the brain or behavior. The approach is quite straightforward—in a double-blind between-subjects design, research participants receive a nasal spray containing either syntocinon (the synthetic equivalent of oxytocin) or a placebo (either the same carrier liquid as the syntocinon spray, but without the neuropeptide, or a standard saline solution). Participants self-administer 24 to 40

International Units (IU) of oxytocin (or the placebo), which comes down to three to five puffs per nostril, and proceed with irrelevant tasks until 40 minutes have passed and the induced neuropeptide has passed the blood–brain barrier (e.g., Baumgartner, Heinrichs, Vonlanthen, Fischbacher, & Fehr, 2008). They perform experimental tasks while brain activation is assessed (e.g., neuro-imaging) or behavioral responses are being registered (e.g., in economic decision-making games). These experimental sessions typically last for another 30–60 minutes. Emerging evidence shows that intranasal administration of oxytocin shows up in elevated levels of saliva oxytocin up to 90 minutes later (Weisman et al., 2012).

The experimental approach may appear rather straightforward, yet we learned a few new things that initially slowed down the project and taxed our resilience. A standard saline solution may function well as a placebo, yet has olfactory elements that render it quite distinct from the oxytocin spray. We set out to manufacture a placebo that contains exactly the same carrier liquid and ingredients as the syntocinon spray, and this required substantial negotiations with the relevant pharmaceutical company to provide us with the preparation protocols and ingredients used for producing the carrier liquid of syntocinon sprays. Ultimately these items were provided to us, which enabled us to contract a pharmacy specializing in producing medical trial medication. Again, this required passing new hurdles as, in fact, we were creating novel trial medication, and thus needed to make sure that indeed the placebo manufactured for our studies was solid and without side-effects. Ultimately, placebos were produced according to the Good Medical Practice and Good Clinical Practice quality standards, and we were ready to proceed.[2]

A next step in the design of our experiments involved a study of the methodological practices in fields other than social psychology. We desired to merge insights from neurobiology, behavioral economics, and social psychology, and tried to do so in a way that would be informative to each of these distinct audiences. Behavioral economics has developed strong views about the need for actual pay for performance; and in many of our experiments, where subjects invested money in in-group and out-group, they received actual money and earned additional money directly depending on their own performance and that of others with whom they were paired. In addition to a €15 show-up fee to compensate them for a medical screening and a two-hour experimental session, they could earn between 0 and €35. This rendered our experiments relatively expensive and taxed our research funds substantially. Relatedly, beyond social psychological science, in which mild forms of experimenter deception are common and deemed appropriate, such practice is adamantly opposed in other fields (e.g., Hertwig & Ortmann, 2001). No form of deception was used in any of the studies we conducted—what we promised was delivered. This rendered the design of the studies sometimes more difficult and seemingly complicated. It taxed our creativity as researchers, yet paid off in our being able to publicize our work beyond our own scientific discipline and to enter into communications with adjacent fields of inquiry.

Why and How Oxytocin Modulates Cooperation

A common assumption in behavioral economics, psychology, and biology is that most group settings contain incentives for their individual members to compete with others so as to defend and promote immediate self-interest (e.g., personal gains and status), and incentives to cooperate with others so as to establish well-functioning long-term relationships that provide greater benefit to all than mutual competition (e.g., Deutsch, 1973; Kelley & Thibaut, 1978; Schelling, [1960] 1980). In essence, families, small groups, communities or work organizations all share two basic properties: (i) each individual member serves personal interests best by opting for non-cooperation; and (ii) when all members opt for non-cooperation, each is worse off than when all had opted for cooperation (Bornstein, 2003; Dawes, 1980; De Dreu, Nijstad, & Van Knippenberg, 2008; Komorita & Parks, 1995). And precisely because of these properties, individuals would be expected to sacrifice immediate self-interest and to cooperate only when they have, first of all, *trust*—the positive expectation that others will reciprocate one's cooperative effort (Pruitt & Kimmel, 1977).[3] Second, they cooperate especially when they have *other-concern*—the care for others' outcomes, interests, and well-being, including a desire for fairness and to benefit the collective rather than oneself (Bohnet & Frey, 1999; Carnevale & Pruitt, 1992; De Dreu et al., 2000; Pruitt & Kimmel, 1977; Weber, Kopelman, & Messick, 2004;).

In general, people have stronger trust in and concern for others they like and feel close to, with whom they share common goals and values, with whom they anticipate future interaction, or with whom they share group membership (e.g., Batson, 1998; De Dreu, 2010; Komorita & Parks, 1995; Pruitt & Kimmel, 1977; Weber et al., 2004). Put differently, human cooperation is *parochial*—people more readily cooperate with members of their in-group than with members of more or less rivaling out-groups (Bernhard, Fischbacher, & Fehr, 2006; Choi & Bowles, 2007; Darwin, 1871; Hammond & Axelrod, 2006; Wildschut, Pinter, Vevea, Insko, & Schopler, 2003; Wit & Kerr, 2002).

Parochial altruism serves individual and group survival both in ancestral and contemporary societies (Darwin, 1871), and may be rooted in evolved neurobiological circuitries (De Dreu, 2012a; Rilling & Sanfey, 2011). One likely candidate is the oxytocinergic circuitry, because of three interrelated possibilities—it modulates:

1. social categorization of others as in-group versus not in-group;
2. enabling trust to develop; and
3. up-regulation of other-concern.

These possible functions are now briefly elaborated upon, and evaluated in light of the existing research evidence.

Oxytocin Modulates Social Categorization

Humans quickly distinguish others as in-group versus out-group (Allport, 1954; Brewer, 1999; Kurzban, Tooby, & Cosmides, 2001; Tajfel & Turner, 1979), and such social discrimination may be modulated by oxytocin. For example, when a bond is formed between sexual partners in monogamous species, or between a mother and her offspring, an olfactory memory is forged in the olfactory bulb—during initial interaction (e.g., following parturition), oxytocin released in the brain helps to establish the olfactory signatures of the protagonist as memorable (Brennan & Kendrick, 2006). Accordingly, male rodents engineered to lack (forebrain) oxytocin receptors no longer discriminate between familiar and unfamiliar females—compared to normal rodents, these knock-out rodents spent just as little time investigating female rodents with whom they had shared a cage for several days, and because they spent significantly less time investigating novel females that they met for the first time, the time spent on familiar versus unfamiliar females no longer differed among knock-out rodents (Ferguson et al., 2000; Ferguson, Young, & Insel, 2002; Macbeth, Lee, Edds, & Young, 2009).

In humans, evidence is less direct, but still suggestive. Rimmele, Hediger, Heinrichs, and Klaver (2009) gave participants intranasal oxytocin or a placebo and showed them a series of pictures of faces. One day later, participants returned to the laboratory and were given a surprise recognition task, where they had to indicate for a series of pictures whether they had seen the picture before. Participants who learned pictures of faces under oxytocin performed better one day later than those who had learned the pictures under the placebo—oxytocin makes a face in memory more familiar. Familiarity, in turn, is a key driver of social categorization (Mateo, 2004; Tang-Martinez, 2001), with familiar others being more likely to be categorized as in-group than unfamiliar others (e.g., Castelli & Zogmeister, 2000).

Humans not only use familiarity as a cue to social categorization. They also, for example, discriminate others on the basis of perceptual and cognitive cues such as physical appearance and attitude similarity, and auditory cues such as native language spoken (Dovidio & Gaertner, 2010; Yzerbyt & Demoulin, 2010). Research is needed to test the proposition that oxytocin guides social discrimination in humans, and to uncover whether and how oxytocin modulates the neural circuitries involved in classifying others into in-group versus out-group on the basis of perceptual and/or auditory cues. Such research is needed also to distinguish clearly between mere categorization, and the motivation to treat familiar/in-group others more positively than unfamiliar/out-group others.

Oxytocin Modulates Fear, Betrayal Aversion, and Trust

For trust to develop, humans need to down-regulate fear and betrayal aversion. The role of oxytocin here is rather well documented. Oxytocin's targets include the amygdala, hippocampus, and regions of the spinal cord that regulate the parasympathetic branch of the autonomic nervous system (Ludwig & Leng, 2006; Neumann, 2008; Roderigues et al., 2009). Oxytocin interacts with the hypothalamic–pituitary–adrenal axis to attenuate stress responses, and this has a pervasive influence throughout both the body and the brain (Neumann, 2008; Roderigues & Sapolsky, 2009). Specifically, oxytocin reduces cortisol levels after exposure to stressors (Heinrichs, Baumgartner, Kirschbaum, & Ehlert, 2003), and inhibits cardiovascular stress responses (Uvnas-Moberg, 1998). A pioneering neuro-imaging study by Kirsch et al. (2005) provided the most direct evidence here. These researchers gave participants either an intranasal dose of oxytocin or a placebo, and showed them fearful or neutral stimuli while brain activity was being registered. Results showed that oxytocin reduced the activation of the amygdala and attenuated its coupling to brainstem centers responsible for autonomic and behavioral components of fear.

Fearful, anxiety-provoking stimuli and situations typically motivate an immediate and automatic fight-or-flight response (LeDoux, 2000; Phelps, 2006). However, because of its anxiolytic effects at both the physiological and neurological levels, oxytocin may allow the individual to consider alternatives to fight or flight, including pro-social approach (Heinrichs et al., 2009; Lim & Young, 2006; Taylor et al., 2000). This was the key hypothesis tested in the study by Kosfeld et al. (2005) referred to earlier. These authors engaged participants in a so-called trust game. The Trust Game involves two participants, each with a US$10 endowment. Player 1 (henceforth Investor) is asked to choose an amount to be transferred to Player 2 (henceforth Trustee). As explained to both participants, the amount transferred by Investor is tripled on the way to Trustee. Trustee is then asked to choose an amount to transfer back to Investor. The back transfer is not tripled, and the final amounts in the two accounts are paid. In this game, the main reason for Investors to transfer is trust: the positive expectation that Trustee will reciprocate and transfer back. For Trustees, the main reason to transfer back is other-concern.

Kosfeld et al. (2005) observed that Investors in the oxytocin group transferred the maximum amount more often than those in the placebo group, and the average transfer was also somewhat higher among participants given oxytocin rather than a placebo. A follow-up study by Baumgartner et al. (2008) considered Investor transfers both before and after they learned about their Trustee's back transfer (set at a constant 50% back transfer in six trials of the Trust Game). Intranasal oxytocin did not influence pre-feedback transfers by Investors. However, after Investors learned that Trustees transferred back in 50% of the trials, Investors given oxytocin continued to make substantial transfers, whereas those given a placebo significantly decreased their transfers. Thus, oxytocin inoculated against betrayal aversion among Investors (see also De Dreu, 2012b; Keri & Kiss, 2011; Mikolajczak et al., 2010; Rilling et al., 2012).

In the previous sub-section, I reviewed evidence that oxytocin may be involved in the social categorization of others as familiar versus unfamiliar, and conjectured that oxytocin facilitates the classification of self and others into in-group and out-group members. Importantly, perception of in-group members links with reduced amygdala activation and betrayal aversion, and greater trust. Perceiving out-group members, in contrast, has been linked to increased activation in the amygdala, increased vigilance, and reduced trust (Beer et al., 2008; Hein, Silani, Preuschoff, Batson, & Singer, 2010; Van Bavel, Packer, & Cunningham, 2008). From this, it thus follows that hypothalamic release (or infusion) of oxytocin has effects on the amygdala and the development of trust that are also expected to happen when perceiving in-group members. The perception of out-group members triggers neural and behavioral tendencies that go against those elicited by oxytocin. Accordingly, effects of oxytocin on trust and cooperation should be stronger when the protagonist is categorized as in-group rather than out-group—oxytocin's effects on trust and cooperation are, in other words, parochial.[4]

The notion that oxytocin's effects are primarily parochial is not inconsistent with the effects on trust reviewed above, where, in fact, effects of oxytocin were stronger when the protagonist was familiar rather than a stranger, or when additional information revealed that the other party was not untrustworthy (De Dreu, 2012a). For example, Declerck, Boone, and Kiyonari (2010) studied cooperation in an Assurance Game where the only reason not to cooperate is distrust; they found that individuals given oxytocin expected more cooperation and cooperated more, but only when they had familiarized themselves with their protagonist. When individuals lacked such positive prior interaction, oxytocin actually led to *less* cooperation than with a placebo.

Oxytocin Up-regulates Empathic Other-concern

The second core pathway through which cooperation develops as a function of oxytocin runs through empathic concern and its neurobiological substrates. Concern for others is mediated by neural circuitries involved in empathy and reward processing, including the inferior frontal gyrus and the inferior parietal lobe necessary for emotional recognition and contagion, as well as the evolutionary and phylogenetically more recent ventromedial prefrontal cortex, and the temporoparietal junction and the medial temporal lobe (Baumgartner, Fischbacher, Feierabend, Lutz, & Fehr, 2009; Decety & Chaminade, 2003; Halko, Hlushchuk, Hari, & Schurmann, 2009; Keuken et al., 2011; Schnell, Bluschke, Konradt, & Walter, 2011; Shamay-Tsoory, 2011). Empathy refers to the individual's capacity to experience affective reactions to the observed or anticipated experiences of another individual, and to take another person's perspective (Batson, 1998; Frith & Singer, 2008). The neural circuitries involved in empathy are activated more when individuals are exposed to in-group rather than out-group members (Harris & Fiske, 2007; Hein et al., 2010; Van Bavel et al., 2008) and, importantly, these circuitries are up-regulated by hypothalamic release (or infusion) of oxytocin.[5] Accordingly, oxytocin promotes the development of other-concern for in-group others (and may counteract the tendency to withhold other-concern from out-group others).

Several studies support the idea that empathy and other-concern are mediated by oxytocin. In women exposed to infant crying, intranasal oxytocin modulates activity in the inferior frontal gyrus (Riem et al., 2011); fathers given oxytocin rather than a placebo are more stimulating of their toddler's exploration and show less hostility (Naber, Van IJzendoorn, Deschamps, Van Engeland, & Bakermans-Kranenburg, 2010); and participants given oxytocin rather than a placebo have increased sensitivity to others' fear (Fischer-Shofty, Shamay-Tsoory, Harari, & Levkovitz, 2010), rate images of neutral faces as more trustworthy and attractive (Theodoridou, Rowe, Pentok-Voak, & Rogers, 2009), and more accurately infer emotions expressed by others (e.g., Domes, Heinrichs, Michel, Berger, & Herpertz, 2007). In the trust game discussed in the previous sub-section, higher levels of blood plasma oxytocin correlated with increased back-transfers, reflecting increased trustworthiness and other-concern (Morhenn et al., 2008; Zak, Kurzban, & Matzner, 2005). Finally, individuals given oxytocin rather than a placebo made more generous offers in two-party ultimatum bargaining (Zak, Stanton, & Ahmadi, 2007).

Taken together, there is evidence that oxytocin activates the neural circuitries that sustain and motivate empathic other-concern, and that oxytocin drives behavioral expressions of other-concern through more fair offers in ultimatum bargaining. There is no evidence reviewed yet that oxytocin up-regulates empathic concern for in-group members more than for out-group members, something that would follow from the above discussion. The next section reviews work speaking to this possibility in particular.

Oxytocin Motivates In-group Favoritism and Parochial Cooperation

It was proposed that the effects of oxytocin on empathy and concomitant cooperation emerge especially when protagonists are categorized as in-group, and not when they are categorized as out-group. This proposition was tested in De Dreu, Greer, Van Kleef, Shalvi, and Handgraaf (2011a). Based on the well-established finding that humans positively value and evaluate in-group members (henceforth "in-group favoritism"), and sometimes negatively value and evaluate out-group members (henceforth "out-group derogation"; Dovidio & Gaertner, 2010; Yzerbyt & Demoulin, 2010), De Dreu and colleagues hypothesized that oxytocin motivates in-group favoritism and, perhaps, out-group derogation. In five experiments, indigenous Dutch males received oxytocin or a placebo in a double-blind, randomized between-subjects design and, after 40 minutes, were exposed to images of in-group targets (Dutch males) or out-group targets (males of Middle-eastern descent or, in other experiments, Germans). Using different methods, De Dreu and colleagues indeed found that oxytocin motivated stronger in-group favoritism—in-group targets elicited stronger positive associations when participants received oxytocin rather than a placebo. Weak and inconclusive effects were found for out-group derogation.

To illustrate the findings reported in De Dreu et al. (2011a), consider the following unpublished data. Healthy males received 24 IU intranasal oxytocin or a placebo in a double-blind, placebo-controlled between-subjects design and, 40 minutes later, were asked to rate the extent to which "a typical Dutch man," or "a typical German man" would be able to experience 12 possible emotions. The order in which targets were shown was controlled for, and had no effects; emotions were shown in randomized order. Emotions were either positive or negative in valence, and were either primary (typically for both humans and non-humans [e.g., fear, joy]) or secondary (typically seen as uniquely human [e.g., contempt, passion]; see Demoulin et al., 2004). Earlier work has shown a so-called infra-humanization effect, in that humans tend to ascribe secondary emotions, but not primary emotions, to their in-group more than to the rivaling out-group (Yzerbyt & Demoulin, 2010). The idea here was that oxytocin might amplify such infra-humanization, and that was precisely what was found—males given oxytocin rather than a placebo rated it more likely that Dutch rather than German males experienced secondary emotions; for primary emotions, no such effect was observed (see also Figure 4.1).

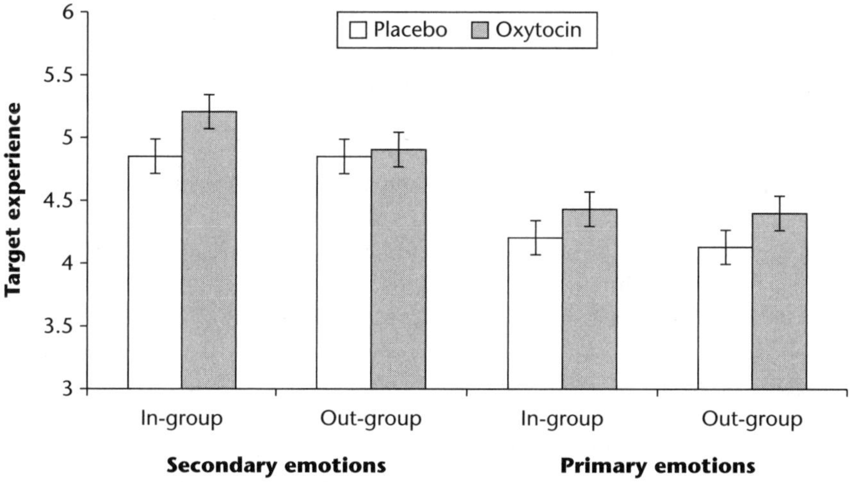

FIGURE 4.1 Oxytocin Motivates Infra-humanization in Intergroup Relations

Notes: A 2 (treatment) × 2 (target) × (emotion) analysis of variance with treatment between-subjects showed that higher ratings for secondary rather than primary emotions, $F(1,70)=250.28$, $p<0.001$, and for in-group than out-group targets, $F(1,70)=10.96$, $p<0.002$, emerged especially among individuals given oxytocin (treatment × group, $F[1,70]=4.66$, $p<0.034$). The significant three-way interaction, $F(1,70)=6.53$, $p<0.013$ showed that among secondary but not primary emotions, participants given oxytocin gave higher ratings for in-group targets (no difference emerged for out-group targets).

These results, and those reported in De Dreu et al. (2011a) suggest that oxytocin up-regulates empathy for in-group members, but does not create more benevolent views of others generally, or expanded social categories. If that had been the case, the data should have shown positive views and evaluations of in-group and out-group members alike. This was never the case (De Dreu et al., 2011a; De Dreu, Greer, Van Kleef, Shalvi, & Handgraaf, 2011b; see also Chen, Kumsta, & Heinrichs, 2011). Furthermore, these results imply that oxytocin motivates parochial cooperation with in-group members, and not with individuals belonging to out-groups. This possibility was the focus of a series of experiments reported in De Dreu et al. (2010).

Males self-administered through nasal spray 24 IU oxytocin or a placebo. Forty minutes later they were, on the basis of a trivial criterion, categorized into two three-person groups, and engaged in the Inter-group Prisoner's Dilemma–Maximizing Differences Game (IPD–MD) explained above (Halevy et al., 2008). Recall that within-group pool investments reflect the motivation to benefit in-group members at a cost to oneself; whereas between-group pool investments reflect out-group hate—the motivation to hurt the out-group at a cost to oneself (Halevy et al., 2008). Results showed three things. First, individuals who received oxytocin rather than a placebo kept less of their endowment to themselves, and invested more—they were less selfish. Second, individuals who invested did so in the within-group pool more than in the between-group pool—individuals clearly engaged in in-group love more than in out-group hate (Allport, 1954; Brewer, 1999; see also Halevy et al., 2008). Third, and finally, this tendency toward in-group love was substantially stronger among individuals given oxytocin, and no effect of oxytocin was found on out-group hate.

Together, these findings indicate that, as predicted, oxytocin motivates in-group favoritism and parochial altruism—under oxytocin, individuals self-sacrifice to benefit their in-group. Parochial altruism not only implies cooperation toward one's in-group, but often comes hand in hand with a competitive approach of rival out-groups (Choi & Bowles, 2007). After all, a competitive approach of rival out-groups indirectly serves the in-group, because it pre-empts out-group attack and diminishes or even neutralizes out-group threat (Arrow, 2007; Bacharach & Lawler, 1981; De Dreu, 1995; Jervis, 1976). Although oxytocin did not influence spiteful out-group hate, it may be that it motivates a defense-motivated competitive approach of rival out-groups. Indeed, lactating Wrister rats selectively bred for high anxiety-related behaviors (HAB) show more maternal aggression against a virgin intruder compared to Wrister rats bred for low anxiety-related behaviors (LAB) (Bosch, Meddle, Beiderbeck, Douglas, & Neumann, 2005). In this study, maternal aggression was positively correlated with oxytocin release in the paraventricular nucleus among HAB rats, and negatively correlated in LAB rats. Finally, a blockade of endogenous oxytocin action by infusion of an oxytocin receptor antagonist reduced maternal aggression among HAB rats, and increased aggression among LAB dams. Bosch et al. (2005) concluded that maternal aggression serves to protect offspring; and from this study, it thus follows that whereas oxytocin reduces defense-motivated aggression toward outsiders among low-anxiety individuals, it up-regulates aggression among high-anxiety individuals.

To test the possibility that oxytocin not only motivates in-group love, but also a defense-motivated competitive approach of rival out-groups, De Dreu et al. (2010) conducted a third experiment. In this experiment, participants played a Prisoner's Dilemma game between their own in-group and a three-person out-group—they decided between cooperation and non-cooperation, with outcomes from their decision accruing to the in-group (the decision maker included). The cardinal payoffs of the Prisoner's Dilemma game were manipulated so that a competitive approach by the out-group (rather than cooperation) had either strong or weak negative effects on in-group outcomes. Thus, to prevent the negative effects of possible competition by their out-group, individuals should choose the non-cooperative rather than the cooperative alternative themselves.

The results of De Dreu et al. (2010, Experiment 3) showed that the variation in out-group threat interacted with whether participants received intranasal oxytocin or a placebo. Under low out-group threat, treatment had no effect on non-cooperation toward the out-group; but when out-group threat was high, group members were significantly more non-cooperative toward the out-group when they had been given oxytocin rather than a placebo. From these results, De Dreu and colleagues concluded that oxytocin motivates humans to approach rivaling out-groups in a competitive manner, especially when the out-group represents a threat to the individual's in-group. In a recent follow-up, De Dreu, Shalvi et al. (2012) replicated these findings. Through further manipulations of cardinal payoffs, this study showed that oxytocin produced similar levels of competitive approach when self-interest was at stake (i.e., one feared the out-group because it threatened personal outcomes). When the out-group threatened the outcomes of one's fellow in-group members, but not one's own outcomes, males given a placebo substantially reduced their competitive approach, whereas those given oxytocin slightly increased their competitive tendencies.

From "Tend-and-Defend" to In-group Maintenance

The studies reviewed above suggest that oxytocin induces a "tend-and-defend" response to intergroup competition and conflict (De Dreu, 2012a; De Dreu et al., 2010). Such tend-and-defend tendencies have since been shown in a variety of other tasks and settings, including alliance seeking. For example, De Dreu, Greer et al. (2012) engaged participants in an intergroup competition and then asked them to select allies into their team. Participants were shown faces of potential allies that were morphed into either high-threat individuals (low on trustworthiness and high on dominance) or low-threat individuals (high on trustworthiness and low on dominance; see e.g., Oosterhof & Todorov, 2008). Males given oxytocin rather than a placebo selected more high- than low-threat allies into their team, and rated these high-threat targets as more useful. The authors concluded that in intergroup competition, oxytocin motivates humans to select allies that have high threat potential and appear aggressive rather than friendly, presumably to make their in-group a stronger and more threatening competitor to rivaling out-groups.

The studies reviewed here suggest that oxytocin plays a particularly important role in sustaining, maintaining, and improving social life within one's in-group. To date, research has focused almost exclusively on either social cognition measures of facial recognition and liking, or behavioral measures of trust and cooperation. However, if indeed oxytocin serves as a single neurohormonal mechanism underlying these and related forms of within-group cooperation, there may be many more social behaviors affected by oxytocin (Madden & Clutton-Brock, 2011). Elsewhere I proposed that oxytocin may modulate a broader variety of group maintenance behaviors, including norm compliance and enforcement, information sharing, and perhaps even group problem solving and creativity (De Dreu, 2012a). New research is needed to substantiate this claim.

Oxytocin's "Dark Side"

The studies reviewed here showed that oxytocin motivates in-group favoritism, and within-group (but not between-group) trust and cooperation. Moreover, oxytocin stimulated competitive approach toward a rival out-group when the out-group constituted an imminent threat to the in-group. Together, these studies suggest that oxytocin may, albeit indirectly, contribute to intergroup competition and conflict. First, in-group favoritism alone, or in combination with out-group derogation, creates intergroup bias: In-group members get relatively better treatment and receive benefits more readily than out-group members (Dovidio & Gaertner, 2010). Such unfair treatment triggers negative emotions, violent protest, and aggression among disfavored and excluded individuals (Hewstone, Rubin, & Willis, 2002). Accordingly, by stimulating in-group favoritism, oxytocin could trigger a chain reaction toward intense between-group conflict. Second, through its effects on within-group cooperation, oxytocin contributes to making the in-group function well and be strong, not only in absolute terms but also relative to rivaling out-groups. Perceiving such effective and relatively strong groups may trigger, among rivaling out-groups, pre-emptive strikes and aggression (Bacharach & Lawler, 1981; De Dreu, 2010; Deutsch, 1973; Jervis, 1976). Again, oxytocin's effects on within-group cooperation may, inadvertently, contribute to elevated levels of intergroup tension, competition, and conflict.

That oxytocin is involved in, and may even indirectly promote, intergroup tension and conflict needs further testing. First, in humans, oxytocin motivated intergroup competition when out-group threat was high and not when out-group threat was low (De Dreu et al., 2010). In nursing rats, oxytocin motivated aggression against intruders when rats were bred for high anxiety; among those bred for low anxiety, oxytocin actually reduced aggression against intruders (Bosch et al., 2005). Thus, there may be important individual differences and contextual factors that temper or even reverse the effects of oxytocin on intergroup competition. A recent study by Israel, Weisman, Ebstein, and Bornstein (2012) showed, for example, that when in-group and out-group interests were aligned rather than opposed, intranasal administration of oxytocin boosted contributions to both the in-group and to the overarching collective of in- and out-group together. This work thus suggests that the competitive structure of the intergroup relation may be a critical moderator of the effects of oxytocin on within-group cooperation and between-group competition. Other work indicates that effects of oxytocin on trust and cooperation are stronger in, for example, individuals scoring low rather than high on attachment security (Bartz et al., 2010; De Dreu, 2012b). It would be interesting to test the hypothesis that oxytocin moderates the impact of attachment security on social cognition and behavior in intergroup relations (for a review, see Mikulincer & Shaver, 2007).

Second, current evidence is limited to one-shot interactions, and although one would expect escalatory spirals to be set in motion by oxytocin-induced in-group favoritism and parochial cooperation, research is needed to test such unfolding of cooperation and aggression over time. It would be particularly informative to see how individuals given oxytocin versus a placebo respond to their protagonist's transgressions—failures to reciprocate cooperative effort. Would oxytocin make people more forgiving, and would such enhanced forgiveness be limited to in-group protagonists?

Third, and finally, we included in our studies healthy male subjects, in whom no side-effects of drug administration have been found. In females, exogenous oxytocin may have negative side-effects, especially in the case of (undetected) pregnancy. Our findings pertain to males only and we need to know whether and to what extent effects generalize to females. This issue is particularly pressing, as social psychological studies of intergroup relations uncovered, for example, that women identify with their in-group more strongly than men (Wilson & Liu, 2003), and that men have stronger social dominance orientation (Sidanius, Levin, Liu, & Pratto, 2000). Moreover, women show in-group favoritism, regardless of whether they are dependent on the in-group; whereas men show in-group bias when they depend on in-group members for outcomes (Gaertner & Insko, 2000). In a sense, it appears almost that when dosed with oxytocin, the males in our studies responded more like females in previous studies. Oxytocin may have similar or stronger effects in female participants, but it is also possible that women would be less affected by oxytocin as their in-group favoritism is already at a relatively high level. Future research involving and comparing male and female subjects is urgently needed to settle this issue.

Conclusion

The research reviewed here shows that hypothalamic release (or infusion) of oxytocin:

1. down-regulates fear and anxiety, and enables trust to develop, especially with protagonists who are familiar and/or categorized as in-group;
2. motivates empathic other-concern, in-group favoritism, and parochial cooperation; and
3. motivates non-cooperation toward potentially threatening out-groups.

These conclusions subscribe to the emerging insight that oxytocin-induced goodwill (empathy, trust, cooperation) is far from indiscriminate and highly contingent upon the perceived features and characteristics of goodwill's targets (Bartz et al., 2010; Chen et al., 2011; De Dreu et al., 2011a, 2011b). These conclusions also fit with the Darwinian insight that human self-sacrifice and cooperation serve in-group functioning and thereby enhance individual prosperity and survival. Accordingly, the oxytocinergic circuitry may have evolved to sustain within-group cooperation, in-group protection, and if needed, competition toward rivaling out-groups. In addition to reproduction and pair-bond formation, the oxytocinergic circuitry serves to "tend-and-defend" the in-group.

Notes

1 In these initial stages, I discussed the puzzles and theoretical solutions with Eric van Dijk, and pharmacological and medical issues with Sander Feith. Both joined what later became the O-team with Shaul Shalvi, Lindy Greer, Michel Handgraaf, Gerben Van Kleef, and myself as core members.

2 We were greatly helped in our endeavors by Novartis Netherlands which provided at no cost the preparation protocols for producing the carrier liquid, by Marleen Kemper from the Academic Medical Center, Amsterdam, who provided clinical/pharmacological expertise, and by the pharmacists from the Stichting Apotheek der Haarlemse Ziekenhuizen who manufactured the placebos.

3 Trust as defined here is inversely related to fear of being exploited by others, and betrayal aversion (Baumgartner et al., 2008; De Dreu, Shalvi et al., 2012).

4 De Dreu (2012a) discusses emerging work suggesting that in-group protection and cooperation beget cooperation, and that displays of trust and cooperation by others—especially familiar others like parents and intimate partners—promote hypothalamic release of oxytocin. This gives reason to assume a spiral of oxytocin-mediated and mutually reinforcing cooperation among the individual and his or her protagonist(s), and that such positive spirals are more likely when protagonists are categorized as in-group rather than out-group members.

5 A distinction has been made between affective and cognitive empathy (e.g., Frith & Singer, 2008), with oxytocin being involved primarily in affective empathy through its influence on the inferior frontal gyrus and the inferior frontal lobe, and indirectly in cognitive empathy through its interaction with dopaminergic circuitries (striatum, ventromedial prefrontal cortex). For a review and discussion, see Shamay-Tsoory (2011).

Suggestions for Further Reading

Bowles, S., & Gintis, H. (2011). *A cooperative species: Human reciprocity and its evolution.* Princeton, NJ: Princeton University Press.

This scholarly work introduces and integrates evolutionary, psychological, biological, and economic theories on human cooperation.

Heinrichs, M., von Dawans, B., & Domes, G. (2009). Oxytocin, vasopressin, and human social behavior. *Frontiers in Neuroendocrinology, 30*, 548–557.

This is an excellent review paper that provides a solid discussion of the methods used in research (e.g., blood plasma, intranasal administration), and discusses a wealth of research on oxytocin and its reproductive, bonding, and stress-regulatory functions.

Kirsch, P., Esslinger, C., Chen, Q., Mier, D., Lis, S., Siddhanti, S. . . . Meyer-Lindenberg, A. (2005). Oxytocin modulates neural circuitry for social cognition and fear in humans. *The Journal of Neuroscience, 25*, 11489–11493.

Pioneering study linking oxytocin to fear-signaling in the human brain.

Zak, P. J., Kurzban, R., & Matzner, W. T. (2005). Oxytocin is associated with human trustworthiness. *Hormones and Behavior, 48*, 522–527.

Prototypical study linking variations in blood plasma oxytocin to behavior in an economic trust game.

References

Allport, G. W. (1954). *The nature of prejudice.* Cambridge, MA: Addison-Wesley.

Apicella, C. L., Cesarini, D., Johannesson, M., Dawes, C. T., Lichtenstein, P., Wallace, B. . . . Westberg, L. (2010). No association between oxytocin receptor (OXTR) gene polymorphisms and experimentally elicited social preferences. *PLos ONE, 5*, e11153.

Arrow, H. (2007). The sharp end of altruism. *Science, 318*, 581–582.

Bacharach, S., & Lawler, J. (1981). *Bargaining power.* San Francisco: Jossey-Bass.

Bartz, J. A., Zaki, J., Bolger, N., Hollander, E., Ludwig, N. N., Kolevzon, A., & Ochsner, K. N. (2010). Oxytocin selectively improves empathic accuracy. *Psychological Science, 21*, 1426–1428.

Batson, C. D. (1998). Altruism and prosocial behavior. In D. Gilbert, S. Fiske & G. Lindzey (Eds.), *The handbook of social psychology* (pp. 282–316). New York: McGraw-Hill.

Baumgartner, T., Fischbacher, U., Feierabend, A., Lutz, K., & Fehr, E. (2009). The neural circuitry of a broken promise. *Neuron, 64*, 756–770.

Baumgartner, T., Heinrichs, M., Vonlanthen, A., Fischbacher, U., & Fehr, E. (2008). Oxytocin shapes the neural circuitry of trust and trust adaptation in humans. *Neuron, 58*, 639–650.

Beer, J. S., Stallen, M., Lombardo, M. V., Gonsalkorale, K., Cunningham, W. A., & Sherman, J. W. (2008). The Quadruple Process model approach to examining the neural underpinnings of prejudice. *NeuroImage, 43*, 775–783.

Bernhard, H., Fischbacher, U., & Fehr, E. (2006). Parochial altruism in humans. *Nature, 442*, 912–915.

Bohnet, I., & Frey, B. S. (1999). Social distance and other-regarding behavior in dictator games: Comment. *American Economic Review, 89*, 335–339.

Bornstein, G. (2003). Intergroup conflict: Individual, group, and collective interests. *Personality and Social Psychology Review*, 7, 129–145.

Bosch, O. J., Meddle, S. L., Beiderbeck, D. I., Douglas, A. J., & Neumann, I. D. (2005). Brain oxytocin correlates with maternal aggression: Link to anxiety. *The Journal of Neuroscience, 25*, 6807–6815.

Bowles, S., & Gintis, H. (2011). *A cooperative species: Human reciprocity and its evolution*. Princeton, NJ: Princeton University Press.

Brennan, P. A., & Kendrick, K. M. (2006). Mammalian social odours: attraction and individual recognition. *Philosophical Transactions of the Royal Society B: Biological Society, 361*, 2061–2078.

Brewer, M. B. (1999). The psychology of prejudice: Ingroup love or outgroup hate? *Journal of Social Issues, 55*, 429–444.

Campbell, A. (2010). Oxytocin and human social behaviour. *Personality and Social Psychology Review, 14*, 281–295.

Carnevale, P. J., & Pruitt, D. G. (1992). Negotiation and mediation. *Annual Review of Psychology, 43*, 531–582.

Carter, C. S., Grippo, A. J., Pournajafi-Nazarloo, H., Ruscio, M. G., & Porges, S. W. (2008). Oxytocin, vasopressin, and sociality. *Progress in Brain Research, 170*, 331–336.

Castelli, L., & Zogmeister, C. (2000). The role of familiarity in implicit memory effects: The case of examplar activation. *European Journal of Social Psychology, 30*, 223–234.

Chen, F. S., Kumsta, R., & Heinrichs, M. (2011). Oxytocin and intergroup relations: Goodwill is not a fixed pie [Letter to the Editor]. *Proceedings of the National Academy of Sciences of the United States of America, 108*, E45.

Choi, J. K., & Bowles S. (2007). The coevolution of parochial altruism and war. *Science, 318*, 5850.

Darwin, C. (1871). *The descent of man*. New York: Appleton.

Dawes, R. M. (1980). Social dilemmas. *Annual Review of Psychology, 31*, 169–193.

Decety, J., & Chaminade, T. (2003). Neural correlates of feeling sympathy. *Neuropsychology, 41*, 127–138.

Declerck, C. H., Boone, C., & Kiyonari, T. (2010). Oxytocin and cooperation under conditions of uncertainty: The modulating role of incentives and social information. *Hormones and Behavior, 57*, 368–374.

De Dreu, C. K. W. (1995). Coercive power and concession making in bilateral negotiation. *Journal of Conflict Resolution, 39*, 646–670.

De Dreu, C. K. W. (2010). Social conflict: The emergence and consequences of struggle and negotiation. In S. T. Fiske, D. T. Gilbert & L. Gartner (Eds.), *The handbook of social psychology* (Vol. 2, pp. 983–1023). New York: Wiley.

De Dreu, C. K. W. (2012a). Oxytocin modulates cooperation within and competition between groups: An integrative review and research agenda. *Hormones and Behavior, 61*, 419–428.

De Dreu, C. K. W. (2012b). Oxytocin modulates the link between adult attachment and cooperation because of betrayal aversion. *Psychoneuroendocrinology, 37*, 871–880.

De Dreu, C. K. W., Greer, L. L., Handgraaf, M. J. J., Shalvi, S., Van Kleef, G. A., Baas, M. . . . Feith, S. W. W. (2010). The neuropeptide oxytocin regulates parochial altruism in intergroup conflict among humans. *Science, 328*, 1408–1411.

De Dreu, C. K. W., Greer, L. L., Shalvi, S., Handgraaf, M. J. J., & Van Kleef, G. A. (2012). Oxytocin modulates the selection of allies in intergroup conflict. *Proceedings of the Royal Society B: Biological Sciences, 279*, 1150–1154.

De Dreu, C. K. W., Greer, L. L., Van Kleef, G. A., Shalvi, S., & Handgraaf, M. J. J. (2011a). Oxytocin promotes human ethnocentrism. *Proceedings of the National Academy of Sciences of the United States of America, 108*, 1262–1266.

De Dreu, C. K. W., Greer, L. L., Van Kleef, G. A., Shalvi, S., & Handgraaf, M. J. J. (2011b). Reply to Chen et al: Perhaps goodwill is unlimited, but oxytocin-induced goodwill is not. *Proceedings of the National Academy of Sciences of the United States of America*, *108*, E46.

De Dreu, C. K. W., Nijstad, B. A., & Van Knippenberg, D. (2008). Motivated information processing in group judgment and decision making. *Personality and Social Psychology Review*, *12*, 22–49.

De Dreu, C. K. W., Shalvi, S., Greer, L. L., Van Kleef, G. A., & Handgraaf, M. J. J. (2012). Oxytocin motivates competitive approach in intergroup relations to protect vulnerable in-group members. *PLos OUR*, 7, e46751. doi:10.1371/journal.pone.0046751

De Dreu, C. K. W., Weingart, L. R., & Kwon, S. (2000). Influence of social motives on integrative negotiation: A meta-analytic review and test of two theories. *Journal of Personality and Social Psychology*, *78*, 889–905.

Demoulin, S., Leyens, J-Ph., Paladino, M-P., Rodriquez-Torres, R., Rodrigues-Peres, A., & Dovidio, J. F. (2004). Dimensions of "uniquely" and "non-uniquely" human emotions. *Cognitive Emotion*, *18*, 71–96.

Deutsch, M. (1973). *The resolution of conflict: Constructive and destructive processes.* New Haven, CT: Yale University Press.

Domes, G., Heinrichs, M., Michel, A., Berger, C., & Herpertz, S. C. (2007). Oxytocin improves "mind-reading" in humans. *Biological Psychiatry*, *61*, 731–733.

Donaldson, Z. R., & Young, L. J. (2008). Oxytocin, vasopressin, and the neurogenetics of sociality. *Science*, *322*, 900–903.

Dovidio, J. F., & Gaertner, S. L. (2010). Intergroup bias. In S. T. Fiske, D. T. Gilbert and L. Gartner (Eds.), *The handbook of social psychology* (pp. 1084–1123). New York: Wiley.

Ellemers, N., De Gilder, D., & Haslam, S. A. (2004). Motivating individuals and groups at work: A social identity perspective on leadership and group performance. *Academy of Management Review*, *29*, 459–478.

Feldman, R., Gordon, I., Schneiderman, I., Weisman, O., & Zagoory-Sharon, O. (2010). Natural variations in maternal and paternal care are associated with systematic changes in oxytocin following parent–infant contact. *Psychoneuroendocrinology*, *35*, 1133–1141.

Feldman, R., Weller, A., Zagoory-Sharon, O., & Levine, A. (2007). Evidence for a neuroendocrinological foundation of human affiliation. *Psychological Science*, *18*, 965–970.

Ferguson, J. N., Young, L. J., Hearn, E. F., Matzuk, M. M., Insel, T. R., & Winslow, J. T. (2000). Social amnesia in mice lacking the oxytocin gene. *Nature Genetics*, *25*, 284–288.

Ferguson, J. N., Young, L. J., & Insel, T. R. (2002). The neuroendocrine basis of social recognition. *Frontiers in Neuroendocrinology*, *23*, 200–224.

Fischer-Shofty, M., Shamay-Tsoory, S. G., Harari, H., & Levkovitz, Y. (2010). The effect of intranasal administration of oxytocin on fear recognition. *Neuropsychology*, *48*, 179–184.

Frith, C. D., & Singer, T. (2008). The role of social cognition in decision making. *Philosophical Transactions of the Royal Society B: Biological Society*, *363*, 3875–3886.

Gaertner, L., & Insko, C. A. (2000). Intergroup discrimination in the minimal group paradigm: Categorization, reciprocation, or fear? *Journal of Personality and Social Psychology*, *79*, 77–94.

Gordon, I., Zagoory-Sharon, O., Leckman, J. F., & Feldman, R. (2010). Prolactin, oxytocin, and the development of paternal behavior across the first six months of fatherhood. *Hormones and Behavior*, *58*, 513–518.

Halevy, N., Bornstein, G., & Sagiv, L. (2008). "In-group love" and "out-group hate" as motives for individual participation in intergroup conflict. *Psychological Science, 19*, 405–411.

Halko, M-L., Hlushchuk, Y., Hari, R., & Schurmann, M. (2009). Competing with peers: Mentalizing-related brain activity reflects what is at stake. *Neuroimage, 46*, 542–548.

Hammond, A., & Axelrod, R. (2006). The evolution of ethnocentrism. *Journal of Conflict Resolution, 50*, 926–936.

Harris. L. T., & Fiske, S. T. (2007). Social groups that elicit disgust are differentially processed in mPFC. *Social Cognitive and Affective Neuroscience, 2*, 45–51.

Hein, G., Silani, G., Preuschoff, K., Batson, C. D., & Singer, T. (2010). Neural responses to ingroup and outgroup members' suffering predict individual differences in costly helping. *Neuron, 68*, 149–160.

Heinrichs, M., Baumgartner, T., Kirschbaum, C., & Ehlert, U. (2003). Social support and oxytocin interact to suppress cortisol and subjective responses to psychological stress. *Biological Psychiatry, 54*, 1389–1398.

Heinrichs, M., von Dawans, B., & Domes, G. (2009). Oxytocin, vasopressin, and human social behavior. *Frontiers in Neuroendocrinology, 30*, 548–557.

Henrich, N., & Henrich, J. (2007). *Why humans cooperate*. Oxford, UK: Oxford University Press.

Hertwig, R., & Ortmann, A. (2001). Experimental practices in economics: A methodological challenge for psychologists? *Behavioral and Brain Sciences, 24*, 383–412.

Hewstone, M., Rubin, M., & Willis, H. (2002). Intergroup bias. *Annual Review of Psychology, 53*, 575–604.

Israel, S., Lerer, E., Shalev, I., Uzefovsky, F., Riebold, M., Laiba, E. . . . Ebstein, R. P. (2009). The oxytocin receptor (OXTR) contributes to prosocial fund allocations in the Dictator Game and the Social Value Orientations Task. *PLos ONE, 4*, e5535.

Israel, S., Weisman, O., Ebstein, R. P., & Bornstein, G. (2012). Oxytocin, but not vasopressin, increases both parochial and universal altruism. *Psychoneuroendocrinology, 37*, 1341–1344.

Jervis, R. (1976). *Perception and misperception in international relations*. Princeton, NJ: Princeton University Press.

Kelley, H. H., & Thibaut, J. W. (1978). *Interpersonal relations: An interdependency theory approach*. New York: Academic Press.

Keri, S., & Kiss, I. (2011). Oxytocin response in a trust game and habituation of arousal. *Physiology and Behavior, 102*, 221–224.

Keuken, M. C., Hardie, A., Dorn, B. T., Dev, S., Paulus, M. P., Jonas, K. J. . . . Pineda, J. A. (2011). The role of the left inferior frontal gyrus in social perception: An rTMS study. *Brain Research, 1383*, 196–205.

Kirsch, P., Esslinger, C., Chen, Q., Mier, D., Lis, S., Siddhanti, S. . . . Meyer-Lindenberg, A. (2005). Oxytocin modulates neural circuitry for social cognition and fear in humans. *The Journal of Neuroscience, 25*, 11489–11493.

Komorita, S. S., & Parks, C. D. (1995). Interpersonal relations: Mixed-motive interaction. *Annual Review of Psychology, 46*, 183–207.

Kosfeld, M., Heinrichs, M., Zak, P. J., Fischbacher, U., & Fehr, E. (2005). Oxytocin increases trust in humans. *Nature, 435*, 673–676.

Kruglanski, A. W., Chen, X., Deschesne, M., Fishman, S., & Orehek, E. (2009). Fully committed: Suicide bombers' motivation and the quest for personal significance. *Political Psychology, 30*, 331–357.

Kurzban, R., Tooby, J., & Cosmides, L. (2001). Can race be erased? Coalitional computation and categorization. *Proceedings of the National Academy of Sciences of the United States of America, 98*, 15387–15392.

LeDoux, J. E. (2000). Emotion circuits in the brain. *Annual Review of Neuroscience, 23*, 155–184.

Lim, M. M., & Young, L. J. (2006). Neuropeptic regulation of affiliative behavior and social bonding in animals. *Hormones and Behavior, 50*, 506–517.

Ludwig, M., & Leng, G. (2006). Dendritic peptide release and peptide-dependent behaviors. *National Review of Neuroscience*, 7, 126–136.

Macbeth, A. H., Lee, H. J., Edds, J., & Young, W. S. III (2009). Oxytocin and the ocytocin receptor underlie intrastrain, but not interstrain, social cognition. *Genes, Brain and Behavior, 8*, 558–567.

Madden, J. R., & Clutton-Brock, T. H. (2011). Experimental peripheral administration of oxytocin elevates a suite of cooperative behaviours in a wild social animal. *Proceedings of the Royal Society B: Biological Sciences, 278*, 1189–1194.

Mateo, J. M. (2004). Recognition systems and biological organization: The perception component of social recognition. *Annales Zoologici Fennici, 41*, 729–745.

Mikolajczak, M., Gross, J. J., Lane, A., Corneille, O., de Timary, Ph., & Luminet, O. (2010). Oxytocin makes people trusting, not gullible. *Psychological Science, 21*, 1072–1075.

Mikulincer, M., & Shaver, Ph. R. (2007). Boosting attachment security to promote mental health, prosocial values, and inter-group tolerance. *Personality and Social Psychology Review, 18*, 139–156.

Morhenn, V. B., Park, J. W., Piper, E., & Zak, P. J. (2008). Monetary sacrifice among strangers is mediated by endogenous oxytocin release after physical contact. *Evolution and Human Behavior, 29*, 375–383.

Naber, F., Van IJzendoorn, M. H., Deschamps, P., Van Engeland, H., & Bakermans-Kranenburg, M. J. (2010). Intranasal oxytocin increases fathers' observed responsiveness during play with their children: A double-blind within-subject experiment. *Psychoneuroendocrinology, 35*, 1583–1586.

Neumann, I. D. (2008). Brain oxytocin: A key regulator of emotional and social behaviours in both females and males. *Journal of Neuroendocrinology, 20*, 858–865.

Oosterhof, N. N., & Todorov, A. (2008). The functional basis of face evaluation. *Proceedings of the National Academy of Sciences, 105*, 11087–11092.

Ostrom, E. (1998). A behavioural approach to the rational choice theory in collective action. *American Political Science Review, 92*, 1–22.

Penner, L. A., Dovidio, J. F., Piliavin, J. A., & Schroeder, D. A. (2005). Prosocial behavior: Multilevel perspectives. *Annual Review of Psychology, 56*, 365–392.

Phelps, E. A. (2006). Emotion and cognition: Insights from studies of the human amygdala. *Annual Review of Psychology, 57*, 27–53.

Pruitt, D. G., & Kimmel, M. J. (1977). Twenty years of experimental gaming: Critique, synthesis, and suggestions for the future. *Annual Review of Psychology, 28*, 363–392.

Riem, M. M. E., Bakermans-Kranenburg, M. J., Pieper, S., Tops, M., Boksem, M. A. S., Vermeiren, R. R. J. M. . . . Rombouts, S. A. R. B. (2011). Oxytocin modulates amygdala, insula, and inferior frontal gyrus responses to infant crying: A randomized controlled trial. *Biological Psychiatry, 70*, 291–297.

Rilling, J. K., DeMarco, A. C., Hackett, P. D., Thompson, R., Ditzen, B., Patel, R., & Pagnoni, G. (2012). Effects of intranasal oxytocin and vasopressin on cooperative behavior and associated brain activity in men. *Psychoneuroendocrinology, 37*, 447–461.

Rilling, J. K., & Sanfey, A. G. (2011). The neuroscience of social decision making. *Annual Review of Psychology, 62*, 23–48.

Rimmele, U., Hediger, K., Heinrichs, M., & Klaver, P. (2009). Oxytocin makes a face in memory more familiar. *The Journal of Neuroscience, 29*, 38–42.

Roderigues, S. M., & Sapolsky, R. M. (2009). Disruption of fear memory through dual-hormone gene therapy. *Biological Psychiatry, 65*, 441–444.

Roderigues, S. M., Saslow, L. R., Garcia, N., John, O. P., & Keltner, D. (2009). Oxytocin receptor genetic variation relates to empathy and stress reactivity in humans. *Proceedings of the National Academy of Sciences of the United States of America, 106*, 21437–21441.

Schelling, T. ([1960] 1980). *The strategy of conflict.* Cambridge, MA: Harvard University Press.

Schnell, K., Bluschke, S., Konradt, B., & Walter, H. (2011). Functional relations of empathy and mentalizing: An fMRI study on the neural basis of cognitive empathy. *Neuroimage, 54*, 1743–1754.

Shamay-Tsoory, S. (2011). The neural bases for empathy. *Neuroscience Update, 17*, 18–24.

Sidanius, J., Levin, S., Liu, J., & Pratto, F. (2000). Social dominance orientation, anti-egalitarianism and the political psychology of gender: An extension and cross-cultural replication. *European Journal of Social Psychology, 30*, 41–67.

Skuse, D. H., & Gallagher, L. (2008). Dopaminergic-neuropeptide interactions in the social brain. *Trends in Cognitive Sciences, 13*, 27–35.

Stouffer, S. A., Lumsdaine, A. A., Lumsdaine, M. H., Williams, R. M., Smith, M. B., & Janis, I. L. (Eds.). (1949). *The American soldier: Combat and its aftermath. Vol. 2. Studies in social psychology.* Princeton, NJ: Princeton University Press.

Tajfel, H., & Turner, J. C. (1979). An integrative theory of intergroup conflict. In W. Austin & S. Worchel (Eds.), *The social psychology of intergroup relations* (pp. 33–48). Pacific Grove, CA: Brooks/Cole.

Tang-Martinez, Z. (2001). The mechanism of kin discrimination and the evolution of kin recognition in vertebrates. *Behavioural Processes, 53*, 21–40.

Taylor, S. E., Klein, L. C., Lewis, B. P., Gruenewald, T. L., Gurung, R. A. R., & Updegraff, J. A. (2000). Biobehavioral responses to stress in females: Tend-and-befriend, not fight-or-flight. *Psychological Review, 107*, 411–429.

Theodoridou, A., Rowe, A. C., Pentok-Voak, I. S., & Rogers, P. J. (2009). Oxytocin and social perception: Oxytocin increases perceived facial trustworthiness and attractiveness. *Hormones and Behavior, 56*, 128–132.

Tobena, A. (2009). Lethal altruists: Itineraries along the dark outskirts of moralistic prosociality. *Annals of the New York Academy of Sciences, 1167*, 5–15.

Tost, H., Kolachana, B., Hakimi, S., Lemaitre, H., Verchininski, B. A., Mattay, V. S. . . . Meyer-Lindenberg, A. (2010). A common allele in the oxytocin receptor gene (OXTR) impacts prosocial temperament and human hypothalamic-limbic structure and function. *Proceedings of the National Academy of Sciences of the United States of America, 107*, 13936–13941.

Uvnas-Moberg, K. (1998). Oxytocin may mediate the benefits of positive social interaction and emotions. *Psychoneuroendocrinology, 23*, 819–835.

Van Bavel, J. J., Packer, D. J., & Cunningham, W. A. (2008). The neural substrates of in-group bias: A functional magnetic resonance imaging investigation. *Psychological Science, 11*, 1131–1139.

Weber, J. M., Kopelman, S., & Messick, D. M. (2004). A conceptual review of decision making in social dilemmas: Applying a logic of appropriateness. *Personality and Social Psychology Review, 8*, 281–307.

Weisman, O., Zagoory-Sharon, O., & Feldman, R. (2012). Intranasal oxytocin administration is reflected in human saliva. *Psychoneuroendocrinology, 37*, 1582–1586.

Wildschut, T., Pinter, B., Vevea, J. L., Insko, C. A., & Schopler, J. (2003). Beyond the group mind: A quantitative review of the interindividual–intergroup discontinuity effect. *Psychological Bulletin, 129*, 698–722.

Wilson, M. S., & Liu, J. H. (2003). Social dominance orientation and gender: The moderating role of gender identity. *British Journal of Social Psychology, 42*, 187–198.

Wit, A., & Kerr, N. (2002). "Me versus just us versus us all" categorization and cooperation in nested social dilemmas. *Journal of Personality and Social Psychology, 83*, 616–637.

Yzerbyt, V., & Demoulin, S. (2010). Intergroup relations. In S. T. Fiske, D. T. Gilbert and L. Gartner (Eds.), *The handbook of social psychology* (pp. 1024–1083). New York: Wiley.

Zak, P. J., Kurzban, R., & Matzner, W. T. (2005). Oxytocin is associated with human trustworthiness. *Hormones and Behavior, 48*, 522–527.

Zak, P. J., Stanton, A. A., & Ahmadi, S. (2007). Oxytocin increases generosity in humans. *PLos ONE, 11*, e1128.

PART II

Person Perception and Stereotyping

5

SCANNING FOR SCHOLARS

How Neuro-imaging the MPFC Provides Converging Evidence for Interpersonal Stratification

Susan T. Fiske, Daniel L. Ames, Mina Cikara, and Lasana T. Harris

Like repotting a houseplant, moving to a new university allows a researcher to grow roots in an expanded, fertile, fresh environment. One then hopes for new foliage as well. In our lab's case, we moved just as social psychology was opening up to neuro-imaging techniques, and our new setting provided ample opportunities for collaboration and facilities for scanning, in particular. I was excited immediately because my prior work on the continuum model of categorization was a decade old, and we had run the course of the obvious conventional lab studies (Fiske, Lin, & Neuberg, 1999). Still, part of me wondered whether the distinction between category-based and individuated processes was a theoretical convenience to organize the literature, or something more. Then I encountered a paper by my prospective colleague Marcia Johnson, suggesting that aging adults over-rely on stereotypes when they cannot retrieve individuating information to remember who said what (Mather, Johnson, & De Leonardis, 1999). What caught my eye was that these impairments correlated with neuropsychological tests implicating distinct brain regions for each process. My first reaction was, "Maybe we aren't making it up," the categorizing–individuating distinction, and I rushed to Marcia's door, only to find it closed. She had left Princeton University for Yale University just as I arrived, so our lab was on our own in this quest.

Still, Princeton had much to offer in the way of exciting collaborations. A student request resulted in a project that did differentiate categorizing from individuating processes, using functional magnetic resonance imaging (fMRI). The empirical context was the publication of two simultaneous articles showing amygdala activation in White people's reactions to Black faces (Hart et al., 2000; Phelps et al., 2000). The amygdala at that point was implicated mainly in fear responses (later, in vigilance more broadly), so the inference was that White people, at least implicitly prejudiced ones, are more or less automatically afraid of

Black people. This seemed problematic to me because White Americans have complicated and malleable reactions to Black Americans. Drawing on the continuum model, we noted that the pioneering race–amygdala studies had put participants in a categorical mindset, according to the continuum model, so we varied participants' social goals while looking at Black and White yearbook photographs of men (Wheeler & Fiske, 2005): a categorical age judgment (over or under 21?), a non-social surface judgment (dot on the photo, yes or no?), and an individuating judgment (would he like the vegetable in the preceding slide, yes or no?). This study required us to run back and forth to an off-campus medical facility's scanner, while the department awaited its own research scanner. Whites' amygdala activation to Black faces replicated under the age judgment (categorical condition), but to our delight, disappeared when participants had to decide what vegetable he liked (individuation), as well as under the dot (superficial) task. We had learned that the categorizing and individuating of a racial out-group member were reflected in distinct neural activations. I was hooked.

What We Wanted to Know: Theoretical Background

Within a couple of years, an incoming graduate student staked his career on daring to plunge into this new area with me. Together, Lasana Harris and I replicated the Wheeler result on the department's own brand-new scanner. We did this because the technique and results were new to us, and we wanted to be sure they were reliable, both in our methods and in the data. In my lab, we do not like to go public with any results that we have not ourselves replicated, to guard against chance, overenthusiasm, and mistakes. Also, Lasana's first neuro-imaging study was a training experience, so the replication was not so much about two different scanners as about different humans.

We then shifted our attention to the textures of social categories, moving from the categorical and individuating processes described by the continuum model to describing distinct types of social categories in the stereotype content model (SCM) (for detailed theoretical context, see Fiske, 2012).

Social cognition turns out to be captured nicely by two fundamental and apparently universal dimensions, now described in the stereotype content model (Fiske, Cuddy, Glick, & Xu, 2002), but also represented in a variety of contemporary and historical person–perception research (see Fiske, Cuddy, & Glick, 2007, for a review; or Abele, Cuddy, Judd, & Yzerbyt, 2008, for a sampling). The convergence is reassuring, as is the cross-cultural validation (Cuddy et al., 2009; Durante et al., 2013).

For the first dimension, people want to know whether another's intentions are good or bad. That is, is the other individual (coming toward me at night) or group (coming to our country) a friend or a foe? Our term for this is the other's apparent warmth, which combines good intentions and trustworthiness (see Leach, Ellemers, & Barreto, 2007, on the correlated dimensions of morality and sociability). Warmth is fundamental, associated with safety from an early age (Williams & Bargh, 2008), inferred faster than other dimensions (Willis & Todorov, 2006), prioritized in impressions of others (Abele & Bruckmüller, 2011), and predictive of active helping versus harming (Cuddy, Fiske, & Glick, 2007). People infer warmth from the other's cooperative or competitive goals (Fiske et al., 2002; Kervyn, Fiske, & Yzerbyt, 2013), as well as from faces that look faintly happy (and not angry; Todorov, Dotsch, Wigboldus, & Said, 2011).

Having inferred another's intent, people want to know if the other can enact that intent; in other words, if the other is competent. The intentions of a competent individual or group matter more than those of one incompetent to act effectively. Competence is inferred from status to a surprising degree (Cuddy et al., 2009; Fiske et al., 2002; Kervyn et al., 2013). In faces, dominance (which we would interpret as status), has facial cues that are more masculine and older than low-dominance (status) faces, which are more feminine and younger (Todorov et al., 2011).

Crossing the warmth and competence dimensions yields a recognizable description of stereotypes in every culture examined so far (more than three dozen; Durante et al., 2013). As Table 5.1 indicates, each combination tends to elicit specific emotions. When we started doing the neuro-imaging work, we had only surveys and experiments using participant questionnaire reports of "society's" predictors, stereotypes, emotions, and behavioral tendencies, but nothing personal and nothing to go beyond deliberate verbal responses. Yet, if warmth/cooperation and competence/status characterize every society as functional adaptations, should we not have typical neural responses to these two dimensions—friends and foes, those up and down the hierarchy—as well as their combinations?

Warmth and competence are respectively predicted by cooperation and status. Example groups appear in each cell in Table 5.1, followed by the emotion typically reported for each cluster of groups.

The next section describes how our neuro-imaging analyses came to focus on the medial prefrontal cortex (MPFC) as a region of interest. The subsequent section describes what we found when we did, focusing on the stereotype content model's dimensions and quadrants. A final section evaluates the costs and benefits of these approaches, in our experience.

TABLE 5.1 Stereotype Content Model Dimensions (Perceived Warmth and Competence)

	Low Competence/ Low Status	*High Competence/ High Status*
High Warmth/ Cooperative	Older, disabled Pity	In-group, allies, middle class Pride
Low Warmth/ Competitive	Poor, immigrant, homeless Disgust	Rich, professionals Envy

How We Aimed to Discover It: Methods

Lasana Harris and I tackled the first of the potential neural signatures of responses in these quadrants. At first, in grant proposals and in our own minds, we speculated about a variety of possible neural patterns: attention directed up and neglect directed down the status–competence hierarchy (maybe reflected in the fusiform face area, for example); reward areas activating for warm groups more than for cold ones; and more. Social neuroscience was young, and our hypotheses were vague and unsubstantiated. The grants were not awarded.

Meanwhile, Lasana had developed elaborate stimulus materials: for two groups in each quadrant, several instantly recognizable photographs of individuals apparently from those groups, with neutral facial expressions and pre-tested on various dimensions (Harris & Fiske, 2009). In most of the SCM studies, we used photographs of people from distinct quadrants of the space because those elicit an immediate response, with pictures pre-tested to depict readily the relevant groups. At first, we went fishing, to see if any patterns were immediately obvious for warm versus cold groups or for competent versus incompetent ones. At that time, with those stimuli, and our emerging skills, nothing was obvious. Maybe the grant reviewers were right.

Nevertheless, we persisted, and one pattern appeared. Some background will explain our focus on a particular neural region of interest. Specifically, we heeded the emergent, consistent finding that the MPFC consistently responds to other people as stimuli (now collected in meta-analyses: Amodio & Frith, 2006; Mitchell, 2008; van Overwalle, 2009). This finding is so reliable that it was obvious even to us as observers of the early literature. Indeed, we conducted a study replicating an earlier social psychology questionnaire paradigm, manipulating verbal descriptions that modulate dispositional attributions (consensus, consistency, and distinctiveness; McArthur, 1972); we showed that in the only two out of eight behavior–description combinations that typically produce a dispositional inference, the MPFC activated above the baseline (Harris, Todorov, & Fiske, 2005). We were excited because to us this implicated the MPFC in people thinking about other people's dispositions. Our skeptical cognitive colleagues asked what made this so particularly social, and we answered by showing that the same actions by non-human objects (e.g., brooms) and by people did not equally evoke the MPFC (Harris & Fiske, 2008).

As an ongoing method, the MPFC became an index in several of our studies, because of its important theoretical and empirical role in social cognition. We return to those studies in a moment, but first, other converging evidence developed simultaneously to focus our analyses on the role of the MPFC.

Despite the agreement among social neuroscientists, one issue with the MPFC is its overlap with the ventral (lower) areas that also reliably activate in non-social reward processing. Indeed, this issue became so contentious at one point, that we formed an "adversarial collaboration" with our close colleagues from an adjoining lab who had a dramatically different take on at least that part of the MPFC's role.

The two labs together designed two simultaneous studies, to compare the disparate interpretations of MPFC responsiveness: as a social-cognition specialist versus a not-especially-social reward specialist.

One study (Harris, McClure, van den Bos, Cohen, & Fiske, 2007) compared reactions to pictures of people and objects, both manipulated along the warmth and competence dimensions to fit the four SCM quadrants. To highlight the reward qualities of these pictures, we contrasted the low–low cell (see Table 5.1) as uniformly negative, and the rest as at least partially positive. We developed two types of judgments, one that we thought would focus people on whether the photographs depicted humans or not: The intrinsically social judgment (person or object?) asked them to distinguish humans from non-humans. In another condition, we asked them to make a judgment that did not require them to distinguish: They made a non-social judgment (good or bad?) that applied to both people and things as origins of reward. The type of task ended up making no difference.

However, an anterior (rostral) region of MPFC (arMPFC; see Figure 5.1) activated more to positive than negative people, consistent with findings reported

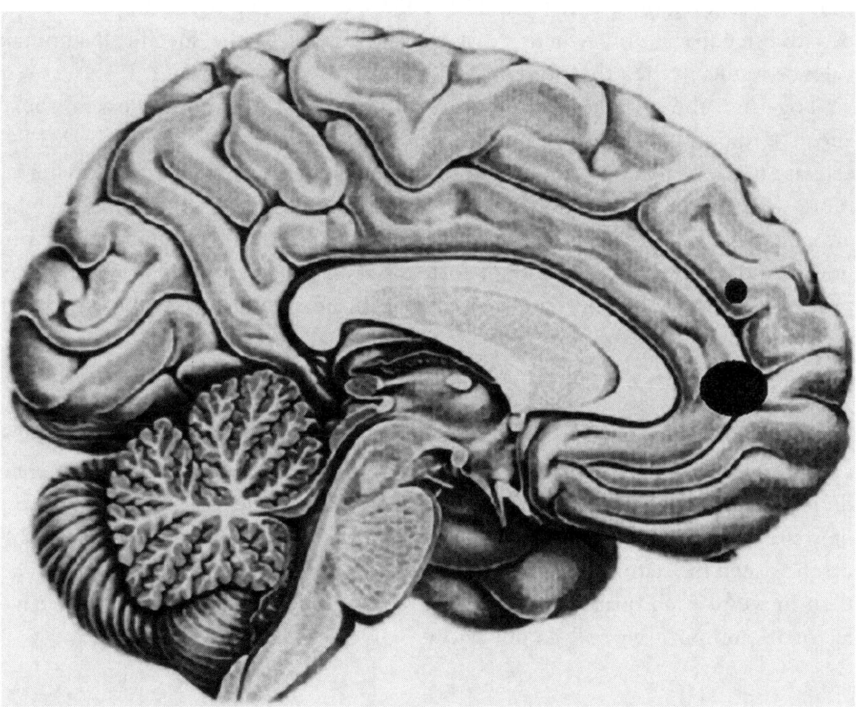

FIGURE 5.1 Schematic showing Anterior Rostral MPFC (Top Dot), Activating more to Positive than Negative People, and Para-ACC (Lower Dot), Activating more to all Positive Rewarding Entities, both People and Objects; see text and Harris et al. (2007).

below that theory-of-mind areas do not activate to the low–low out-groups. The same valence effect did not occur for objects, suggesting that it is the out-groups' sociality, not the valence alone, that drives this effect. (This pattern, however, held regardless of the social/non-social judgment task.) A nearby but more ventral MPFC area, para-anterior cingulate cortex (pACC), previously implicated in reward processing, activated *only* to stimulus valence, again regardless of task (social/non-social), but this time also regardless of target (human/non-human). Thus, one MPFC area differentiated extremely negative people from (at least partially) positive people, but did not show the same response to objects. Another nearby area differentiated rewarding entities (both people and objects) from unrewarding ones.

The companion study (van den Bos, McClure, Harris, Fiske, & Cohen, 2007) tackled the same MPFC problem, focusing also on the pACC. Participants performed a time-estimation task, for which they were reinforced by either a human experimenter or a computer program, receiving either positive (juice) or negative (quinine) feedback. Consistent with the other study, the pACC responded to feedback valence, regardless of human or non-human delivery. However, two other adjacent regions of the ventral MPFC responded respectively to positive and negative feedback only when it was social, parallel to the specifically human valence results in the other study.

Together, these studies separate reward processing from more general social affective responses. Other research indicates that more ventral areas of the MPFC activate to the self and people similar to oneself, who are generally liked, whereas more dorsal regions activate to groups perceived as different from oneself, who are generally disliked (Mitchell, Macrae, & Banaji, 2006). But much remains to understand about this large swath of cortex, which of course is implicated in many processes, social and otherwise. Nevertheless, these studies contributed to our methodological focus on the MPFC as a region of interest.

What We Have Learned: Results

Continuing our inquiry into neural signatures of social cognition's fundamental dimensions, but now armed with methods that focus on the MPFC, several studies illustrate some potential patterns of how people regard friends and foes (warmth), up and down the status hierarchy (competence). We start with the dimensions and then move to each quadrant in turn. Throughout, we reflect on what we have learned—and what we still do not know.

Wrong Side of the Tracks? The Trolley Study of Status Effects

An undergraduate's brainstorm triggered one study focused on the SCM status/competence dimension, exploring whether people value the lives of high-status/competent others over low-status/incompetent others (Cikara, Farnsworth, Harris,

& Fiske, 2010). Status divides people from each other in numerous ways (Fiske, 2011), and society values people of higher prestige. We wondered whether individuals also would favor them in simulated life-or-death decisions.

Experimental philosophers interrogate people's judged value by using various versions of the Trolley Dilemma, the footbridge version of which (Thomson, 1986) essentially goes like this :

> An empty runaway streetcar speeds down the tracks toward five people. Joe, from an overpass, sees this accident unfolding. If Joe chooses, he can shove a bystander off the overpass to block the streetcar, saving the five people. How morally acceptable is it for Joe to push the bystander off the overpass?
>
> *(Cikara et al., 2010, p. 205)*

We were interested in using this dilemma to see whether moral acceptability depends on the stereotypes associated with the person being sacrificed and the people being saved. Our neuroscience question was whether the MPFC and related social decision-making networks would be implicated in these trade-offs, consistent with the idea of valuing people by valuing their minds.

Normally, trading off one human life against others is morally unacceptable: Most people (80–90%) reject sacrificing one person to save five; at least, that is the result in this scenario, which entails actually pushing the one person (e.g., Hauser, Cushman, Young, Jin, & Mikhail, 2007). Our data reverse that, with 69% accepting the sacrifice of a low-status/low-competence person, 77% endorsing the rescue of the high-status/high-competence groups, and fully 84% approving the extreme combination of sacrificing one low–low (e.g., homeless) person to save five high–high people (e.g., in-group, middle-class Americans).

It is the neuro-imaging data that most clearly suggest the processes at work here. We wondered whether the decision would be easy and automatic, or whether overriding a normally unacceptable choice would require the engagement of complex social decision-making networks involving frontal cortex areas such as the MPFC, dorsolateral PFC, and orbitofrontal cortex (as well as social-cognition network areas such as the precuneus and posterior cingulate cortex) (see Figure 5.2). The imaging data were consistent with the moral override being a reluctant decision (i.e., saying it is acceptable to do something a vast majority of people do not condone); the social-cognition network indeed activated most for the most acceptable (least unacceptable) trade-off (sacrifice one homeless or drug-addicted person to save five students or middle-class Americans).

Producing these results presented some challenges distinct from a purely questionnaire version of the same study. First, to obtain reliable imaging data, we had to create a massive repeated-measures design; in this case, 128 ratings per participant (4 SCM quadrants $\times$ 2 groups from each quadrant $\times$ 16 instances of each group). One thus worries about fatigue effects and routinized responding, but to detect those, one can always examine the data for order effects.

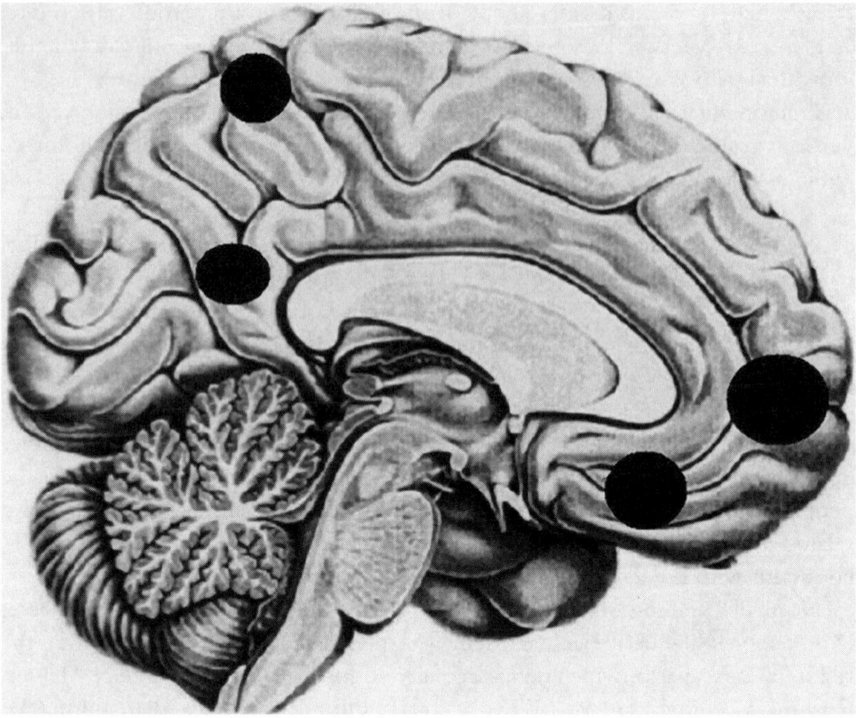

FIGURE 5.2 Social Cognition Network (Clockwise from Right: MPFC, Orbitofrontal Cortex, Posterior Cingulate Cortex, Precuneus) Activating to a Difficult Social Decision; see Cikara et al. (2010) and text.

Second, the key correlation was between our Likert scales of acceptability and average activity in neural regions previously defined by contrasting the most extreme combination (sacrifice one low–low person to save five high–high people) against all the other conditions). However, we had a relatively small sample size (n = 18), so correlations, though medium in magnitude (about .3), were nonsignificant.

Third, in keeping with basic psychometric principles, one wants to prevent violations of independence—using an effect both to identify relevant brain areas and then to correlate them with behavior. Identifying the brain areas a priori by contrasts and then correlating responses in these regions with self-reported acceptability ratings avoids the Vul, Harris, Winkielman, and Pashler (2009) critique of calculating correlations based on the same data already used to identify a region of interest.

Fourth, to avoid reverse inference from activation to cognition, one must admit that the cognitive process was still unclear—whether the apparent activation of a difficult social-choice network instead indicated that the participants were exercising cognitive control to override moral aversion to taboo trade-offs (an especially plausible interpretation of the dorsolateral PFC). Also unclear was whether responses in the trade-off of one low–low out-group member versus five high–high in-group members depended more on the one person sacrificed or the five people saved. Even though the images appeared sequentially (one sacrificed, then all five saved), the temporal sensitivity of fMRI is not sufficiently acute to distinguish between these two adjacent events in activating the relevant network.

Finally, the argument that this activation pattern fits difficult social decision making, while having precedents (e.g., Greene, Sommerville, Nystrom, Darley, & Cohen, 2004), is not by any means the only interpretation one might make. Well-practiced routines versus less predictable events might show a similar contrast (Wood & Grafman, 2003).

We found the neuro-imaging results to be helpful in understanding both the utilitarian valuing of high-status people over low-status ones, and that the decision involved a neural network previously implicated in complex social decisions, as if people were not making this trade-off lightly. We were disappointed to get no main effects for warmth, but the interaction effects did emphasize the extreme out-groups versus the unambiguous in-groups.

Beyond the Pale? MPFC Suggests Dehumanizing Homeless and Addicts

Among the most socially relevant neural areas, recall that early studies and meta-analyses had immediately implicated the MPFC in social cognition. We leveraged this initial discovery in our first search for neural signatures of SCM space, focusing on the worst-case scenario of the low–low extreme out-groups, derogated on both primary dimensions (Harris & Fiske, 2006). We predicted MPFC activation for most groups, any even partially positive ones, including mixed out-groups (pitied and envied ones), and certainly in-groups and allies (eliciting pride and admiration), but not low–low ("disgusting") groups.

As noted, the first challenge had been to find dozens of photographs of out-group members from each quadrant; in this study, participants saw only 48 images (4 SCM quadrants × 2 groups each × 6 instances each). We also had to think of a task to occupy participants and direct their attention to the pictures, without interfering with their spontaneous responses. At first, we did not know whether individual perceivers would report the same emotions that previous survey respondents had reported as being "society's" responses. But we took the risk—and besides we needed to know anyway—so we asked scanning participants to indicate which of the four emotions they felt in response to the picture. Reassuringly, the vast majority reported the pre-tested emotion for each picture.

What is more, in three quadrants, MPFC activation was significantly above a fixation-cross baseline, but as predicted, not in the fourth, low–low quadrant, for reactions to homeless and addicted people (Figure 5.3). Effect sizes in that one cell

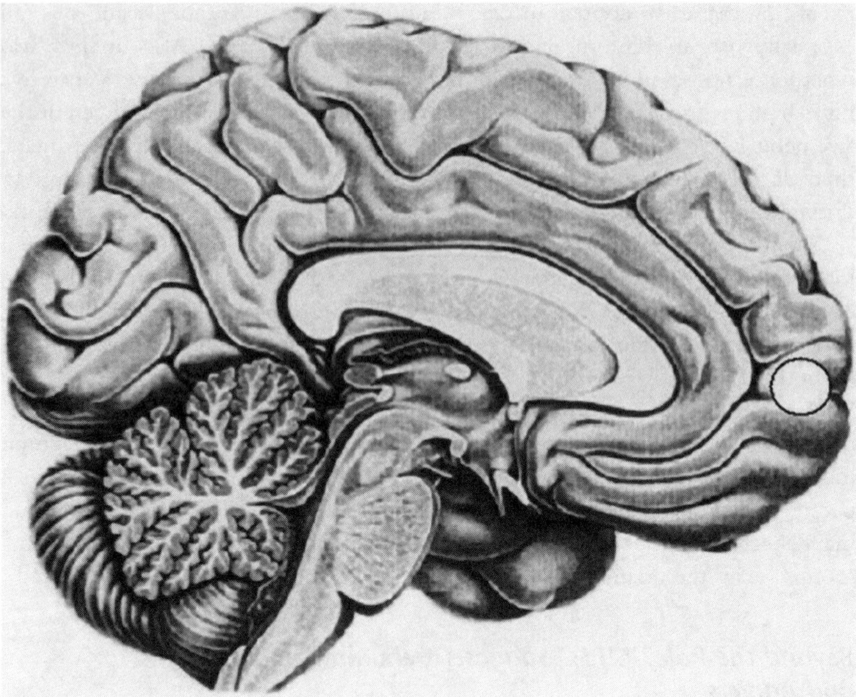

FIGURE 5.3 The Area of MPFC that Failed to Activate to Photos of Homeless People and Drug Addicts, Relative to all other Social-Group Members; see Harris and Fiske (2006) or text.

were half that of the remaining cells. Also, the low–low cell did show some amygdala and insula activation, consistent with that cell reportedly evoking a disgust response, as in other reactions to other highly stigmatized individuals (Krendl, Macrae, Kelley, Fugelsang, & Heatherton, 2006).

These results sparked questionnaire study follow-ups, which examined other self-report patterns associated with extreme out-groups (Harris & Fiske, 2011, Study 1). Consistent with the depressed MPFC, participants reported dehumanizing reactions, such as difficulty attributing a mind and lower likelihood of interacting. Moreover, open-ended descriptions of people from this cluster, as compared to the other three, used fewer mental-state verbs, and rating scale data reported less ease of attributing a mind to them and less ease in inferring their dispositions (all indicating less mentalizing of them). Ratings also indicated that they seemed less competent, articulate, and intelligent, as well as less warm, familiar, and similar; overall, people rated them as seeming marginally less typically human, but having more ups-and-downs in life.

A subsequent neuro-imaging study (Harris & Fiske, 2011, Study 2) combined the prior imaging and questionnaire studies, in order to examine the correlated features of dehumanizing perception. Out of 10 questionnaire ratings, seven correlated with neural activations in each participant's individual viewing of 60 photographs across SCM space.

From the SCM perspective, two neural areas correlated most with the two fundamental ratings dimensions: (Right anterior) insula correlated negatively with warmth ratings, and (left posterior) insula correlated positively with competence. To the extent that the anterior insula is implicated in interoceptive processes (that is, experiencing bodily reactions; Craig, 2009), then one might speculate that low-warmth reactions fitting SCM disgust and envy quadrants are more bodily in nature (respectively, "yuk!" or "cringe!"), but perhaps also that high-status/competence people merit attention, as other data would indicate (Fiske, 2010). Caution requires acknowledging that:

1. One should not make a reverse inference from activation to mental processes.
2. These neural areas were identified by their association (parametric modulation) with ratings on all 10 dimensions, so the correlation with an individual dimension is not entirely independent (see Vul et al., 2009), but the areas had to pass a $p < .001$ and 10-contiguous-voxel threshold.
3. The insula is a large, complex area associated with many neural and psychological processes.

Still, the results were provocative and heuristic for future research.

Finally, the bilateral superior temporal sulcus (STS) correlated with:

1. mentalizing—how easy it was to think about the person's mind;
2. how similar the person seemed to the self; and
3. how familiar the person seemed.

These were reassuring results on two counts: The STS is a well-established part of a broader social cognition network, and it correlated with three of the humanizing indicators. And the anterior cingulate cortex (conflict monitoring; Botvinick, Braver, Barch, Carter, & Cohen, 2001) correlated with typical humanity ratings. Curiously, the MPFC did not.

Still, we pursued the role of the MPFC in dehumanized perception (Harris & Fiske, 2007). Drawing on the earlier result (Wheeler & Fiske, 2005) that individuating tasks overcome amygdala (vigilance) responses by Whites to Black faces, we applied the same task to people viewing images of homeless people and drug addicts. A baseline condition replicated our previous de-activation of MPFC to low–low out-groups, in this case under a superficial categorization task (under/over 35 years old), but the individuating vegetable preference task reversed that, showing that neural signatures of mentalizing (not just vigilance) respond to social goals.

Our work on dehumanization and the MPFC taught us another lesson about this emerging field: The brain works in networks, not modules. The entire social cognition network's activation, not just the MPFC, reduces when perceiving dehumanized targets, and checks of all fMRI dehumanized perception studies confirm this fact. Indeed, in the same year in which our first dehumanized perception paper was published, another imaging paper by a group in communication studies had participants play combat video games in the scanner. This group found that just before participants pulled the trigger in this virtual environment to destroy an enemy, there was similar reduction in the MPFC, the social-cognition neural network, and many other parts of the brain since implicated in dehumanized perception (Mathiak & Weber, 2006). This converging evidence provided by another lab in another field, using a different paradigm but the same technology, is comforting for validating our early efforts, but disquieting because of the level of complexity it reveals.

Envy's Delight: Schadenfreude's Smiles and Rewards

After our lab team had examined dehumanization in the most extreme, low–low quadrant of SCM space, another enterprising graduate student, Mina Cikara, turned to another potentially controversial quadrant, the competent-but-cold envied out-groups that include rich people and outsider entrepreneurs. These high-status, allegedly competitive/exploitative out-groups uniquely elicit a peculiar reaction, namely malicious glee or Schadenfreude. Envied out-groups do not in the least elicit pity, sympathy, or empathy, the humane reactions that sometimes comfort even the low–low out-groups. When a designer-clad investment banker sprints to his waiting limousine, but on the way steps in dog faeces, observers cannot help but smile a little at his comeuppance. We captured that Schadenfreude to dirtied bankers, splashed yacht owners, and gummed-up CEOs by measuring participants' subtle smiles, using electromyography (EMG; Cikara & Fiske, 2012). We chose this method because certain expressions, such as smiles, are relatively reliable (Brown & Schwartz, 1980) and intuitively valid indicators of pleasure. What is more, social desirability probably constrains self-reported Schadenfreude, so more indirect measures (in this case, subtle muscle activations related to smiling) are apt.

Applying EMG to our study of Schadenfreude was not a trivial challenge. First, we had to learn how to do it, but at least we had the facilities, and facial EMG is considerably easier to learn, analyze, and interpret than fMRI data. Second, we created all-new SCM images that removed background and minimized differences in face size, expression, and orientation. Armed with these methods and materials, we found that three of the SCM quadrants elicited more smiling to mundane positive events ("ate a good sandwich") than to mundane negative ones ("sat on chewing gum"). As predicted, envied out-group members uniquely elicited more smiles when paired with negative than positive events (Figure 5.4). These were the field's first studies to show that stereotyped group membership alone is sufficient to elicit Schadenfreude. But establishing the neural correlates of this

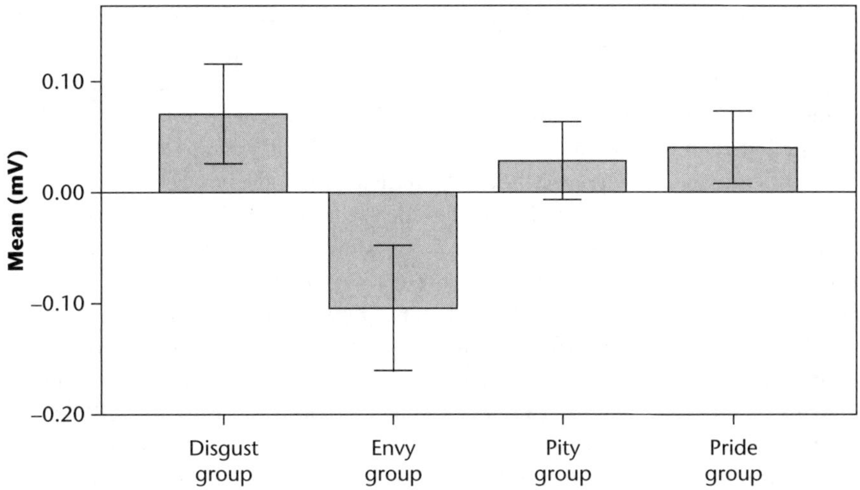

FIGURE 5.4 Zygomaticus Major (ZM) Activation is Greater for Positive than Negative Events, Except for Envy Group Members, where ZM Activates More to Negative than Positive Events; see Cikara and Fiske (2011) or text.

phenomenon had to be undertaken separately, because we lacked the equipment to record EMG in the scanner.

A paired fMRI study (Cikara & Fiske, 2011) observed increased anterior insula activation to positive events overall, and the magnitude of this response correlated with reported willingness to harm envied targets, but decreased willingness to harm in-group targets. Although open to multiple interpretations, this insula correlation with in-group loyalty versus envied-out-group harm might echo the Harris and Fiske (2011) main effects for the insula activating as a function of both (high) competence and (low) warmth. That is, a common mechanism might be the insula activating to competent and cold groups, who are the most threatening. But we are reluctant to over-interpret anatomically approximate patterns in different paradigms with different stimuli. Still, the insula is worth watching.

A related fMRI study refocused on Schadenfreude per se, though not drawing so directly on the SCM stereotypes, but still studying envy. Instead, we (Cikara, Botvinick, & Fiske, 2011) examined own-team versus envied-team baseball plays. Avid fans of the Boston Red Sox and New York Yankees watched simulated winning and losing plays involving these two teams (to assess pleasure at own wins and opponent losses). The pure Schadenfreude condition examined reactions to opponents losing to a third team. Negative outcomes activated the insula, perhaps consistent with the insula/threat reactions just mentioned. In addition, the anterior cingulate cortex (possibly discrepancy monitoring) also activated to one's own failures and a rival's successes. More to the point here, positive outcomes (own success and rival's failure, even to a third team) activated the ventral striatum (VS), a region consistently implicated in processing rewards, consistent with subjective pleasure. Ventral striatum activity (during these subjectively pleasurable plays) correlated with self-reported willingness to harm a rival fan. We construed these results as relevant to envy because each side reported envying the high-threat rivals, and the behavioral implications are especially disturbing.

A critic might ask, why no MPFC? Admittedly, for conceptual coherence, some MPFC activity might have linked to the idea that people like to understand the minds of competitors (e.g., Ruscher & Fiske, 1990) and of powerful people (e.g., people who control one's outcomes: Erber & Fiske, 1984). Yet people do not bother to make sense of those minds when their own potential control is minimal; for example, if they report lack of control (Ruscher & Fiske, 1990) or confront a monolithic, unilateral out-group (Dépret & Fiske, 1999). So far, at least, the envied out-groups are the exception to our persistent findings that implicate MPFC.

Before departing from the Schadenfreude response to envied out-groups, note that even envied out-groups can elicit other more positive emotions when they act more like admirable in-groups or pitied out-groups, and even more negative emotions when they act like disgusting out-groups (Cikara & Fiske, 2011). Because these are priming and questionnaire results, they are less relevant to our main charge here, focusing on neuroscience evidence. But we mention it for the record and to finish this sub-section on a less cynical note about reactions to envied out-groups.

Mindless Mascots? Sexual Hostility Diminishes Other Minds

Sexualization disrupts social cognition. Sexualized women might most often be viewed as mere mindless chicks (Eckes, 2002), pitiful in the SCM space, not competent, but nice enough. Former postdoctoral fellow Jennifer Eberhardt suggested that neuro-imaging would provide an angle on this issue (Cikara, Eberhardt, & Fiske, 2011). To start, images of bikini-clad and fully clothed men and women were matched for facial attractiveness, and standardized on posture, size, orientation, expression, and background.

Initial implicit-association data indicated that men (but not women) viewing pictures of bikini-clad women (but not bikini-clad or fully clothed men or fully clothed women) apparently denied them one critical element of mind: agency. Agency, autonomy, and independence fit the SCM competence dimension, which would describe sexualized women as low-status/competence in SCM space. Implicit associations were weak between sexualized women and third-person, agentic verbs ([she] handles, controls, uses, manages. . .), but higher between those women and first-person verbs ([I] handle, control, use, manage. . .). We interpreted this as denying agency to sexualized women by (at least some) men; women did not show this effect, and men associated more agency with professionally clad women. Hostile sexism predicted men's tendency to assign differential agency to sexualized and non-sexualized women. To the extent that attributed agency is a crucial element of attributing a mind (Gray, H., Gray, K., & Wegner, 2007), denying agency denies a mind to bikini-clad women.

Neuro-imaging allowed us a converging indicator of mind denial by focusing on the MPFC. The MPFC and other social-cognition network areas (temporal poles and posterior cingulate cortex) correlated negatively with hostile sexism in men viewing bikini-clad women (Figure 5.5). The finding that mind-perception areas simultaneously deactivated to sexualized women also fits other behavioral data: In a pre-test, men rated the sexualized women as less in control of their lives; and in Study 2, they recalled the bodies, but not the faces of the women in bikinis, another measure of sexualizing, but not individuating them.

One of the main challenges in this research was selecting the relevant brain areas. We compared neural responses to sexualized men and sexualized women, the most conservative test because those men were second to those women in ratings of having least control over their lives. That contrast, at conservative thresholds for probabilities and minimum contiguous voxels to make a cluster, was combined with using half the data for each participant (alternate runs) to identify the relevant brain areas, and then the other half to test their correlation with hostile sexism, thereby again avoiding the Vul et al. (2009) non-independence critique.

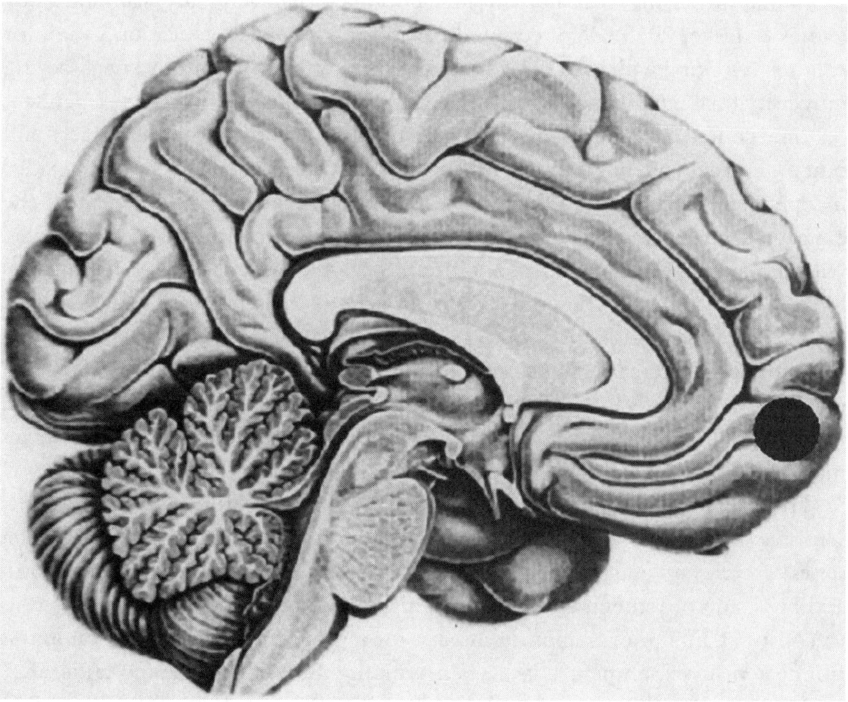

FIGURE 5.5 MPFC Activation Correlates Negatively with Hostile Sexism in Men Viewing Pictures of Bikini-clad Women; see text or Cikara, Eberhardt, & Fiske (2011).

Search for Solutions: People Who Need People

On a more upbeat note, we have pursued several solutions to understanding how other people might move from disrespected (low-status, low-competence) out-groups or disliked (competitive/exploitative, low-warmth) out-groups, to richer, more individuated, fully human others. The warmth dimension is predicted by interdependence, with cooperation increasing warmth and competition decreasing warmth. One study illustrates how one might leverage social interdependence to warm up reactions to an unfamiliar other.

Drawing on outcome dependency's motivating force, prior research had shown that people think harder when they depend on another person for a valued outcome. If the default form of social cognition is categorical, the more individuating form increases attention, encourages dispositional inferences (trait attributions that go beyond the category information), and generates impressions more idiosyncratic to the particular perceiver (Fiske et al., 1999). We wanted to go beyond verbal evidence, to generate even more specific converging evidence. We were also excited about the possibility of adding to the support for the MPFC as an indicator of considering other minds.

All this required incoming graduate student Daniel Ames to create an ecologically valid, involving, real, human interaction, a challenge in any scanning study (Ames & Fiske, 2013). We recruited participants to create education games for children, working with each of two supposed education majors from a neighboring university (making them out-groupers). Each participant was outcome-dependent on one confederate (for a monetary prize based on their joint product) and outcome-independent from the other (another prize depended on participants' independent work). When they met their two partners, one indicated that she expected to do well on the task, and the other not so well (fully counterbalanced with the outcome-dependency condition). Regardless of expectancy, participants looked more often at the confederate on whom their outcomes depended, a converging indicator of attention.

In the scanner, participants viewed a randomized series of slides, each one picturing a confederate and a positive or negative sentence from her alleged teaching evaluations. Thus, each slide reaffirmed or disconfirmed the expectancies introduced earlier.

To provide an individually tailored estimate of each participant's impression-sensitive area of the MPFC, at the end of these runs, each participant viewed a series of irrelevant person-information pairs, under two sets of instructions—form an impression or remember the order. This social–non-social contrast allowed us to construct an a priori region of interest for impression formation, autonomous from the main experiment's data, again avoiding the Vul et al. (2009) criticism.

As predicted, this separately defined region of the MPFC (see Figure 5.6) activated to expectancy-consistent information in the outcome-independent condition, consistent with the default process of confirming one's expectancies.

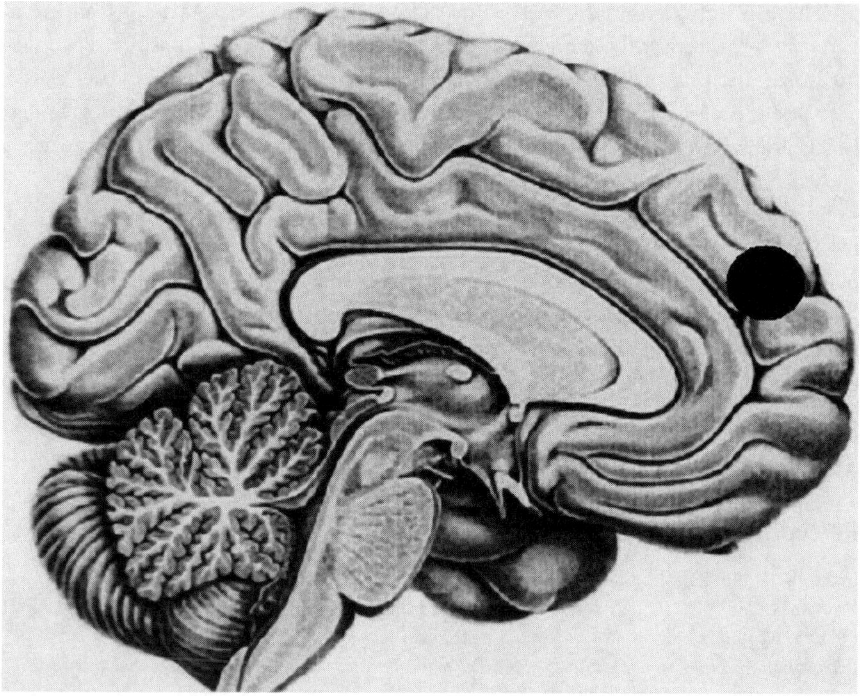

FIGURE 5.6 MPFC Area Activates Differentially to Expectancy-Consistent and Inconsistent Information, under Outcome Dependence and Independence; see text or Ames and Fiske (2013).

However, outcome dependency reversed that, activating the MPFC selectively to inconsistent information, which would be more diagnostic and individuating.

Added Value: Worth the Effort?

We believe these findings collectively inform theory and benefit research on intergroup relations. Without these expensive measures, we would not have developed the concept of dehumanization applied to the stereotype content model's low–low quadrant, which then sparked a series of questionnaire studies designed to interrogate dehumanized perceptions. Without these measures, we would not have linked sexual objectification to these other, superficially different dehumanizing processes (low agency/competence, but not low warmth). Without them, we would not have understood the lesser personhood of envied targets, whose misfortunes uniquely elicit glee. And without them, we would not have such a handle on whether dual processes of social cognition—categorical and individuating—are merely a convenient theoretical fiction, for the moment, or a process that is deeper and more enduring.

These insights emerged with some acknowledged costs: Start-up training is extensive, scanning time expensive, data analysis intensive, and massive repeated-measures designs are required. Social neuroscience has complex and costly dependent measures, but complex and inexpensive independent variables. Social psychologists are better at the independent variable side, and this is our relative advantage.

Neuro-imaging is not a cure-all. Social psychology is enjoying its heyday of media attention, mostly admiring. Students find our mainstream appealing; policy-makers occasionally listen to us. What neuro-imaging adds is merely another converging scientific tool, one among many. More than 10 years ago, as social neuroscience was first appearing, the senior author remarked to an equally senior colleague: "This is what will make us obsolete." She was wrong.

Suggestions for Further Reading

Relevant Social Psychology Theory

Fiske, S. T. (2012). The continuum model and the stereotype content model. In P. A. M. Van Lange, A. W. Kruglanski & E. T. Higgins (Eds.), *Handbook of theories of social psychology* (Vol. 1, pp. 267–288). Thousand Oaks, CA: Sage.

This summarizes the senior author's two social cognition theories, relevant research, and their links.

Fiske, S. T., Cuddy, A. J. C., & Glick, P. (2007). Universal dimensions of social perception: Warmth and competence. *Trends in Cognitive Science*, *11*, 77–83.

This briefly summarizes the stereotype content model, related models, and evidence, as of that date.

Harris, L. T., & Fiske, S. T. (2009). Dehumanized perception: The social neuroscience of thinking (or not thinking) about disgusting people. In M. Hewstone & W. Stroebe (Eds.), *European review of social psychology* (Vol. 20, pp. 192–231). London, UK: Wiley.

This summarizes the Harris–Fiske research program on dehumanized perception.

Relevant Social Neuroscience Methods

Ames, D. L., Fiske, S. T., & Todorov, A. (2011). Impression formation: A focus on others' intents. In J. Decety & J. Cacioppo (Eds.), *The Oxford handbook of social neuroscience* (pp. 419–433). New York: Oxford University Press.

This summarizes findings across neuroscience and social psychology focused on mind perception as the central future of individuating social cognition.

Lieberman, M. (in press). Methods for social neuroscience. In H. T. Reis & C. M. Judd (Eds.), *Handbook of research methods in social and personality psychology* (2nd ed). Cambridge, UK: Cambridge University Press.

This summarizes up-to-date methods of neuroscience, focused on neuro-imaging.

Todorov, A., Fiske, S. T., & Prentice, D. (Eds.). (2011). *Social neuroscience: Toward understanding the underpinnings of the social mind*. New York: Oxford University Press.

This collects examples of research programs from a variety of social neuroscience labs.

References

Abele, A. E., & Bruckmüller, S. (2011). The bigger one of the "Big Two"? Preferential processing of communal information. *Journal of Experimental Social Psychology, 47*, 935–948.

Abele, A. E., Cuddy, A. J. C., Judd, C. M., & Yzerbyt, V. Y. (2008). Fundamental dimensions of social judgment. *European Journal of Social Psychology, 38*, 1063–1065.

Ames, D. L., & Fiske, S. T. (2013). Outcome dependency alters the neural substrates of impression formation. Manuscript under review.

Amodio, D. M., & Frith, C. D. (2006). Meeting of minds: The medial frontal cortex and social cognition. *Nature Reviews Neuroscience*, 7, 268–277.

Botvinick, M., Braver, T., Barch, D., Carter, C., & Cohen, J. (2001). Conflict monitoring and cognitive control. *Psychological Review, 108*, 624–652.

Brown, S., & Schwartz, G. E. (1980). Relationships between facial electromyography and subjective experience during affective imagery. *Biological Psychology, 11*, 49–62.

Cikara, M., Botvinick, M. M., & Fiske, S. T. (2011). Us versus them: Social identity shapes neural responses to intergroup competition and harm. *Psychological Science, 22*, 306–313.

Cikara, M., Eberhardt, J. L., & Fiske, S. T. (2011). From agents to objects: Sexist attitudes and neural responses to sexualized targets. *Journal of Cognitive Neuroscience, 23*, 540–551.

Cikara, M., Farnsworth, R. A., Harris, L. T., & Fiske, S. T. (2010). On the wrong side of the trolley track: Neural correlates of relative social valuation. *Social Cognitive and Affective Neuroscience, 5*, 404–413. doi: 10.1093/scan/nsq011

Cikara, M., & Fiske, S. T. (2011). Bounded empathy: Neural responses to out-group targets' (mis)fortunes. *Journal of Cognitive Neuroscience, 23*, 3791–3803.

Cikara, M., & Fiske, S. T. (2012). Stereotypes and Schadenfreude: Behavioral and physiological markers of pleasure at others' misfortunes. *Social Psychological and Personality Science, 3*, 63–71.

Craig, A. D. (2009). How do you feel—now? The anterior insula and human empathy awareness. *Nature Reviews Neuroscience, 10*, 59–70.

Cuddy, A. J. C., Fiske, S. T., & Glick, P. (2007). The BIAS map: Behaviors from intergroup affect and stereotypes. *Journal of Personality and Social Psychology, 92*, 631–648.

Cuddy, A. J. C., Fiske, S. T., Kwan, V. S. Y., Glick, P., Demoulin, S., Leyens, J-Ph., Ziegler, R. (2009). Is the stereotype content model culture-bound? A cross-cultural comparison reveals systematic similarities and differences. *British Journal of Social Psychology, 48*, 1–33.

Dépret, E. F., & Fiske, S. T. (1999). Perceiving the powerful: Intriguing individuals versus threatening groups. *Journal of Experimental Social Psychology, 35*, 461–480.

Durante, F., Fiske, S. T., Kervyn, N., Cuddy, A. J. C., Akande, A., Adetoun, B. E., Storari, C. C. (2013). Nations' income inequality predicts ambivalence in stereotype content: How societies mind the gap. *British Journal of Social Psychology.*

Eckes, T. (2002). Paternalistic and envious gender stereotypes: Testing predictions from the stereotype content model. *Sex Roles*, 47, 99–114.

Erber, R., & Fiske, S. T. (1984). Outcome dependency and attention to inconsistent information. *Journal of Personality and Social Psychology*, 47, 709–726.

Fiske, S. T. (2010). Interpersonal stratification: Status, power, and subordination. In S. T. Fiske, D. T. Gilbert & G. Lindzey (Eds.), *The handbook of social psychology* (5th ed., pp. 941–982). New York: Wiley.

Fiske, S. T. (2011). *Envy up, scorn down: How status divides us*. New York: Russell Sage Foundation.

Fiske, S. T. (2012). The continuum model and the stereotype content model. In P. A. M. Van Lange, A. W. Kruglanski & E. T. Higgins (Eds.), *Handbook of theories of social psychology* (Vol. 1, pp. 267–288). Thousand Oaks, CA: Sage.

Fiske, S. T., Cuddy, A. J. C., & Glick, P. (2007). Universal dimensions of social perception: Warmth and competence. *Trends in Cognitive Science*, *11*, 77–83.

Fiske, S. T., Cuddy, A. J. C., Glick, P., & Xu, J. (2002). A model of (often mixed) stereotype content: Competence and warmth respectively follow from perceived status and competition. *Journal of Personality and Social Psychology*, *82*, 878–902.

Fiske, S. T., Lin, M. H., & Neuberg, S. L. (1999). The continuum model: Ten years later. In S. Chaiken & Y. Trope (Eds.), *Dual process theories in social psychology* (pp. 231–254). New York: Guilford Press.

Gray, H. M., Gray, K., & Wegner, D. M. (2007). Dimensions of mind perception. *Science*, *315*, 619.

Greene, J. D., Sommerville, R. B., Nystrom, L. E., Darley, J. M., & Cohen, J. D. (2004). The neural basis of cognitive conflict in moral judgment. *Neuron*, *44*, 389–400.

Harris, L. T., & Fiske, S. T. (2006). Dehumanizing the lowest of the low: Neuro-imaging responses to extreme outgroups. *Psychological Science*, *17*, 847–853.

Harris, L. T., & Fiske, S. T. (2007). Social groups that elicit disgust are differentially processed in mPFC. *Social Cognitive and Affective Neuroscience*, *2*, 45–51.

Harris, L. T., & Fiske, S. T. (2008). The brooms in Fantasia: Neural correlates of anthropomorphizing objects. *Social Cognition*, *26*, 210–223.

Harris, L. T., & Fiske, S. T. (2009). Dehumanized perception: The social neuroscience of thinking (or not thinking) about disgusting people. In M. Hewstone & W. Stroebe (Eds.), *European review of social psychology* (Vol. 20, pp. 192–231). London, UK: Wiley.

Harris, L. T., & Fiske, S. T. (2011). Dehumanized perception: A psychological means to facilitate atrocities, torture, and genocide? *Zeitschrift für Psychologie/Journal of Psychology*, *21*(3) [Topical Issue on Torture], 175–181.

Harris, L. T., McClure, S. M., van den Bos, W., Cohen, J. D., & Fiske, S. T. (2007). Regions of MPFC differentially tuned to social and non-social affective evaluation. *Cognitive and Behavioral Neuroscience*, 7, 309–316.

Harris, L. T., Todorov, A., & Fiske, S. T. (2005). Attributions on the brain: Neuro-imaging dispositional inferences, beyond theory of mind. *NeuroImage*, *28*, 763–769.

Hart, A. J., Whalen, P. J., Shin, L. M., McInerney, S. C., Fischer, H., & Rauch, S. L. (2000). Differential response in the human amygdala to racial outgroup vs ingroup face stimuli. *Neuroreport: For Rapid Communication of Neuroscience Research*, *11*, 2351–2355.

Hauser, M., Cushman, F., Young, L., Jin, R. K., & Mikhail, J. (2007). A dissociation between moral judgments and justifications. *Mind and Language*, *22*, 1–21.

Kervyn, N., Fiske, S. T., & Yzerbyt, V. (2013). Why is the primary dimension of social cognition so hard to predict? Symbolic and realistic threats together predict warmth in the stereotype content model. Manuscript under review.

Krendl, A. C., Macrae, C. N., Kelley, W. M., Fugelsang, J. A., & Heatherton, T. F. (2006). The good, the bad, and the ugly: An fMRI investigation of the functional anatomic correlates of stigma. *Social Neuroscience*, *1*(1), 5–15.

Leach, C. W., Ellemers, N., & Barreto, M. (2007). Group virtue: The importance of morality (vs. competence and sociability) in the positive evaluation of in-groups. *Journal of Personality and Social Psychology*, *93*, 234–249.

McArthur, L. A. (1972). The how and what of why: Some determinants and consequences of causal attribution. *Journal of Personality and Social Psychology, 22*, 171–193.

Mather, M., Johnson, M. K., & De Leonardis, D. M. (1999). Stereotype reliance in source monitoring: Age differences and neurophysiological test correlates. *Cognitive Neuropsychology, 16*, 437–458.

Mathiak, K., & Weber, R. (2006). Towards brain correlates of natural behavior: fMRI during violent video games. *Human Brain Mapping, 27*, 948–956.

Mitchell, J. P. (2008). Contributions of functional neuroimaging to the study of social cognition. *Current Directions in Psychological Science, 17*, 142–146.

Mitchell, J. P., Macrae, C. N., & Banaji, M. R. (2006). Dissociable medial prefrontal contributions to judgments of similar and dissimilar others. *Neuron, 50*, 655–663.

Phelps, E. A., O'Connor, K. J., Cunningham, W. A., Funayama, E. S., Gatenby, J. C., Gore, J. C., & Banaji, M. R. (2000). Performance on indirect measures of race evaluation predicts amygdala activation. *Journal of Cognitive Neuroscience, 12*, 729–738.

Ruscher, J. B., & Fiske, S. T. (1990). Interpersonal competition can cause individuating processes. *Journal of Personality and Social Psychology, 58*, 832–843.

Thomson, J. J. (1986). *Rights, restitution, and risk: Essays in moral theory*. Cambridge, MA: Harvard University Press.

Todorov, A., Dotsch, R., Wigboldus, D., & Said, C. P. (2011). Data-driven methods for modeling social perception. *Social and Personality Psychology Compass, 5*, 775–791.

van den Bos, W., McClure, S. M., Harris, L. T., Fiske, S. T., & Cohen, J. D. (2007). Dissociating affective evaluation and social cognitive processes in ventral medial prefrontal cortex. *Cognitive and Behavioral Neuroscience, 7*, 337–346.

van Overwalle, F. (2009). Social cognition and the brain: A meta-analysis. *Human Brain Mapping, 30*, 829–858.

Vul, E., Harris, C., Winkielman, P., & Pashler, H. (2009). Puzzlingly high correlations in fMRI studies of emotion, personality, and social cognition. *Perspectives in Psychological Science, 4*, 274–290.

Wheeler, M. E., & Fiske, S. T. (2005). Controlling racial prejudice and stereotyping: Social cognitive goals affect amygdala and stereotype activation. *Psychological Science, 16*, 56–63.

Williams, L. E., & Bargh, J. A. (2008). Experiencing physical warmth promotes interpersonal warmth. *Science, 322*, 606–607.

Willis, J., & Todorov, A. (2006). First impressions: Making up your mind after a 100-ms exposure to a face. *Psychological Science, 17*, 592–598.

Wood, J. N., & Grafman, J. (2003). Human prefrontal cortex: Processing and representational perspectives. *Nature Reviews Neuroscience, 4*, 139–147.

6

SOCIAL IDENTITY SHAPES SOCIAL PERCEPTION AND EVALUATION

Using Neuroimaging to Look Inside the Social Brain

Jay J. Van Bavel, Y. Jenny Xiao, and Leor M. Hackel

On February 26, 2012, 17-year-old Trayvon Martin was shot and killed by neighborhood watch captain George Zimmerman—who thought the teenager looked suspicious—sparking a national outrage and discussion about race in the United States. Why had an innocent, unarmed African American teenager died on his way home? In the aftermath, some sought answers not in explicit racism and prejudice, but rather in the domain of implicit racism—arguing, as two social psychologists wrote in an op-ed (newspaper opinion piece), that "our minds are colored by race" (Goff & Richardson, 2012).

According to a standard dual process account developed over the last 25 years, the implicit effects of race that colored George Zimmerman's mind that night were sadly predictable. Seeing a Black face would likely trigger automatic negative evaluations and associations for Mr Zimmerman, regardless of his explicit beliefs about race (see Devine, 1989). Moreover, due to his inexperience with Black faces, he may have failed to see Mr Martin as an individual, but rather merely as an interchangeable member of the social category "African American." According to the standard dual process account, these events took place automatically and inevitably, triggered in a reflex-like manner by the salient visual cue of race. However, in this chapter, we provide evidence that the psychological and neural processes underlying person perception are context-dependent, dynamic, and shaped by motivational states, rather than inevitable, reflexive, and predictable. Specifically, we review research showing that *self-categorization* and *social identity* structure social perception and evaluation. Self-categorization involves the activation of psychological connections between the self and some class of stimuli at the personal (i.e., defining oneself as unique from others) or collective (i.e., defining oneself in terms of similar characteristics to one's social group) level (Turner, Hogg, Oakes, Reicher, & Wetherell, 1987; Turner, Oakes, Haslam, & McGarty, 1994).

In the following pages, we will describe and critique the dual process account of racial bias, and present an alternative, dynamic systems approach. We will then review a series of experiments supporting this perspective by showing how self-categorization and social identity shape social perception and evaluation (Van Bavel & Cunningham, 2011). Specifically, we will present studies showing that social identities emerge very rapidly under minimal conditions, and can override biases in social perception and evaluation toward groups with much greater historical and cultural baggage, such as race. Importantly, however, we will present research showing that race is not merely "erased" from the visual system, suggesting that currently salient identities may guide low-level aspects of perception and evaluation even when aspects of race (e.g., physiognomic features) are still represented in the brain (see Kaul, Ratner, & Van Bavel, 2012; Ratner, Kaul, & Van Bavel, 2012). Finally, we will discuss the methodologies we used and comment on the value, costs, and benefits of these methods. Specifically, to examine the impact of self-categorization on ostensibly automatic components of the person perception network, we took a multi-level approach—termed social neuroscience (Cacioppo, Berntson, Sheridan, & McClintock, 2000). This approach is based on the assumption that complex phenomena, like prejudice, are best understood by combining social and biological theories and methods (see also Cunningham & Van Bavel, 2009; Van Bavel & Cunningham, 2009b). By understanding more about underlying neural processes, we hope to develop a functional understanding of the biological systems that underlie social perception and evaluation, and help to build the foundation for future research and intervention.

The Dual Process Approach: Race as a Bottom-up Cue in Social Perception

Our approach differs in a number of important ways from dual process and systems models, which have emerged and dominated various topic areas in psychology—including conceptualizations of stereotypes and prejudice (e.g., Devine, 1989; Fazio, Jackson, Dunton, & Williams, 1995)—since the 1980s. According to some dual process models, people have two systems for attitudinal representations: an *implicit* representational system, which gives rise to automatic attitudes; and an *explicit* system, which gives rise to controlled attitudes (e.g., Wilson, Samuel, & Schooler, 2000). This dual process approach was influential in early social neuroscience research on prejudice, as researchers sought to distinguish automatic (e.g., fast, unintentional, outside of awareness) and controlled (e.g., slow, deliberate) processes in the brain. For instance, research on this topic revealed that racial biases emerged within the first few hundred milliseconds of perceptual processing (Ito & Urland, 2003) and were linked to relatively early components in the face-processing network (Golby, Gabrieli, Chiao, & Eberhardt, 2001). Moreover, the neural substrates of the dual processes appeared to be separable, such that the amygdala, for example, was strongly correlated with *implicit*, but not *explicit*, measures of racial bias (Phelps et al., 2000). Meanwhile, people with sufficient motivation and opportunity to control biased responses appeared to employ the anterior cingulate cortex (Amodio et al., 2004) and lateral prefrontal cortex (Cunningham et al., 2004), regions that have both been implicated in cognitive control and self-regulation, to help control these biases (see Figure 6.1).

A Dynamic Systems Approach: Identity Shapes Social Perception and Evaluation

Although dual process models serve as a useful heuristic for the human mind and have sparked numerous studies, advances in social and cognitive neuroscience suggest that the human evaluative system is more widely distributed and dynamic than that proposed by traditional dual process models (see Van Bavel, Xiao, & Cunningham, 2012). As such, we argue that there is unlikely to be a clear dissociation between explicit and implicit representations in the brain (see Cunningham, Zelazo, Packer, & Van Bavel, 2007; Van Bavel, Xiao et al., 2012). Rather, we argue that representations are constructed from the dynamic interaction of multiple brain systems, and feature the recursive interaction between bottom-up cues (e.g., skin color or hair length) and top-down cues (e.g., attention or motivations) that interact in cycles until the evaluative system settles on a

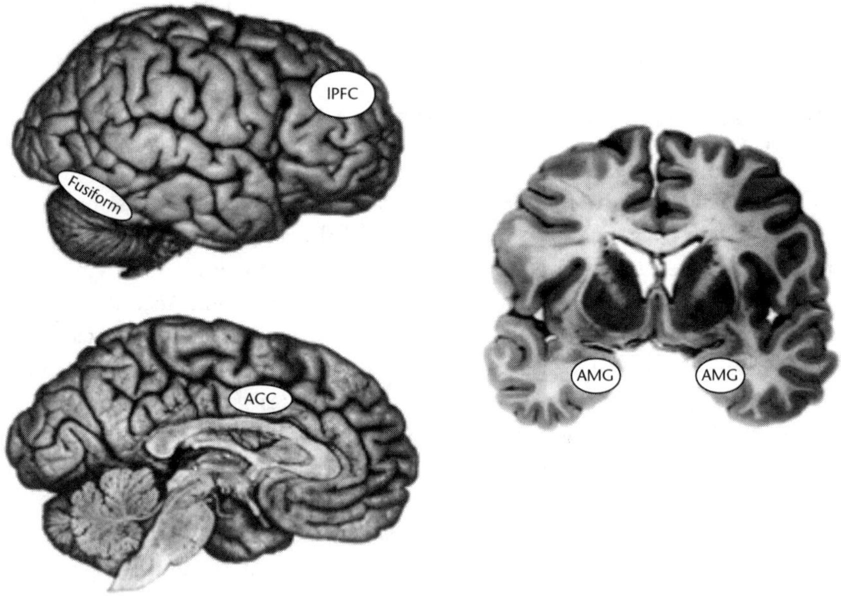

FIGURE 6.1 Anatomical Images of the Human Brain Identifying the Approximate Spatial Locations of the (A) Fusiform Gyri and Lateral Prefrontal Cortext (lPFC), (B) Anterior Cingulate Cortex (ACC), and (C) Left and Right Amygdala (AMG)

representation of a target (Cunningham et al., 2007; Freeman & Ambady, 2011). Further, the dynamic approach suggests that virtually all aspects of evaluation and perception states are influenced by the context and motivation of the perceiver. As such, we assume that early aspects of social perception depend on social identity more than on the ostensibly automatic processes assumed by most dual process models of personal perception and evaluation (see Van Bavel & Cunningham, 2011 for an extended discussion).

Therefore, we argue that salient social identities—rather than race, per se—can shape person perception processes in this dynamic fashion. It is true that biases and stereotypes about certain social groups, such as racial and gender groups, are often built upon years of exposure to cultural stereotypes and personal experience. However, the more general process of categorizing one's self as a group member (i.e., self-categorization) can have an early influence on intergroup perception, evaluation, and behaviors (Turner et al., 1994; Van Bavel & Cunningham, 2011). As social beings, humans are remarkably adept at dividing up the world into *us* and *them*, even in the absence of any factors typically posited to account for intergroup bias, such as intergroup competition or conflict (Tajfel, Billig, Bundy, & Flament, 1971). In a classic minimal groups paradigm used by social identity researchers, people form groups quickly and favor members of their own group, even when groups are formed under rather arbitrary premises (e.g., the flip of a coin). Such evidence points to the value humans place on social identities and the context-dependent nature of identification. In fact, self-categorization with a group may occur in a reflexive and automatic manner, and guide automatic evaluations and behaviors (e.g., Otten & Wentura, 1999). Building on self-categorization theory, we assume that self-categorization is "inherently variable, fluid, and context dependent" (Turner et al., 1994, p. 454). In other words, the social context can heighten the accessibility of a particular social identity (e.g., a racial identity, a national identity, etc.), which in turn elicits perceptions and evaluations consistent with the activated aspects of this identity.

Furthermore, in terms of the time course of these processes, we argue that prior states of the evaluative system set the stage for automatic construals of subsequent stimuli. Higher-order processes—mediated by top-down control signals from the frontal and parietal networks—can incorporate expectations, goals, bodily states, and context into representations that are deemed most relevant in a given context (see Miller & Cohen, 2001), which can then lead to different patterns of self- and social categorization. When a given self-categorization becomes salient, one may in turn be more likely to see others in light of their membership vis-à-vis this salient category. As a consequence, this may pre-empt ostensibly automatic racial bias (or other visually salient social categories) that some researchers have characterized as inevitable. Thus, the preceding context and motivational state of an organism informs subsequent evaluative processes (and vice versa) in a continually dynamic manner. Based on this model, we reasoned that self-categorization with a group should shape people's perceptions and evaluations of others in terms of their current and most salient self-categorization, not necessarily race. We used a social neuroscience approach to test whether the neural underpinnings of social categorization more closely match the proposed dynamic model of the human mind than dual process models (see Van Bavel & Cunningham, 2011, for a more detailed discussion).

How We Tested Our Perspective

With only behavioral measures, it may be difficult to illustrate *when* in the processing stream higher-order constructs such as self-categorization and social identity exert influence on perception and evaluation, and it would be impossible to demonstrate *where* in the brain these influences occur. By tapping into neural activity with functional magnetic resonance imaging (fMRI) and electroencephalography (EEG), we could more precisely compare different models of social processing, and delineate differences between alternative models of perception and evaluation.

Functional magnetic resonance imaging provides an indirect index of neural activity by measuring changes in cerebral blood flow, which correlates with neural activity (see Huettel, Song, & McCarthy, 2004, for a review). Compared to other non-invasive measures of neural activity, fMRI offers superior *spatial* resolution (on the order of cubic millimeters), although it has inferior *temporal* resolution to methods like EEG (on the order of 4–6 seconds for fMRI, as opposed to milliseconds [ms] for EEG; see Cunningham, Packer, Kesek, & Van Bavel, 2009, for a discussion). As such, fMRI is better suited to answer questions about the localization of cognitive processes, whereas EEG is better suited to delineate their time course. Examining the neural substrates of person perception allowed us to test the impact of social categorization along the neuroaxis. Does self-categorization merely alter activity in brain regions involved in reflective, controlled processing (e.g., lateral prefrontal cortex)? Or does it also affect activity in brain regions implicated in ostensibly bottom-up, automatic processing (e.g., the amygdala)?

Empirical Evidence: Dissociating the Effects of Race and Self-categorization

Behavioral Investigations

To examine the relationship between self-categorization and intergroup perception and evaluation, we conducted several experiments in which we assigned participants to one of two mixed-race groups (e.g., the Lions or Tigers) with an equal number of Black and White males in each group. Participants then had a few minutes to memorize the group membership of these faces before we assessed dependent measures of their attitudes (Van Bavel & Cunningham, 2009a), memory

(Van Bavel & Cunningham, 2012b; Van Bavel, Swencionis, O'Connor, & Cunningham, 2012), and/or brain activity (Van Bavel, Packer, & Cunningham, 2008, 2011). Importantly, the faces on each team were fully counterbalanced across participants to ensure that any effects of group membership were due to self-categorization and identification, and *not* the visual properties of different faces (e.g., attractiveness, luminance, symmetry, etc). Moreover, race was orthogonal to team membership and there were no visual cues to group membership during the administration of the dependent measures. This mixed-race paradigm allowed us to examine whether individuals automatically and inevitably categorize others according to a visually salient category like race (e.g., Devine, 1989; Ito & Urland, 2005). We predicted that a salient self-categorization—however minimal—would override or even pre-empt racial bias.

In a pair of initial studies, we examined the influence of a shared social identity on ostensibly automatic evaluations (Van Bavel & Cunningham, 2009a). We measured automatic evaluations of the faces described above using a computerized response-window priming task (Draine & Greenwald, 1998). During this task, participants were instructed to categorize a word rapidly on each trial as "good/liked" or "bad/disliked." On each trial, a face appeared at the center of the computer monitor for 150 ms (followed by a blank screen for 50 ms) before an unambiguously positive (e.g., love) or negative (e.g., hatred) target word appeared. We assumed that faces with positive associations would increase accuracy to positive words and decrease accuracy to negative words. By the same logic, faces with negative associations would decrease accuracy to positive words and increase accuracy to negative words.

Replicating previous research, participants who merely saw two mixed-race groups without being assigned to one of them showed the standard pattern of automatic racial bias: more positive evaluations of White compared to Black faces (Fazio et al., 1995). In other words, mere exposure to a racially diverse environment was not sufficient to override racial bias. In contrast, participants who were actually part of a mixed-race group had positive automatic evaluations of White *and* Black in-group members, and these evaluative preferences were driven by in-group favoritism and not out-group derogation. That is, group membership increased relative positivity toward Black in-group members relative to Black out-group members, eliminating the standard pattern of automatic racial bias among in-group members. Thus, participants' evaluations reflected their current salient self-categorization even when the groups had no history of contact or conflict, and when there was an orthogonal, visually salient social category cue (i.e., race) with strong existing evaluative connotations. In short, "automatic" racial bias was *not* automatic.

Neuroimaging Investigations

Although the neural networks involved in evaluation are widely distributed (Cunningham et al., 2007), initial research focused on a small structure in the temporal lobe called the amygdala (see Figure 6.1 above), a region in the extended face network that plays an important role in social evaluation (see Macrae & Quadflieg, 2010). The amygdala has been implicated in a host of social and affective processes, including fear conditioning and processing of negative stimuli (for a review, see Phelps, 2006). More strikingly, the amygdala is activated even during rapid subliminal presentations of affectively significant faces (Whalen et al., 1998). Several studies have found that individual differences in amygdala activity for Black compared to White faces correlate with implicit measures of racial bias including the Implicit Association Test (IAT) (Cunningham et al., 2004; Phelps et al., 2000). These correlations with racial bias led some researchers to interpret differences in amygdala activation in intergroup contexts as evidence of negativity toward stigmatized groups. However, differences in amygdala activity to faces of different racial groups are generally uncorrelated with explicit measures of prejudice (Phelps et al., 2000).

The dissociation between implicit and explicit measures of racial bias, along with the dissociation between explicit measures of racial bias and amygdala activity, are consistent with numerous dual process models of prejudice. Indeed, several studies have examined the control of automatic racial biases. For example, one fMRI study examined both automatic and controlled responses to Black and White faces (Cunningham et al., 2004). Several White participants were presented with Black and White faces for 30 ms or 525 ms. Consistent with the assumption that rapid subliminal presentation (i.e., 30 ms) would elicit automatic racial biases, participants had greater amygdala activity following exposure to subliminal Black faces than following subliminal White faces. Moreover, differential amygdala activity in response to Black versus White faces was correlated with individual differences in racial bias on the IAT (Greenwald, McGhee, & Schwartz, 1998). In contrast, when the faces were presented supraliminally (i.e., 525 ms), this differential amygdala activity was significantly reduced, and brain regions involved in conflict detection and regulatory control (i.e., the anterior cingulate cortex [ACC] and lateral prefrontal cortex [PFC]) showed greater activity for Black compared to White faces. Such findings suggest that participants were controlling their automatic racial bias.

Based on our research on the malleability of automatic evaluations (Van Bavel & Cunningham, 2009a), however, we reasoned that flexibly construing people as in-group members might provide a powerful alternative to the traditional dual process models of control evident in previous research (see also Wheeler & Fiske, 2005). We had previously shown that the amygdala may play a role in processing any motivationally relevant stimuli, regardless of valence (Cunningham, Van Bavel, & Johnsen, 2008). Thus, we reasoned that when race is the most salient social category, the amygdala may indeed be responsive to members of groups who are stereotypically associated with threat or novelty—explaining some previous findings from neuroimaging studies of prejudice. In contrast, when race is not the most salient social category, groups that are currently motivationally relevant would be associated with greater amygdala activity. In minimal group contexts, in-group members tend to be motivationally relevant and afford group members the opportunity to meet belongingness goals (Van Bavel & Cunningham, 2012b; Van Bavel, Swencionis et al., 2012).

Similar to our other experiments, we randomly assigned White participants to a minimal mixed-race group, asked them to learn the faces of each group, and then presented them with in-group and out-group faces during neuroimaging (Van Bavel et al., 2008). Crossing race and group membership allowed us to examine the role of self-categorization in neural processing: Would membership in a new group lead participants to process targets in terms of this salient group membership rather than race? Importantly, assigning people to mixed-race groups equated in-group and out-group members in familiarity and novelty. Participants in previous neuroimaging studies on race had different experiences and associations with Black versus White social categories, making it possible that novelty with Black faces may have elicited differences in amygdala activity (Dubois et al., 1999).

In our experiment, during neuroimaging, participants categorized each face according to either team membership (Leopards or Tigers) or skin color (Black or White). As predicted, participants had greater amygdala activity to in-group (i.e., same-team) than out-group (i.e., other-team) faces, regardless of task. Importantly, this in-group bias in neural processing occurred within minutes of group assignment, in the absence of explicit team-based rewards or punishments, *and* independent of pre-existing attitudes, stereotypes, or familiarity. In-group bias in neural activity was not moderated by target race or categorization task (i.e., categorizing by team or by skin color), suggesting that it did not require explicit attention to group membership and may have occurred relatively automatically. Again, this suggests that self-categorization can shape relatively automatic aspects of social perception and evaluation.

Whereas earlier studies often interpreted amygdala activity to out-group faces as reflecting negativity or fear toward stigmatized group members, participants in our experiment (Van Bavel et al., 2008) had greater amygdala activity to in-group members. These results support the idea that the amygdala may be involved in segregating relevant from irrelevant stimuli to enhance perception of *important* stimuli (Anderson & Phelps, 2001; Vuilleumier, 2005; Whalen, 1998). Importantly, the relevance of different social categories varies according to social context (Turner et al., 1987). In contexts where race provides the most salient group distinction, racial attitudes, cultural stereotypes, and personal values (e.g., egalitarianism) may provide the most relevant motivational guides. Most neuroimaging studies in this literature make race the only salient difference between faces by showing 50% White faces and 50% Black faces. This departure from the population base-rate where most participants live may artificially make race and racial stereotypes more salient than usual and therefore heighten racial biases reflected in amygdala activity in those studies. However, assigning people to mixed-race groups may change the way people construe race and other social categories, and sensitize perceptual and evaluative processes to other contextually relevant group memberships. Indeed, people categorize others according to race when it is the most salient social category, but categorize according to other group memberships (and ignore race) when they are part of a mixed-race group (Kurzban, Tooby, & Cosmides, 2001). The heightened amygdala activity to in-group members in the current study may stem from their motivational relevance and salience in the current group context.

Empirical Evidence: Own Race Bias or Own "Group" Bias?

Next, to obtain more evidence that self-categorization, rather than more bottom-up aspects of race, drives social perception, we extended our research to one of the most robust and widely replicated phenomena in social perception—the own-race bias. Extensive research has shown that people appear to be better at remembering people from their own race than from other races (Malpass & Kravitz, 1969)—an effect that has been variably termed the cross-race effect, same-race bias or own-race bias (ORB). This simple psychological phenomenon has caused countless individuals to exclaim that members of another race or ethnicity "all look the same to me," providing fodder for cartoonists, comedians, and satirical websites (e.g., http://www.alllooksame.com). Although the ORB may appear to be relatively innocuous, it can lead an eyewitness in a criminal case to misidentify a suspect from another race, leading to the conviction of an innocent person. Indeed, approximately 36% of wrongful convictions are due to erroneous cross-race eyewitness identification in which Caucasian witnesses misidentify minority defendants (Scheck, Neufeld, & Dwyer, 2000).

For the past several decades, perceptual expertise—a bottom-up mechanism—has been widely accepted as the primary psychological explanation for ORB. According to this account, people become expert at identifying individuals within their own race by virtue of greater exposure to own-race individuals, including family, friends, and acquaintances, relative to members of another race. This increased exposure produces a specific expertise for encoding and/or recalling own-race faces. Over the course of a lifetime of interactions with people from the same race, experience in making both within- and between-race distinctions tunes the perceptual system to make finer distinctions among exemplars within own-race faces than within other-race faces (Malpass & Kravitz, 1969).

At the neural level, one fMRI study (Golby et al., 2001) examined the relationship between the ORB and activation in the fusiform face area (FFA), a sub-region of the fusiform gyrus (see Figure 6.1, above) located on the ventral surfaces of the temporal lobe (Kanwisher, McDermott, & Chun, 1997). Building on research showing that FFA activity increases with visual expertise (see Palmeri & Gauthier, 2004, for a review), Golby and colleagues (Golby et al., 2001) presented Black and White participants with pictures of Black and White faces during fMRI. Activity in FFA was greater to own-race than other-race faces for both Black and White participants (see also Lieberman, Hariri, Jarcho, Eisenberger, & Bookheimer, 2005). Moreover, on a subsequent memory test, the degree of same-race bias (i.e., superior memory for same-race over other-race faces) was predicted by fusiform gyrus activation to racial in-group members. Consistent with the "perceptual expertise" hypothesis, these experiments suggest that extensive visual experience with faces or other stimulus categories, including one's race, may gradually tune neurons in the FFA to encode stimuli at the subordinate/individual level—that is, to make fine-grained discriminations between exemplars within a stimulus category (Tarr & Gauthier, 2000).

More recently, social categorization approaches have challenged the perceptual expertise model of ORB (Hugenberg, Young, Bernstein, & Sacco, 2010; Sporer, 2001). According to these approaches, categorizing others as in-group or out-group members may alter the depth or type of processing they receive, such that own-race faces are processed as individuals by default and other-race faces as interchangeable representatives of a social category, leading to superior recognition memory for own-race faces (Bernstein, Young, & Hugenberg, 2007; Sporer, 2001). Moreover, activity in the fusiform may not be fully contingent on expertise with specific categories, but rather may be sensitive to top-down factors, such as the level of categorization that perceivers use to recognize stimuli (Gauthier, Anderson, Tarr, Skudlarski, & Gore, 1997).

In line with this top-down approach and our previous work, we predicted that people assigned to a minimal group would encode in-group members at a subordinate level and out-group members at a superordinate level, and that this differential encoding would be reflected in differences in fusiform activity (in-group > out-group), despite participants' limited exposure to members of both categories. We reasoned that if the fusiform is merely processing expert stimuli, then White participants would show greater fusiform activity to White faces relative to Black faces, regardless of contextual information. In contrast, if the fusiform is flexibly involved in individuating stimuli—which is subject to the influence of categorizing another person as an in-group member (Bernstein et al., 2007; Brewer, 1988)—participants would show greater activity to in-group relative to out-group faces, regardless of race. Indeed, consistent with the latter hypothesis, we found greater activation within the bilateral fusiform gyri for in-group than out-group faces (Van Bavel et al., 2008). These results provide convergent evidence that the fusiform is sensitive to shifts in social contexts, responding selectively to face stimuli imbued with psychological significance by virtue of their group membership and encoding the more motivationally relevant in-group faces at the subordinate level. Moreover, these effects were not moderated by race (nor was there a main effect of race; see also Hehman, Maniab, & Gaertner, 2010; Van Bavel & Cunningham, 2012b).

We have recently replicated this pattern of in-group bias in the face-sensitive sub-region of the fusiform (i.e., the FFA) and shown that relatively greater activity in this region mediates the effects of group membership on recognition memory—

a behavioral index of individuation (Van Bavel et al., 2011). Specifically, we found a positive correlation between the FFA differences for in-group versus out-group faces and recognition memory differences for in-group versus out-group faces. These findings imply that in-group members are more likely to be processed as individuals or exemplars in a non-categorical fashion than out-group members, consistent with social cognitive models of person perception (Brewer, 1988; Fiske & Neuberg, 1990; Sporer, 2001). Thus, our research provides evidence that the motivational relevance of categories, like group membership, can affect fusiform activity in a flexible and dynamic fashion even in the absence of explicit task instructions or long-term experience with the category.

We believe our study suggests that the fusiform may play a key role in processing in-group members in greater depth than out-group members—placing in-group biases in perception firmly within the realm of motivated social perception (Balcetis & Dunning, 2006). As such, these studies on the fusiform, along with a series of recent behavioral studies (see Hugenberg et al., 2010, for a review), led us to predict that social memory would also be sensitive to the motivational aspects of social identity. In a series of studies, we examined the influence of social identification, social roles, and belonging needs on recognition memory (Van Bavel & Cunningham, 2012b; Van Bavel, Swencionis et al., 2012). As predicted, we found that the motivational aspects of the perceiver's social identity shape social attention and memory over and above mere categorization into groups. For instance, participants who reported a strong need to belong or were highly identified with their minimal in-group had a memory advantage for in-group versus out-group faces. However, social affordances attenuated in-group bias: Memory for out-group faces was heightened among participants who were assigned to a role (i.e., spy) that required attention toward out-group members. This research suggests that many aspects of social identity play a role in shaping social perception.

Although we have now accumulated extensive evidence that social identity can override or pre-empt racial bias in a number of domains, we are not suggesting that people are becoming colorblind to race. It seems likely that race may be represented in the brain, even when it is silent on a specific mental process or task, simply because it generally co-varies with visually salient features (e.g., color). Indeed, we have recent evidence that race may be encoded in the visual system, even when it does not produce racial biases (Ratner et al., 2012). As we noted above, our previous work suggests that overall activation levels in the fusiform vary as a function of salient group membership, not necessarily race. However, we re-analyzed the data using multivariate pattern analysis—a technique that can identify *patterns* of neural activity representing a type of stimulus even in the absence of greater *mean activation* in response to that stimulus category. We found that patterns of neural activity within the early visual cortex and a face-sensitive sub-region of the fusiform gyri (FG) represented the race of faces, even though the FG showed similar *overall* levels of activation to White and Black faces (Kaul et al., 2012;

Ratner et al., 2012). Moreover, race was represented in the fusiform to a greater extent than in the early visual cortex, suggesting that fusiform activity did not merely reflect low-level perceptual information (e.g., color) from the early visual cortex. The results indicate that *patterns* of activation within the FG encode race even when the *mean level* of fusiform activation is driven by other group memberships. Therefore, the human visual system may still encode color and physiognomic features that allow them to distinguish between Black and White faces, even when other more motivationally salient social categorizations pre-empt or override their influence on specific processes or tasks like face individuation. This also suggests that racial bias may (re-)emerge relatively quickly if race is made psychologically salient (see Van Bavel & Cunningham, 2011, for a discussion).

Implications

Over the past few decades, dual process and systems models have emerged as the dominant perspective in psychology (Chaiken & Trope, 1999). In particular, the development and widespread utilization of implicit measures, buttressed by research on automaticity, have suggested that stereotypes and prejudice can operate automatically and efficiently below conscious control and despite intentions to the contrary (Greenwald & Banaji, 1995). Take racial attitudes, for example: Many dual process models would predict that exposure to a Black target *automatically* gives rise to negative associations in White participants, and non-prejudiced perceivers are motivated to then exert *control* over these attitudes (Devine, 1989).

In contrast to dual process models, our research takes a dynamic systems approach to social perception and evaluation (Van Bavel, Xiao et al., 2012). Using a combination of social cognitive and neuroscience techniques, we show the flexible nature of ostensibly automatic social perception, as well as the mechanism through which these influences occur. We demonstrate that even rapidly and arbitrarily created social categories, which people have no prior knowledge about or contact with, can pre-empt or override the influence of existing social categories that often carry a great deal of societal and historical baggage, such as racial groups. Specifically, we show that automatic evaluations of and neural responses to other people are shaped by salient social and self-categorization.

Perhaps the most arresting aspect of this research is that very brief exposure to arbitrary intergroup alliances was sufficient to elicit categorization and identification according to minimal group membership, making this a more potent social category than race—a category marked by years of exposure and associated with relatively stable stereotypes and attitudes. Therefore, racial categorization may be malleable in certain contexts, including ones in which race is irrelevant to another psychologically salient social identity. Further, mere membership in an arbitrary group is sufficient to increase evaluative and behavioral preferences for in-group members; people who are actually *assigned* to one of the groups used group membership as a cue for categorization rather than race, and revealed a preference for in-group members, regardless of race, relative to those who were exposed to the groups but not made a member of either one.

Thus, while visually salient categories like race may trigger bottom-up, perceptual processes due to low-level visual features (e.g., physiognomic features), a psychologically salient social identity can trigger top-down perceptual and evaluative processing, which can attenuate the ostensibly automatic effects of race. Moreover, the top-down aspects of identity can alter relatively early aspects of perceptual and evaluative processing. This is potentially important, because it introduces the possibility that transient aspects of self-categorization can pre-empt or override the effects of visually salient and socially important categories—including categories with which people have extensive experience—perhaps before these social categories even begin to influence the perceptual and evaluative system.

This chapter is a sample of a broader research program examining the influence of social identity on perceptions of the *social* world, including rapid attention orientation (Brosch & Van Bavel, 2012), person memory (Van Bavel & Cunningham, 2012b; Van Bavel, Swencionis et al., 2012), automatic evaluation (Van Bavel & Cunningham, 2009a), and empathy (Cikara, Bruneau, Van Bavel, & Saxe, 2013). More strikingly, we have also recently found that social identity can shape group members' perception of the *physical* world, including their perception and representation of physical distance (Xiao & Van Bavel, 2012) and olfaction (Coppin et al., 2013).

Costs and Benefits

In a similar way to other chapters in this volume, we use a combination of techniques and methodologies from social psychology and cognitive neuroscience. Were it not for the development and utilization of techniques like fMRI, investigation of the flexibility of perception and evaluation would have remained on the level of behavior. For instance, in a behavioral study, we could manipulate presentation durations of our stimuli and use reaction-time measures to look at the time course of processing social stimuli. However, with neuroscience techniques, we can examine the sensitivity of different brain regions to top-down influences, challenging hard-wired views of brain function. Taking social perception, for example, we now know that the evaluative function of the amygdala is not driven purely in a bottom-up fashion by low-level features of stimuli, but also by higher-order motives (Cunningham et al., 2008; Van Bavel et al., 2008). This empirical evidence would have been impossible to establish without utilization of neuroimaging techniques to supplement our existing body of literature, which has a long history of behavioral investigations.

Moreover, our approach offers new perspectives to social psychology and cognitive neuroscience. By bringing a neuroscience approach to social psychology, we can explore concrete mechanisms underlying the abstract process of self-categorization and the construct of social identity. Additionally, considering a neuroscience perspective may help us to re-evaluate classic psychological theories, challenging the distinction between automatic and controlled processing, and suggesting that a strict dissociation is unlikely to be a natural kind grounded in brain structure or function. Meanwhile, by bringing social psychological theories to cognitive neuroscience, this research demonstrates the flexibility of the person perception network and the power of "top-down" systems to alter supposedly "bottom-up" processing. The social cognitive neuroscience perspective highlights the power of the situation to influence cognition, suggesting that even basic cognitive processes must be considered within the social identity contexts in which they are situated (Turner et al., 1994). For example, processing in the fusiform gyri—part of the ventral visual stream—appears to depend on the current motivational relevance of the stimulus being processed.

On a broader level, this work implies a model of human cognition in which a currently active mindset determines supposedly "automatic" reactions. Top-down influences on cognition and emotion need not emerge only after unavoidable automatic responses, with these responses being controlled through strategies like suppression or reappraisal. Rather, top-down influences before the fact—what we term "pre-appraisal" (Van Bavel, Xiao et al., 2012)—can determine which responses become active in the first place, thus influencing cognitive, evaluative, and affective outcomes.

Although we focus on fMRI research in this chapter, it is not well suited for addressing all types of research questions (see Cunningham et al., 2009 for a discussion). For instance, fMRI is usually more suited to delineating activities in distinct brain regions and different sub-areas, while EEG may directly examine the time course of psychological processes such as automaticity (Cunningham et al., 2009). Therefore, we have conducted several studies using EEG to examine the time course—a central feature of automaticity—of these processes. These studies suggest that social identity and other motivational factors can shape responses to faces within 100 ms of face presentation (Cunningham, Van Bavel, Arbuckle, Packer, & Waggoner, 2012; Van Bavel & Cunningham, 2012a). In other words, social identity concerns may *pre-empt* the activation of ostensibly automatic racial bias (see also Van Bavel & Cunningham, 2009a).

On a theoretical level, it is important to recognize that as useful as these methodologies are, they are by no means suited to all types of psychological research questions. In our lab, we spend a considerable amount of time trying to identify the optimal methodology and level of analysis for each research question. Ultimately, understanding social perception and evaluation with multiple methods and across multiple levels of analysis offers the greatest promise of generating integrative, process-oriented theories of self- and social categorization, and subsequently developing interventions for social issues like prejudice and discrimination.

Despite these numerous benefits, it is crucial to recognize the potential pitfalls of applying neuroscience techniques to investigations of social psychological questions, as well as some potential disadvantages in comparison with behavioral assessments. On the theoretical level, the use of reverse inference, in which one infers the presence of a cognitive process based solely on brain activity, is potentially problematic—as one brain region may be involved in a number of psychological processes—and researchers should exercise caution in this respect (see Poldrack, 2006). Researchers may want to include behavioral manipulations and measures that link cognitive processes to brain activity, and consider whether prior research has established a clear and sufficiently exclusive link between a particular psychological process and a particular brain region to support a reverse inference.

Furthermore, in terms of practicality, there are several potential disadvantages of using neuroscience techniques in social psychology research. First, research involving neuroscience techniques is usually much more costly than behavioral investigations. Second, facilities suited to conducting neuroscience research may be less readily accessible. Moreover, it usually requires prolonged training to acquire sufficient expertise to utilize neuroscience techniques and conduct appropriate data analysis. Although these practical concerns may be more mundane than potential theoretical pitfalls, it is nevertheless important to keep these in mind when conducting social neuroscience research.

Conclusion

In a complex and dynamic social world, a central challenge for adaptive human behavior is the flexible and appropriate categorization and evaluation of others. In this chapter, we discuss a social neuroscience approach to self- and social categorization, linking the effects of self-categorization and social identity on perception and evaluation to brain function. Our research illustrates that self-categorization with a social group can dramatically shape social perception and evaluation, and can pre-empt or override ostensibly pervasive racial biases. Although the effects of social categories such as race are relatively robust, our research shows that self-categorization can alter the effects of race on variables ranging from perception to evaluation, including underlying brain function. Using a social neuroscience approach not only helps to elucidate the neural substrates that underlie self- and social categorization, but also suggests that even putatively hard-wired aspects of brain function are sensitive to the top-down influence of contextual and motivational factors. As such, our responses to someone like Trayvon Martin may be influenced by seemingly trivial features of our salient social identities—such as whether we define ourselves as members of a neighborhood watch or a community association—-rather than being hard-wired and predetermined.

Suggestions for Further Reading

Theoretical

Cunningham, W. A., Zelazo, P. D., Packer, D. J., & Van Bavel, J. J. (2007). The Iterative Reprocessing Model: A multi-level framework for attitudes and evaluation. *Social Cognition, 25*, 736–760.

This article provides a theoretical framework for understanding how attitudes and evaluations can emerge dynamically over time, with contextual and motivational influences, as an alternative to strict dual process frameworks.

Turner, J. C., Hogg, M. A., Oakes, P. J., Reicher, S. D., & Wetherell, M. S. (1987). *Rediscovering the social group: A self-categorization theory*. Oxford, UK: Basil Blackwell.

This article outlines self-categorization theory, according to which representations of self- and social categorization are flexibly constructed and determine social perception.

Van Bavel, J. J., & Cunningham, W. A. (2011). A social neuroscience approach to self and social categorisation: A new look at an old issue. *European Review of Social Psychology, 21*, 237–284.

This article integrates research on self-categorization and social neuroscience.

Methodological

Amodio, D. M. (2010). Can neuroscience advance social psychological theory? Social neuroscience for the behavioral social psychologist. *Social Cognition, 28*, 695–716.

This article discusses the use of neuroscience methods in social psychological research, including when it can be useful and, importantly, how to avoid mistakes and pitfalls in social neuroscience.

Huettel, S. A., Song, A. W., & McCarthy, G. (2004). *Functional magnetic resonance imaging*. Sunderland, MA: Sinauer Associates.

A guidebook to theory and practice in the use of fMRI.

Ochsner, K. N. (2007). Social cognitive neuroscience: Historical development, core principles, and future promise. In A. Kruglanski & E. T. Higgins (Eds.), *Social psychology: A handbook of basic principles* (Vol. 2, pp. 39–66). New York: Guilford Press.

This chapter presents a history, rationale, and statement of aims of social cognitive neuroscience.

References

Amodio, D. M., Harmon-Jones, E., Devine, P. G., Curtin, J. J., Hartley, S. L., & Covert, A. E. (2004). Neural signals for the detection of unintentional race bias. *Psychological Science, 15*, 88–93.

Anderson, A. K., & Phelps, E. A. (2001). Lesions of the human amygdala impair enhanced perception of emotionally salient events. *Nature, 411*, 305–309.

Balcetis, E., & Dunning, D. (2006). See what you want to see: Motivational influences on visual perception. *Journal of Personality and Social Psychology, 91*, 612–625.

Bernstein, M. J., Young, S. G., & Hugenberg, K. (2007). The cross-category effect: Mere social categorization is sufficient to elicit an own-group bias in face recognition. *Psychological Science, 18*, 709–712.

Brewer, M. B. (1988). A dual process model of impression formation. In T. K. Srull & R. S. Wyer (Eds.), *Advances in social cognition* (Vol. 1, pp. 1–36). Hillsdale, NJ: Lawrence Erlbaum Associates, Inc.

Brosch, T., & Van Bavel, J. J. (2012). The flexibility of emotional attention: Accessible social identities guide rapid attentional orienting. *Cognition, 125*, 309–316.

Cacioppo, J. T., Berntson, G. G., Sheridan, J. F., & McClintock, M. K. (2000). Multilevel integrative analyses of human behavior: Social neuroscience and the complementing nature of social and biological approaches. *Psychological Bulletin, 126*, 829–843.

Chaiken, S., & Trope, Y. (1999). *Dual-process theories in social psychology*. New York: Guilford Press.

Cikara, M., Bruneau, E., Van Bavel, J. J., & Saxe, R. (2013). The roots of intergroup empathy bias: Intergroup competition and differentiation shape empathy toward in-group and out-group members. Unpublished manuscript, MIT.

Coppin, G., Delplanque, S., Oud, B., Margot, C., Sander, D., & Van Bavel, J. J. (2013). Swiss identity smells like chocolate: Social identity shapes olfactory experience. Unpublished manuscript, University of Geneva.

Cunningham, W. A., Johnson, M. K., Raye, C. L., Gatenby, J. C., Gore, J. C., & Banaji, M. R. (2004). Separable neural components in the processing of Black and White faces. *Psychological Science, 15*, 806–813.

Cunningham, W. A., Packer, D. J., Kesek, A., & Van Bavel, J. J. (2009). Implicit measures of attitudes: A physiological approach. In R. E. Petty, R. H. Fazio & P. Brinol (Eds.), *Attitudes: Insights from the new implicit measures* (pp. 485–512). New York: Psychology Press.

Cunningham, W. A., & Van Bavel, J. J. (2009). A neural analysis of intergroup perception and evaluation. In G. G. Berntson & J. T. Cacioppo (Eds.), *Handbook of neuroscience for the behavioral sciences* (pp. 975–984). Hoboken, NJ: Wiley.

Cunningham, W. A., Van Bavel, J. J., Arbuckle, N. L., Packer, D. J., & Waggoner, A. S. (2012). Rapid social perception is flexible: Approach and avoidance motivational states shape P100 responses to other-race faces. *Frontiers in Human Neuroscience, 6*, 140.

Cunningham, W. A., Van Bavel, J. J., & Johnsen, I. R. (2008). Affective flexibility: Evaluative processing goals shape amygdala activity. *Psychological Science, 19*, 152–160.

Cunningham, W. A., Zelazo, P. D., Packer, D. J., & Van Bavel, J. J. (2007). The iterative reprocessing model: A multi-level framework for attitudes and evaluation. *Social Cognition, 25*, 736–760.

Devine, P. G. (1989). Stereotypes and prejudice: Their automatic and controlled components. *Journal of Personality and Social Psychology, 56*, 5–18.

Draine, S. C., & Greenwald, A. G. (1998). Replicable unconscious semantic priming. *Journal of Experimental Psychology: General, 127*, 286–303.

Dubois, S., Rossion, B., Schiltz, C., Bodart, J. M., Michel, C., Bruyer, R. et al. (1999). Effect of familiarity on the processing of human faces. *NeuroImage, 9*, 278–289.

Fazio, R. H., Jackson, J. R., Dunton, B. C., & Williams, C. J. (1995). Variability in automatic activation as an unobtrusive measure of racial attitudes: A bona fide pipeline? *Journal of Personality and Social Psychology, 69*, 1013–1027.

Fiske, S. T., & Neuberg, S. L. (1990). A continuum of impression formation, from category-based to individuating processes: Influences of information and motivation on attention and interpretation. In M. P. Zanna (Ed.), *Advances in experimental social psychology* (Vol. 23, pp. 1–74). New York: Academic Press.

Freeman, J. B., & Ambady, N. (2011). A dynamic interactive theory of person construal. *Psychological Review, 118*, 247–279.

Gauthier, I., Anderson, A. W., Tarr, M. J., Skudlarski, P., & Gore, J. C. (1997). Levels of categorization in visual recognition studied with functional MRI. *Current Biology*, *7*, 645–651.

Goff, P. A., & Richardson, L. S. (2012). Running from race in our minds. *Huffington Post.* Retrieved from http://www.huffingtonpost.com/phillip-atiba-goff/trayvon-martin-race_b_1376621.html

Golby, A. J., Gabrieli, J. D. E., Chiao, J. Y., & Eberhardt, J. L. (2001). Differential fusiform responses to same- and other-race faces. *Nature Neuroscience, 4*, 845–850.

Greenwald, A. G., & Banaji, M. R. (1995). Implicit social cognition: Attitudes, self-esteem, and stereotypes. *Psychological Review, 102*, 4–27.

Greenwald, A. G., McGhee, D. E., & Schwartz, J. L. K. (1998). Measuring individual differences in implicit cognition: The Implicit Association Test. *Journal of Personality and Social Psychology, 74*, 1464–1480.

Hehman, E., Maniab, E. W., & Gaertner, S. L. (2010). Where the division lies: Common ingroup identity moderates the cross-race facial-recognition effect. *Journal of Experimental Social Psychology, 46*, 445–448.

Huettel, S. A., Song, A. W., & McCarthy, G. (2004). *Functional magnetic resonance imaging.* Sunderland, MA: Sinauer Associates.

Hugenberg, K., Young, S. G., Bernstein, M. J., & Sacco, D. F. (2010). The Categorization-Individuation Model: An integrative account of the other-race recognition deficit. *Psychological Review, 117*, 1168–1187.

Ito, T. A., & Urland, G. R. (2003). Race and gender on the brain: Electrocortical measures of attention to the race and gender of multiply categorizable individuals. *Journal of Personality and Social Psychology, 85*, 616–626.

Ito, T. A., & Urland, G. R. (2005). The influence of processing objectives on the perception of faces: An ERP study of race and gender perception. *Cognitive, Affective, and Behavioral Neuroscience, 5*, 21–36.

Kanwisher, N., McDermott, J., & Chun, M. (1997). The fusiform face area: A module in human extrastriate cortex specialized for the perception of faces. *Journal of Neuroscience, 17*, 4302–4311.

Kaul, C., Ratner, K. G., & Van Bavel, J. J. (2012). Dynamic representation of race: Processing goals shape race encoding in the fusiform gyri. *Social Cognitive and Affective Neuroscience.* doi: 10.1093/scan/nss138

Kurzban, R., Tooby, J., & Cosmides, L. (2001). Can race be erased? Coalitional computation and social categorization. *Proceedings of the National Academy of Sciences of the United States of America, 98*, 15387–15392.

Lieberman, M. D., Hariri, A., Jarcho, J. M., Eisenberger, N. I., & Bookheimer, S. Y. (2005). An fMRI investigation of race-related amygdala activity in African-American and Caucasian-American individuals. *Nature Neuroscience, 8*, 720–722.

Macrae, C. N., & Quadflieg, S. (2010). Perceiving people. In D. T. Gilbert, S. T. Fiske & G. Lindzey (Eds.), *The handbook of social psychology* (5th ed., pp. 428–463). New York: McGraw-Hill.

Malpass, R. S., & Kravitz, J. (1969). Recognition for faces of own and other "race". *Journal of Personality and Social Psychology, 13*, 330–334.

Miller, E. K., & Cohen, J. D. (2001). An integrative theory of prefrontal cortex function. *Annual Review of Neuroscience, 24*, 167–202.

Otten, S., & Wentura, D. (1999). About the impact of automaticity in the Minimal Group Paradigm: Evidence from affective priming tasks. *European Journal of Social Psychology, 29*, 1049–1071.

Palmeri, T. J., & Gauthier, I. (2004). Visual object understanding. *Nature Reviews Neuroscience*, *5*, 291–303.

Phelps, E. A. (2006). Emotion and cognition: Insights from studies of the human amygdala. *Annual Review of Psychology*, *24*, 27–53.

Phelps, E. A., O'Connor, K. J., Cunningham, W. A., Funayama, E. S., Gatenby, J. C., Gore, J. C., & Banaji, M. R. (2000). Performance on indirect measures of race evaluation predicts amygdala activation. *Journal of Cognitive Neuroscience*, *12*, 729–738.

Poldrack, R. A. (2006). Can cognitive processes be inferred from neuroimaging data? *Trends in Cognitive Sciences*, *10*, 59–63.

Ratner, K. G., Kaul, C., & Van Bavel, J. J. (2012). Is race erased? Decoding race from patterns of neural activity when skin color is not diagnostic of group boundaries. *Social Cognitive and Affective Neuroscience*. doi: 10.1093/scan/nss063

Scheck, B., Neufeld, P., & Dwyer, J. (2000). *Actual innocence: Five days to execution and other dispatches from the wrongly convicted*. New York: Doubleday.

Sporer, S. L. (2001). Recognizing faces of other ethnic groups: An integration of theories. *Psychology, Public Policy, and Law*, 7, 36–97.

Tajfel, H., Billig, M., Bundy, R., & Flament, C. (1971). Social categorization and intergroup behaviour. *European Journal of Social Psychology*, *1*, 149–178.

Tarr, M. J., & Gauthier, I. (2000). FFA: A flexible fusiform area for subordinate-level processing automized by expertise. *Nature Neuroscience*, *3*, 764–769.

Turner, J. C., Hogg, M. A., Oakes, P. J., Reicher, S. D., & Wetherell, M. S. (1987). *Rediscovering the social group: A self-categorization theory*. Oxford, UK: Basil Blackwell.

Turner, J. C., Oakes, P. J., Haslam, S. A., & McGarty, C. (1994). Self and collective: Cognition and social context. *Personality and Social Psychology Bulletin*, *20*, 454–463.

Van Bavel, J. J., & Cunningham, W. A. (2009a). Self-categorization with a novel mixed-race group moderates automatic social and racial biases. *Personality and Social Psychology Bulletin*, *35*, 321–335.

Van Bavel, J. J., & Cunningham, W. A. (2009b). A social cognitive neuroscience approach to intergroup perception and evaluation. In W. P. Banks (Ed.), *Encyclopedia of consciousness* (pp. 379–388). New York: Academic Press.

Van Bavel, J. J., & Cunningham, W. A. (2011). A social neuroscience approach to self and social categorisation: A new look at an old issue. *European Review of Social Psychology*, *21*, 237–284.

Van Bavel, J. J., & Cunningham, W. A. (2012a). The dynamic nature of social perception in the human brain. Unpublished manuscript, New York University.

Van Bavel, J. J., & Cunningham, W. A. (2012b). A social identity approach to person memory: Group membership, collective identification, and social role shape attention and memory. *Personality and Social Psychology Bulletin*, *38*, 1566–1578.

Van Bavel, J. J., Packer, D. J., & Cunningham, W. A. (2008). The neural substrates of in-group bias: A functional magnetic resonance imaging investigation. *Psychological Science*, *19*, 1131–1139.

Van Bavel, J. J., Packer, D. J., & Cunningham, W. A. (2011). Modulation of the fusiform face area following minimal exposure to motivationally relevant faces: Evidence of in-group enhancement (not out-group disregard). *Journal of Cognitive Neuroscience*, *23*, 3343–3354.

Van Bavel, J. J., Swencionis, J. K., O'Connor, R. C., & Cunningham, W. A. (2012). Motivated social memory: Belonging needs moderate the own-group bias in face recognition. *Journal of Experimental Social Psychology*, *48*, 707–713.

Van Bavel, J. J., Xiao, Y. J., & Cunningham, W. A. (2012). Evaluation as a dynamic process: Moving beyond dual system models. *Social and Personality Psychology Compass*, *6*, 438–454.

Vuilleumier, P. (2005). How brains beware: Neural mechanisms of emotional attention. *Trends in Cognitive Sciences*, *9*, 585–594.

Whalen, P. J. (1998). Fear, vigilance and ambiguity: Initial neuroimaging studies of the human amygdala. *Current Directions in Psychological Science*, 7, 177–188.

Whalen, P. J., Rauch, S. L., Etcoff, N. L., McInerney, S. C., Lee, M., & Jenike, M. A. (1998). Masked presentations of emotional facial expressions modulate amygdala activity without explicit knowledge. *Journal of Neuroscience*, *18*, 411–418.

Wheeler, M. E., & Fiske, S. T. (2005). Controlling racial prejudice: Social-cognitive goals affect amygdala and stereotype activation. *Psychological Science*, *16*, 56–63.

Wilson, T. D., Samuel, L., & Schooler, T. Y. (2000). A model of dual attitudes. *Psychological Review*, *107*, 101–126.

Xiao, Y. J., & Van Bavel, J. J. (2012). See your friends close, and your enemies closer: Social identity and identity threat shape the representation of physical distance. *Personality and Social Psychology Bulletin*, *38*, 959–972.

7

THE ROLE OF MEMORY CONSOLIDATION DURING SLEEP IN SOCIAL PERCEPTION AND STEREOTYPING

Michael A. Zárate and Luke R. Enge

Overview

The research presented in this chapter addresses important theoretical and methodological concerns regarding the propensity to use and act upon stereotypes within social perception. Traditionally, research has demonstrated that social perceivers, almost without fail, categorize people into domains such as "race" and "gender," and then apply stereotypes to those people (e.g., "All men are aggressive"). While stereotyping is an efficient mechanism to determine how to act around novel people (e.g., people you have never met before), the research surrounding it has generally painted the social perceiver as uncontrollably racist and sexist. We argue that this perspective may be overstated and is due largely to an over-reliance on "one shot" studies. Thus, we propose that social psychologists should explore the effects of social perception as it occurs over time. This contention requires minimal changes to the traditional social–cognitive methodology. In our studies, participants are required to learn information about various social targets and are then tested on that information over the course of a few days. Here, we will show how naturally occurring processes—primarily memory consolidation and sleep—promote one's ability to integrate and access recently learned personal information about social targets which in turn reduces our propensity to stereotype those social targets. Rather than *measuring* neural indicators of memory consolidation, which is not particularly feasible, in this chapter we present a line of work where we *manipulate* the time between tests to measure the physiological process that we propose underlies our effects. As such, this chapter presents a cost-effective way of studying psychophysiological processes, without needing imaging tools to test more directly the distinct pathways that consolidated versus non-consolidated memories use to produce coherent social representations.

Why Study Memory Consolidation?

Social cognition is more than the study of applied cognition as it relates to social perception (Ostrom, 1984). Rather, social cognition, and social perception more specifically, entail processes unique to humans as social beings. Faces appear to have characteristics that provoke unique types of processing. Social information tends to be much more complex than non-social information. For the typical English reader, the word "tree" will be perceived as the word "tree," and there are few other meanings. The associated semantic networks are then activated, which might facilitate reading the word "shade" and other semantically related words. Social perception, however, is generally much more complex. Any one person will be categorized by race, gender, and approximate age. Under some circumstances, attractiveness, honesty, and a host of other judgments are also used. All of these judgments occur rapidly, often, and without much conscious effort. Moreover, there are associated stereotypes for all of the presented categorizations, and people often then act on those stereotypes.

Our research, however, highlights that there are common and theoretically important limitations to the propensity to use stereotypes. Stereotypes are a convenience that many people abhor using. Moreover, people often recognize that stereotypes are simply wrong, either in content (the stereotype is wrong), or in principle (it is wrong to use stereotypes). Finally, people prefer to use personal or individuated information (information one learns about a person) over stereotypes when interacting with known individuals.

The research we present in this chapter demonstrates that personal information is indeed more readily accessed, but that it simply takes time—and more specifically, consolidation during sleep—once personal information has been learned, for it to be automatically activated. Here, we identify the theoretical processes and the social context for predicting when and how personal information becomes the primary source of information in making inferences about particular people. To do so, we rely on memory consolidation models to make unique predictions about person perception processes. Our research is contrasted with traditional social psychological models that conceptualize stereotyping as an automatic consequence of person perception.

Previous Research on Stereotyping

Social psychological experimental methods have produced a somewhat limited view of human nature regarding stereotyping and prejudice, leading to the conclusion that person perception is dominated by stereotyping rather than person-based processes. In a typical type of experiment, participants are shown photographs of others. The task might be to identify, as quickly as possible, if this person is male or female, their race, etc. (Dovidio, Evans, & Tyler, 1986; Fazio, Jackson, Dunton, & Williams, 1995; Zárate & Smith, 1990). In a different set of studies, faces of men and women might be shown, for example, and participants are given lexical decision tasks. In these tasks, participants often respond faster to words like "warm" after the presentation of a female face than after being primed with a male face. More recently, the Implicit Association Test (IAT) showed that people more easily associate negative words with Black faces than with White faces, and more easily associate positive words with White faces than with Black faces (Greenwald, McGhee, & Schwartz, 1998). From that literature, one infers that humans automatically stereotype others based on social categories such as race or gender. For instance, Breckler, Olson, and Wiggins (2006) contend that,

> Categorization of people also occurs automatically. We perceive instantly and effortlessly various characteristics of others, including their age, sex, racial group, and physical appearance. The relevant schemas, or stereotypes, are activated automatically, whether we "want" them to be activated or not. Like other schemas, stereotypes allow us to draw inferences about individuals based on the assumption that they possess the central features of the category.
>
> *(Breckler et al., 2006, p. 77)*

They are not alone. Banaji, Lemm, and Carpenter (2004) argue that,

> A rich literature on social categorization processes indicates that such processes are automatically prompted by the mere presence of a stimulus target. And in the mind of the social perceiver, stereotypes that accompany a particular category automatically accrue to its members.
>
> *(Banaji et al., 2004, p. 35)*

Those quotes represent the general view that individuals automatically categorize and stereotype others along a few basic dimensions. One might argue, however, that it is almost logical to stereotype others, given the way in which our studies are conducted. All of those paradigms present to participants pictures of novel targets (persons never seen before), and reactions to these novel targets are tested. That offers experimental control, but it also limits the inferences one can make about a person one has never seen before. If all one knows about a person is their gender or race, for instance, it might make sense to make gender- or race-based inferences. Stereotypes provide some value when other forms of information are limited. Our research highlights how learning personal details about various individuals takes time to be reflected in a response-time (RT) type measure. We show that over time, when people have been able to consolidate new individuating information into memory, personal or non-stereotypic information becomes the most easily accessed information. The next section discusses how our research paradigm, in which we use sleep as a proxy for memory consolidation, contrasts with the traditional research paradigm and proposes a cost-effective methodological modification to traditional social cognitive paradigms. We hope this will provide a foundation for developing distinct lines of research regarding not only social perception over time, but the dynamic nature of other important social cognitive variables such as persuasion and other paradigms which include new information. This new information can then be contrasted with reactions based on more well-developed memory structures.

Beyond the Single-session Experiment

Within most social categorization and social perception studies, what drives stereotypic processing? Most stereotyping measures, especially RT-type measures, rely on a semantic network model of some sort. Response-time measures are based on the idea that without conscious effort, information travels between semantically related networks and is made more accessible through priming mechanisms. That semantic network, however, needs to be established for information to travel over it. For most people we just meet, however, what are the available semantic networks? They are the associated stereotypes. Even if I just learn that "Judy is humorous," that information is not well represented in memory, and as such, RT measures should be slow to respond to that newly learned information. Because stereotypes are well-established memory structures, however, saying that "Judy is warm" might come easily for those who have the stereotype that women are warm.

Independent of any methodological or theoretical perspective, the view that everyone is racist and sexist all the time makes life difficult. As a member of an ethnic minority, am I to believe that every White person who is being nice to me is merely suppressing their negative stereotypes (Devine, 1989)? Is it true that it is cognitively taxing for them to talk with me because they are expending extra cognitive energy trying to appear non-racist (Richeson & Trawalter, 2008)? On the other side of the coin, are majority-group members then supposed to avoid intergroup interactions for fear of appearing racist and sexist? No doubt racism is alive and well, but it seems plausible that racism and sexism appear ubiquitous because researchers seek out instances of it. Memory consolidation processes suggest that after time and sleep, person memory processes should become primary. Inferences learned through direct experience are often more vivid than inferences based on general stereotypes. Because individuating information was learned first-hand, perceivers are more likely to believe it to be true. Acting on known traits about people is preferable to acting on stereotypes you believe may be true, or acting on stereotypes you try to avoid using. Person traits are something you have inferred from previous interactions, even when they are stereotype-consistent. Non-stereotypic traits are also less visually salient (Rothbart & John, 1985), and therefore less accessible upon first sight of an individual. Therefore, person information, once consolidated, should become primary when judging the social target to whom that information refers. Thus, the goal of this research is to address a slightly more nuanced question. When are stereotypes applied, and how do stereotypes interact with our daily experiences to influence social perception more broadly?

Many of our daily interactions occur with people over time. One might see the same receptionist at the gym every day at 5 AM. Just as importantly, the receptionist is always cheerful and busy—at 5 AM. One sees the same people during lunch breaks. One need not talk to them to infer that one person always seems nice, whereas another always seems busy. Maybe a third person seems rude, despite the minimal interactions. Many social interactions occur over time. Some interactions are minimal, and other interactions, as with colleagues, for instance, are more extensive. All of these interactions, however, are memorable enough to produce coherent representations of the individuals. These acquaintances represent a large proportion of our daily interactions and possibly represent a limited example of the contact hypothesis (Pettigrew, 1998).

How do interactions with others across time influence how those individuals are perceived? Here, we will integrate memory consolidation constructs with social cognitive models of social perception. Before discussing the importance of memory consolidation constructs for social cognitive models of social perception, we will outline a neurocognitive model of social perception that suggests the parallel and recurrent nature of person and group representations. This model represents a novel theoretical contribution to the traditional discrete, stage-based models of social perception (e.g., Brewer, 1988; Fiske & Neuberg, 1990). Broadly, our work suggests that person and group information are distinct neurological processes that act in parallel. Each process relies on a different set of semantic memory structures that in turn lead to dissociations in social behavior (e.g., individuation versus stereotyping).

Social Neurocognitive Model of Social Perception

This research extends our prior work (Rivera, Arms-Chavez, & Zárate, 2009, 2012; Sanders, McClure, & Zárate, 2004; Zárate, Sanders, & Garza, 2000; Zárate, Stoever, MacLin, & Chavez, 2008) to focus on how person- and group-level representations become integrated over time. Person-level representations are defined as representations that rely on learned individuating traits. Individuating traits represent information about a person, and are irrelevant to any particular stereotype. At times, person-level information may be stereotype-consistent. Thus, one might learn that a particular woman is warm, which would be consistent with the stereotype that women are warm. It may be the case that learning that a person is warm via first-hand observation leads to different consequences from inferring that a person is warm because of a stereotype. Group-level information, however, entails known stereotypes or/and in-group favoritism effects. Group-level stereotypes are generally well learned and tend to be the focus of most social psychological studies.

Our interest is in understanding how person-level and group-level representations co-exist and work together to form coherent social representations. Our previous research investigating cerebral asymmetries in social perception revealed that person- and group-level representations appear to be stored separately. This neurocognitive model (Zárate et al., 2000) contends that processing asymmetries in our visual systems produces distinct social representations. The left cerebral hemisphere, which is known to store semantic information, shows evidence of stereotyping, while the right hemisphere shows evidence of person memory (Zárate et al., 2000). These asymmetries are driven by distinct visual processing sub-systems (Marsolek, 1999). In particular, the left hemisphere processes abstract, or category, membership information more readily because it identifies specific features salient to a given category. Feature identification is consistent with prototype-like processing that produces general group representations. The right hemisphere, on the other hand, is more adept at identifying combinations of features. Feature combinations are a signature of exemplar theories and produce person-based representations (Smith & Zárate, 1992). Across a series of studies, we demonstrated that category targets prime the stereotype only in the left hemisphere. In contrast, the right hemisphere produces mere exposure effects, demonstrating implicit memory for specific items. Rivera et al. (2009) further demonstrated that sex categorization processes are inhibited when left-hemisphere resources are being used. Rivera et al. (2012) demonstrated again that at fast intervals (within about 80 milliseconds [ms]), sex categorization processes are inhibited when distractor faces were presented to the left hemisphere, but fame judgments are inhibited when distractor faces were presented to the right hemisphere. In essence, the research demonstrates that the cerebral hemispheres are suited for distinct person versus group judgments, and that both sets of processes rely on unique working memory resources.

Together, these studies suggest a departure from previous models which contend that person and group processes are either a serial process (one is the default and the other is used under certain circumstances) or some stage-based, discretely represented continuum (Brewer, 1988; Fiske & Neuberg, 1990). Rather, our research demonstrates that there are at least two distinct processes that work in parallel. These systems represent and produce person and group information in parallel, albeit separately, and use distinct resources to accomplish these tasks. Thus, the two processing systems appear to use distinct working memory resources, operating in different cerebral hemispheres (Rivera et al., 2012). They appear to utilize social information differently and rely on different memory structures (Zárate et al., 2000). Our working hypothesis, then, is that these two systems produce a "horse-race" model of sorts. The model is then subject to various top-down (social, contextual, and individual differences variables) and bottom-up (asymmetries in the processing of specific and abstract visual information) constraints where one process wins the race and produces a conscious output. Moreover, once a process wins the race, it seems plausible that it then works to inhibit alternative processes. That research leads to the hypothesis that under some circumstances, stereotyping need not occur at all. These processes, therefore, may also be dynamic in nature—person and group representations change over time—suggesting that humans may not always rely on group information to guide social behavior.

The Social Context of Recognizing Personal Information

Our basic hypothesis is that learning experiences require time and sleep for those experiences to be well integrated in memory and available for automatic activation. Because our experimental methods are often limited to one experimental session, that may have produced an over-reliance on stereotypes to the exclusion of individuating information. To test this general hypothesis, we use traditional semantic priming paradigms and manipulate the number of experimental sessions.

In the next section, we outline how sleep, in particular, should enhance learning experiences and change how individuals are perceived. We then outline a model regarding the distinct effects that memory consolidation produces for social perception along with some supporting research for that model. Finally, we discuss some novel hypotheses regarding memory consolidation and social perception.

The Role of Sleep in Memory Consolidation

Memory consolidation refers to relatively new, unstable memories becoming stable representations through enhanced network connections between regions of the medial temporal lobe and the neocortex (McGaugh, 2004). Memory consolidation occurs primarily during sleep and integrates or solidifies daily learned events. Within the medial temporal lobe, the hippocampus is involved in the storage and retrieval of recently experienced events (i.e., short-term memory). Conversely, the neocortex is the cognitive controller through its overarching ability to implement executive functions and its role in long-term memory storage. Research has explored the role of the hippocampus in relation to areas of the neocortex as the underlying physical substrates that implement memory consolidation during sleep.

Memory consolidation has been explored in terms of two anatomically separate and complementary learning systems. The general principles behind the complementary learning systems are based on models of human learning and memory, wherein the hippocampus rapidly encodes new information and stores this information as a series of independent and orthogonal cognitive representations. Later, through consolidation processes that occur during offline states (e.g., sleep), the neocortex works with the hippocampus in order to integrate that new information with already existing overlapping cortical representations (Lewis & Durrant, 2011; McClelland, McNaughton, & O'Reilly, 1995; O'Reilly & Rudy, 2000). These general principles reflect the anatomy and function of these two distinct learning systems.

One might ask, "Why sleep and not simply time?" From a purely social psychological perspective, it may not matter. Over time, people sleep, meaning time and sleep are naturally confounded. As such, at the beginning stages of this research program, our team has focused on how experiences influence perceptions over time (and therefore sleep) and how those influence person-perception processing. From a broader theoretical perspective, though, distinguishing the processes can illuminate how person and group perception interact. The research suggests, however, a clear role of sleep in memory consolidation. Gais and Born (2004a, 2004b) reviewed multiple studies which suggested that an increase in spindle activity and slow oscillation EEG (slow cortical potentials) in the medial prefrontal cortex after a period of sleep was associated with overall better recall of declarative information, whereas a similar but opposite pattern (a decrease in spindle activity) was found at the initial encoding of the event before sleep. Furthermore, sleep (versus awake states) facilitates functional connections between the hippocampus and neocortex during a behavioral recognition task of emotional material (Payne & Kensinger, 2010, 2011; Sterpenich et al., 2009).

From a neuro-chemical perspective, slow-wave sleep reduces the release of various hormones that play a central role in learning and memory. For example, communication between the medial temporal lobe system and the medial prefrontal cortex that occurs during sleep is regulated by acetylcholine. Hasselmo (1999) and Gais and Born (2004b) demonstrated that reduced cholinergic (reduction in release of acetylcholine) activation during slow-wave sleep increases recall of recently learned word pairs, compared to an increase in activation during awake states. Furthermore, sleep provides an environment that is more resistant to interference. During sleep, sensory input is reduced, providing less interference with previously learned information. During awake states, however, sensory information is constant which can interfere with learning new material. The unique combination of slow-wave sleep, a decrease in acetylcholine, and a reduced level of sensory input all facilitate consolidation and, more specifically, the integration and generalization of recently learned material (Lewis & Durrant, 2011). Future research will make specific comparisons regarding sleep versus time, partly to develop a full understanding of how memory consolidation processes influence social perception, and partly to integrate the social psychological literature with the neurocognitive literature regarding memory consolidation.

Memory Consolidation and Social Psychology: Integration, Accessibility, and Generalization

Memory consolidation processes are uniquely suited for social psychological research. The complementary learning systems framework predicts that sleep will integrate episodic memories with already existing semantic memories. This process maps on well to social memory phenomena. For example, multiple episodic person memories (e.g., small slices of trait information), after consolidation, should bind together to form a coherent semantic memory or impression of a person. After sleep, newly integrated trait information (Carlos is honest) will allow for a more complete impression of a person (Carlos is an honest man) and for novel behavioral inferences about that individual (I trust that Carlos will not steal my wallet when visiting my home). Once information is easily accessible, it is hypothesized also to generalize more readily to similar instances (Smith & Zárate, 1992).

Furthermore, after learning negative emotional information and after that information becomes consolidated during sleep, that negative emotional information is more readily retrieved (e.g., Payne & Kensinger, 2011), suggesting that memory consolidation benefits memory for negative emotional information more than for other emotional material. Given the role of emotion-laden events in person memory, this connection is particularly important in the development of negative stereotypes and prejudice. Collectively, memory consolidation changes how individuals are represented in memory (e.g., the integration of new information changes the representation associated with that person), and how information is accessed from those memories over the course of time (e.g., after sleep, a memory should be more accessible than before sleep, and this should be reflected in behavioral recall tasks).

Collectively, this research suggests that memory consolidation produces three distinct effects: integration of new information into existing knowledge structures; increased accessibility to those newly formed memories; and generalization of that newly integrated information to novel situations (e.g., inference).

Supporting Research

Recent research in our lab demonstrates the utility of the memory consolidation approach. In this work, rather than measuring memory consolidation as the mediating process explaining how personal information becomes solidified in impression formation, we manipulate memory consolidation through within-subject designs in which participants are tested before and after sleep. Future studies will, however, have to test specifically the effects of time versus sleep on consolidation and social perception. Arms-Chavez, Enge, Rivera, and Zárate (2012) presented participants with a set of 12 targets. Participants were asked to form impressions of the 12 targets. Each target was presented as a photo, a trait descriptor, and two sentences that describe that trait. For example, participants would view a photograph of a female named "Edith" who is "active" ("I am Edith, I am active"). Next to the photograph and below the trait, two sentences describing behaviors consistent with "active" would appear. For example, "I enjoy running in the park," and "I always exercise before I go to work in the morning." The target list was split into two equal groups of six (three female, three male), labeled groups A and B, which were used to counterbalance the stimuli. The targets were presented repeatedly to participants to model repeated interactions and to provide distributed practice. Learning time totaled 1 minute per target, which is not particularly extensive, and thus provides a strong test of the basic hypothesis.

Participants then returned for two testing sessions, where they completed a lexical decision task. The short-delay testing session took place approximately 4 hours later that same day. At that session, participants were tested on one half of the photos. Participants also returned 48 hours later and were tested on the other half of the targets. This test included learned personal information from the training session (e.g., "active"), or stereotypic information (e.g., "warm") paired with learned versus novel photographs. All stereotypic traits were pre-tested to ensure that they were perceived as stereotypically male or female for our sample. Thus, this design allows for one learning session, but two testing sessions, and each item is then tested once. The 4-hour delay is intended to provide some memory interference and to reduce short-term memory effects. The long delay (48 hours) is intended to provide for mechanisms acting during sleep to consolidate the recently learned individuating information.

Across two separate studies, participants consistently responded faster to the personal information than to the stereotypic traits, but only after the 2 nights of sleep. This suggests that once person information is consolidated—or made more accessible in memory— that information then becomes primary when judging that social target. From a cognitive perspective, it seems that person information becomes primary simply because it has been integrated into a more accessible semantic memory structure. From a social psychological perspective, this information becomes primary because of the utility it provides for the social interaction—it is no longer profitable to stereotype that individuated target, but rather to interact with them based on the personal traits they possess. For instance, in Experiment 1, participants responded significantly faster to the individuated traits within the long-delay test than within the short-delay test, whereas the responses to the stereotypic traits did not differ across time. Without the delay, participants responded faster to the stereotypic information than to the personal information. Just as importantly, after sleep (in the long-delay test), participants responded faster to the individuating traits than to the stereotypic traits for learned, or individuated, targets. For novel targets, participants still responded faster to stereotypic traits than for non-stereotypic traits, regardless of whether or not they had slept. Experiment 2 replicated that set of effects quite nicely.

Thus, in order for personal information to be reflected in automatic RT measures, time must be allowed for the information to be consolidated into existing memory structures, presumably through sleep. Sleep or time appears to increase the relative accessibility to recently learned information. The strengthening in semantic relation between a learned target and individuating information therefore changes how individuals are perceived and interacted with at a later time (i.e., individuation for those targets becomes facilitated coupled with a decrease in stereotype activation). In addition, included in those studies were measures that mirrored and replicated many social categorization effects from previous studies. For instance, novel targets (not the 12 targets that participants learned about in the first session) still produced faster responses to the stereotypic terms than they did to the neutral traits. That is a replication of the classic stereotype priming effect.

Costs and Benefits of this Approach

This research differs from the normal paradigm in a few basic ways. The primary difference is that to test how social perception changes over time, one must return to the lab after the participants have had an opportunity to sleep. In the normal social perception study, participants are tested in one experimental session. Further, if one wants to test the effects of a short delay versus a long delay, one can test on only one half of the items during the first delay. If the short-delay test included all the items, then it would become another learning session. Thus, participants are asked to participate in at least three different experimental sessions. It is also the case, however, that to use within-participant designs, individuals need to participate at least three times over at least 2 days. Such demands can produce participant loss and may provide participants with the opportunity to develop expectations for the second testing session. Research in our lab is specifically contrasting the use of within-participant designs with between-participant experimental designs to rule out specific forms of practice effects.

The primary benefit of this approach, however, is that time delays can be incorporated into almost any research design and be used by researchers across multiple paradigms, be it social perception, persuasion, and any study using memory. An obvious drawback of our approach, however, is that we do not directly measure consolidation, but rather assume that consolidation occurs with sleep over time. Our goal right now is to focus on how consolidation affects different types of social perception instead of focusing on how different manipulations of consolidation affect social perception.

We also propose that neurophysiological measures are not necessary to yield and test novel hypotheses regarding social perception. Our own research (e.g., Arms-Chavez et al., 2012; Rivera et al., 2009, 2012; Zárate et al., 2008) has demonstrated the utility of modifying basic semantic priming paradigms to understand how person and group information are differentially represented and accessed from memory, how those representations change over time, and how each process can recursively influence the other. As such, we suggest that basic social categorization tasks can be used to test novel predictions regarding social cognitive neuroscience.

Eventually, as behavioral data are accrued, one could implement this paradigm concurrently with neurophysiological measures of cognitive function, attention, and motor activity (EEG/ERP; fMRI; eye-tracking; mouse-tracking). Recent research using neurophysiological measures, specifically EEG/ERPs and fMRI has been able to dissociate important underlying cognitive mechanisms in social perception. Through the use of event-related brain potentials (ERPs), researchers have investigated not only the time course of face perception itself, but the motivational and perceptual constraints involved with sex and race perception and intergroup relations (see Ito & Bartholow, 2009, for a nice review; Amodio, 2010 and Ofan, Rubin, & Amodio, 2011 for experimental evidence; see also the chapters by Amodio [9], Bartholow [10], and Derks [3] in this volume). While these measures have provided an important level of analysis to the research surrounding social perception processes, they can be quite costly to implement. Furthermore, before implementing neurophysiological measures in social perception research, we argue that one should first rely on traditional measures to study social perception. Without strong support from these traditional behavioral methods, the benefits of neurophysiological measures are less clear, and in some cases may become a case of looking at "what lights up," instead of understanding how those underlying mechanisms lead to social behavior.

Memory consolidation effects are important for social psychological research on a number of levels. The research has the potential to change how we think about others. In particular, memory consolidation constructs provide an avenue to test how daily learning experiences influence group perception processes. Our basic hypothesis is that learning about a particular person might produce the "remember" type of conscious responses, but should not produce the "know" type of responses more consistent with automatic processing (Hirshman & Henzler, 1998). After consolidation, "remember" judgments might appear more like "know" judgments, without the conscious effortful recall. Indeed, our supporting research has provided evidence that after consolidation, recently learned personal information is more readily accessed and used than relevant group stereotypes.

Just as importantly, this research ties together multiple areas of psychology. Our daily interactions are more than either close friendships, or a series of in-group and out-group interactions. Rather, individuals are constantly learning personal information about others. Sometimes those facts confirm well-established stereotypes. That woman you just met is warm. At other times, however, that information is irrelevant to known stereotypes. That woman is also funny. At other times still, information contradicts known stereotypes. She does not appear to be comfortable around children. Social psychological theories will benefit from a clear understanding of how that information is integrated with existing knowledge structures.

Extending the Memory Consolidation Paradigm to Study Group Perception

From a behavioral standpoint, the effects produced by memory consolidation underscore the idea that memory is reorganized during sleep to integrate new information in order to use that information to make inferences and generalize those experiences to novel situations. The ability to integrate and generalize recently learned information to guide judgments and behavior in future novel situations is the rationale for our line of research in social cognition. Thus far, our data have demonstrated that learning that "Joe is active" and enjoys "running in the park" affects how Joe is perceived at a later time. Joe may be a male, but after integrating and retrieving the fact that he is active, it changes how one responds to Joe on an implicit memory task. A novel question for future research might be whether this information generalizes to other members of Joe's group? Thus, when given information about Joe and three of his friends, over time, does one start to generalize to say "they are all like Joe"? If I meet a friend of Joe's, do I infer that s/he is also active? Indeed, this may be important for understanding how person perception processes (individuation) inform group perception processes (entitativity and stereotyping; e.g., Hamilton & Sherman, 1996).

Future research should build on this principle in testing how learning about one group member can change perceptions of other members of that same group. First, this would yield novel social psychological hypotheses on how stereotypes form and how various bottom-up (e.g., the presentation of a face) and top-down (e.g., individuating information) processes work to influence not only how a particular person is perceived, but how similar others are perceived. Second, this methodology also allows us to understand the nuances of social learning. Thus far we have used a paired-associate learning task in our research. This explicit encoding mechanism reflects only a part of our social interactions. Often, we are bombarded with social information in an implicit fashion—we are sometimes not aware of the amount and content of information we learn about on a daily basis. Disentangling the differential effects of explicit and implicit social learning on later social perception will also further dissociate the functional capabilities of memory consolidation within social psychology.

Conclusions

Memory consolidation effects (accessibility, integration, and generalizability) are involved in most social psychological constructs, and from a theoretical perspective, can add a great deal to our understanding of social behavior. In addition, our research has only begun to test the boundaries of the effects of memory consolidation on person perception. Right now, we can say confidently that over time, representations change. It takes time for learning experiences to be reflected in implicit-memory-type measures. Thus, stereotyping does occur, but maybe it is less salient once one knows any information about others. Our studies have not, however, directly tested if it is time, per se, or sleep that directly yields these effects. New studies are being designed to test that specific effect. The literature suggests that emotional events take on greater emphasis through memory consolidation than do non-emotional stimuli. Social stimuli are, obviously, often emotion-laden, and moreover, negative interactions intuitively seem more easily learned and generalized to an entire group. Memory consolidation constructs predict that type of process. Thus, future work will contrast how emotional stimuli change over time compared to less emotion-laden stimuli. All of the directions are both theoretically relevant and are easily applied to normal social interactions. Moreover, this research is potentially important because it starts to shed light on one basic issue. Are individuals always and uncontrollably racist? Our research suggests not.

Suggestions for Further Reading

Lewis, P. A., & Durrant, S. J. (2011). Overlapping memory replay during sleep builds cognitive schemata. *Trends in Cognitive Sciences, 15*, 343–351.

This article provides a succinct overview on the contribution of various physiological mechanisms, during sleep, to memory consolidation. These mechanisms are hypothesized to promote a "memory replay," that in turn binds information into a coherent cognitive representation or "schema." The discussion then extrapolates on the functionality of these processes to more abstract behaviors such as inference and creativity.

McClelland, J. L., McNaughton, B. L., & O'Reilly, R. C. (1995). Why there are complementary learning systems in the hippocampus and neocortex: Insights from the successes and failures of connectionist models of learning and memory. *Psychological Review, 102*, 419–457.

A seminal overview on the distinct underlying mechanisms of human learning and memory. The paper builds on research from human lesion studies (e.g., HM; Scoville & Milner, 1957) and decades of research about various subdivisions of memory (e.g., short-term versus long-term memory). The authors argue that memory is subdivided into two anatomically dissociable systems, each with distinct neuro-computational principles and goals. Offline states are hypothesized to play a critical role in the communication and transfer of information between these two systems.

Payne, J. D., & Kensinger, E. A. (2011). Sleep leads to changes in the emotional memory trace: Evidence from fMRI. *Journal of Cognitive Neuroscience, 23*, 1285–1297.

An empirical article exploring how sleep affects emotional memory at both a behavioral and neural level. Sleep is hypothesized to facilitate functional connections between regions of the brain associated with emotional processing (the amygdala) and memory retrieval (areas of the prefrontal cortex). Given the often highly emotional nature of social interactions, we find this connection particularly important for extending the paradigm we presented here to study further the social neuroscience of how intergroup relations, prejudice, and stereotyping change over time.

References

Amodio, D. M. (2010). Coordinated roles of motivation and perception in the regulation of intergroup responses: Frontal cortical asymmetry effects on the P2 event-related potential and behavior. *Journal of Cognitive Neuroscience, 22*, 2609–2617.

Arms-Chavez, C. J., Enge, L. R., Rivera, L. O., & Zárate, M. A. (2012). On the malleability of social perception: Facilitating individuation with time. Unpublished raw data.

Banaji, M., Lemm, K. M., & Carpenter, S. J. (2004). The social unconscious. In M. B. Brewer & M. Hewstone (Eds.), *Social cognition* (pp. 28–53). Malden, MA: Blackwell Publishing.

Breckler, S. J., Olson, J. M., & Wiggins, E. C. (2006). Social cognition: Thinking about people. In S. J. Breckler, J. M. Olson, & E. C. Wiggins, *Social Psychology Alive* (pp. 65–108). Belmont, CA: Thomas Higher Education.

Brewer, M. B. (1988). A dual process model of impression formation. In R. S. Wyer and T. K. Srull (Eds.), *Advances in social cognition* (Vol. 1, pp. 1–36). Hillsdale, NJ: Lawrence Erlbaum Associates.

Devine, P. G. (1989). Stereotypes and prejudice: Their automatic and controlled components. *Journal of Personality and Social Psychology, 56*, 5–18.

Dovidio, J. F., Evans, N., & Tyler, R. B. (1986). Racial stereotypes: The content of their cognitive representations. *Journal of Experimental Social Psychology, 22*, 22–37.

Fazio, R. H., Jackson, J. R., Dunton, B. C., & Williams, C. J. (1995). Variability in automatic activation as an unobtrusive measure of racial attitudes: A bona fide pipeline? *Journal of Personality and Social Psychology, 69*, 1013–1027.

Fiske, S. T., & Neuberg, S. L. (1990). A continuum of impression formation, from category-based to individuating processes: Influences of information and motivation on attention and interpretation. In M. P. Zanna (Ed.), *Advances in experimental social psychology* (Vol. 23, pp. 1–74). New York: Academic Press.

Gais, S., & Born, J. (2004a). Declarative memory consolidation: Mechanisms acting during human sleep. *Learning and Memory, 11*, 679–685.

Gais, S., & Born, J. (2004b). Low acetylcholine during slow wave sleep is critical for declarative memory consolidation. *Proceedings of the National Academy of Sciences of the United States of America, 101*, 2140–2144.

Greenwald, A. G., McGhee, D. E., & Schwartz, J. L. K. (1998). Measuring individual differences in implicit cognition: The Implicit Association Test. *Journal of Personality and Social Psychology, 74*, 1464–1480.

Hamilton, D. L., & Sherman, S. J. (1996). Perceiving persons and groups. *Psychological Review, 103*, 336–355.

Hasselmo, M. E. (1999). Neuromodulation: Acetylcholine and memory consolidation. *Trends in Cognitive Sciences, 3*, 351–359.

Hirshman, E., & Henzler, A. (1998). The role of decision processes in conscious recollection. *Psychological Science, 9*, 61–65. doi: 10.1111/1467-9280.00011

Ito, T. A., & Bartholow, B. D. (2009). The neural correlates of race. *Trends in Cognitive Sciences, 13*, 524–531.

Lewis, P. A., & Durrant, S. J. (2011). Overlapping memory replay during sleep builds cognitive schemata. *Trends in Cognitive Sciences, 15*, 343–351.

McClelland, J. L., McNaughton, B. L., & O'Reilly, R. C. (1995). Why there are complementary learning systems in the hippocampus and neocortex: Insights from the successes and failures of connectionist models of learning and memory. *Psychological Review, 102*, 419–457.

McGaugh, J. L. (2004). The amygdala modulates the consolidation of memories of emotionally arousing experiences. *Annual Review of Neuroscience, 27*, 1–28.

Marsolek, C. J. (1999). Dissociable neural subsystems underlie abstract and specific object recognition. *Psychological Science, 10*, 111–118.

Ofan, R. H., Rubin, N., & Amodio, D. M. (2011). Seeing race: N170 responses to race and their relation to automatic racial attitudes and controlled processing. *Journal of Cognitive Neuroscience, 23*, 3153–3161.

O'Reilly, R. C., & Rudy, J. W. (2000). Computational principles of learning in the neocortex and hippocampus. *Hippocampus, 10*, 389–397.

Ostrom, T. (1984). The sovereignty of social cognition. In R. S. Wyers & T. R. Srull (Eds.), *Handbook of social cognition* (Vol 1., pp. 1–38). Hillsdale, NJ: Lawrence Erlbaum.

Payne, J. D., & Kensinger, E. A. (2010). Sleep's role in the consolidation of emotional episodic memories. *Psychological Science, 19*, 290–295.

Payne, J. D., & Kensinger, E. A. (2011). Sleep leads to changes in the emotional memory trace: Evidence from fMRI. *Journal of Cognitive Neuroscience*, *23*, 1285–1297.

Pettigrew, T. F. (1998). Intergroup contact theory. *Annual Review of Psychology*, *49*, 65–85.

Richeson, J. A., & Trawalter, S. (2008). The threat of appearing prejudiced and race-based attentional bias. *Psychological Science*, *19*, 98–102.

Rivera, L. O., Arms-Chavez, C. J., & Zárate, M. A. (2009). Resource dependent effects during sex categorization. *Journal of Experimental Social Psychology*, *45*, 908–912.

Rivera, L. O., Arms-Chavez, C. J., & Zárate, M. A. (2012). Hemispheric resource availability influences face perception: A multiple resource approach to social perception. *Laterality: Asymmetries of Body, Brain and Cognition*, *17*, 369–383.

Rothbart, M., & John, O. P. (1985). Social categorization and behavioral episodes: A cognitive analysis of the effects of intergroup contact. *Journal of Social Issues*, *41*, 81–104.

Sanders, J. D., McClure, K. A., & Zárate, M. A. (2004). Cerebral hemispheric asymmetries in social perception: Perceiving and responding to the individual and the group. *Social Cognition*, *22*, 279–291.

Scoville, W. B., & Milner, B. (1957). Loss of recent memory after bilateral hippocampal lesions. *Journal of Neurology, Neurosurgery and Psychiatry*, *20*, 11–21.

Smith, E. R., & Zárate, M. A. (1992). Exemplar-based model of social judgment. *Psychological Review*, *99*, 3–21.

Sterpenich, V., Albouy, G., Darsaud, A., Schmidt, C., Vandewalle, G., Dang-Vu, T. T. . . . Maquet, P. (2009). Sleep promotes the neural reorganization of remote emotional memory. *The Journal of Neuroscience*, *29*, 5143–5152.

Zárate, M. A., Sanders, J. D., & Garza, A. A. (2000). Neurological disassociations of social perception processes. *Social Cognition*, *18*, 223–251.

Zárate, M. A., & Smith, E. R. (1990). Person categorization and stereotyping. *Social Cognition*, *8*, 161–185.

Zárate, M. A., Stoever, C. J., MacLin, M. K., & Chavez, C. J. (2008). Neurocognitive underpinnings of face perception: Further evidence of distinct person and group perception processes. *Journal of Personality and Social Psychology*, *94*, 108–115.

8

HEART RATE AND HEART RATE VARIABILITY RESPONSES TO STEREOTYPE ACTIVATION AMONG NON-STEREOTYPED INDIVIDUALS

Stereotype Lift in the Motor Domain

Aïna Chalabaev, Philippe Sarrazin, and Jeanick Brisswalter

Understanding how social beliefs (e.g., stereotypes) create social reality is a major topic in social psychology. We (Philippe Sarrazin and Aïna Chalabaev) investigate this question in the domain of human movement. Our goal is to identify more effectively to what extent motor behaviors reflect internalization of cultural beliefs, and notably gender stereotypes (e.g., Chalabaev, Sarrazin, & Fontayne, 2009; Chalabaev, Sarrazin, Stone, & Cury, 2008; Chalabaev, Sarrazin, Trouilloud, & Jussim, 2009). In this chapter, we present a research program in which we employed cardiovascular (CV) measures to uncover the processes underlying the phenomenon of stereotype lift (i.e., the performance-boosting consequences of salient negative out-group stereotypes) among males performing in the motor domain.

The motor area is a domain where gender differences are among the largest (e.g., Eagly, 1987). These differences are notably observed in activities that require strength, speed, or endurance (e.g., Knisel, Opitz, Wossmann, & Keteihuf, 2009; Olds, Tomkinson, Léger, & Cazorla, 2006). As an illustration, male world records are systematically higher than female ones in athletics (e.g., 100m: 9.58 seconds for men versus 10.49 seconds for women), and swimming (e.g., 100m freestyle: 46.91 seconds for men versus 52.07 seconds for women). How can these gender differences be explained? Usually they are perceived as resulting from natural biological factors: During puberty, physical abilities develop more among males than females, which results in males performing better in motor tasks that require strength or speed (e.g., Thomas & French, 1985).[1] However, in addition to biological factors, may cultural beliefs also partly shape gender differences in the motor domain?

Given the biological factors that lie at the basis of performance differences in sports, it may seem counterintuitive to hypothesize that these differences are due to cultural beliefs. However, it is precisely because people strongly believe in the natural origin of gender differences in the athletic domain that culture may play a role in these differences. Indeed, this highly shared stereotype contrasts with evidence that although in relative terms, gender differences observed in the motor domain are among the largest, in absolute terms, they are not as important as they appear: Gender has been shown to predict only 5% of the variance in physical abilities (e.g., Eagly, 1987), suggesting that gender differences in strength, coordination, or velocity cannot be uniquely explained by biological factors. The goal of this chapter is to examine how gender stereotypes may shape these gender differences. Specifically, we present research that investigates whether activation of gender stereotypes may lead to gender gaps in motor performance through motivational processes (i.e., effort).

The impact of gender stereotypes on motor performance was examined using physiological measures of motivation, within a human movement approach. Research on human movement has a long tradition in exploring the links between the mind and the body, by examining how cognitive, affective, or motivational variables shape motor behaviors (e.g., DeCaro, Thomas, Albert, & Beilock, 2011); and conversely, how motor actions may affect the psychological functioning of individuals (e.g., Cacioppo, Priester, & Berntson, 1993; Cretenet & Dru, 2004). As a consequence, this domain of research has a long tradition in using a multidisciplinary approach. Combining theories and methodologies from psychology, neurosciences, physiology, and biomechanics is indeed necessary to have a more accurate picture of the complex interconnections that exist between the mind and the body. Following this reasoning, scientists have implemented different types of measures in their studies, including psychological variables, muscular activity (e.g., Aarts, Custers, & Marien, 2008; Coombes, Higgins, Gamble, Cauraugh, & Janelle, 2009; Payen et al., 2011); movement kinematics (e.g., Gray, 2004); cardiovascular (CV) responses (e.g., Filaire, Alix, Ferrand, & Verger, 2009); and cortical activity (e.g., Farmer et al., 2004; Omlor, Patino, Hepp-Reymond, & Kristeva, 2007).

Thus, as human movement scientists, our work is grounded in a multidisciplinary tradition. This approach allows us to go beyond the measures commonly used in social psychology when we believe this may add to the literature, which was the case when we started investigating the effects of gender stereotypes on motivational processes and motor behaviors. More particularly, we investigated this topic on the basis of theories of stereotype threat (Steele, 1997) and stereotype lift (Walton & Cohen, 2003). Although a great deal of research has investigated the more general role of gender stereotypes in predicting sport involvement (e.g., Fredricks & Eccles, 2005), very few studies have examined the more specific effects of stereotypes on motor performance. Stereotype-threat and stereotype-lift frameworks seem therefore particularly well-suited for investigating this question as they focus on stereotype effects on performance. If activation of negative stereotypes about females (e.g., females have poor athletic abilities) impairs their performance while simultaneously increasing that of males, this would reinforce in turn the natural differences in athletic abilities existing between males and females. At the time we began our research on this topic (around 2004), many studies using self-reports failed to identify the hypothesized mechanisms of stereotype threat. The small amount of research that found interesting results used indirect measures, including CV responses (e.g., Blascovich, Spencer, Quinn, & Steele, 2001; Croizet et al., 2004), non-verbal behaviors (e.g., Bosson, Haymovitz, & Pinel, 2004), and cognitive measures (e.g., Schmader & Johns, 2003). This chapter describes two studies that build on this previous research to test the effects of gender stereotypes on motor performance.

In what follows we first elaborate on stereotype-threat and stereotype-lift theories and explain why it is necessary to use physiological measures within these research areas. Then we turn to a description of the CV measures used to index motivation in the current research: heart rate (HR) and heart rate variability (HRV). After describing the more technical aspects related to these measures, we present two studies that investigated the effects of activating negative out-group stereotypes on physiological and self-reported indices of motivation and motor performance. Finally, we conclude by discussing the potential costs and benefits of incorporating physiological measures in psychological research.

Stereotype Threat and Stereotype Lift: Two Inequality-producing Mechanisms

Stereotype-threat and stereotype-lift theories model the situational influence that stereotypes may have on stereotyped individuals (stereotype threat) and non-stereotyped individuals (stereotype lift), when they are activated in achievement contexts (i.e., where competence is at stake). More specifically, stereotype threat is a situationally induced identity threat "that can arise when one knows that he or she can possibly be judged or treated negatively on the basis of a negative stereotype about one's group" (Goff, Steele, & Davies, 2008, p. 92). This threat has been shown to hamper cognitive and motor performance through motivational (e.g., motivation to avoid failure), affective (e.g., increased anxiety), cognitive (e.g., reductions in working memory capacity), and physiological (e.g., stress) processes (see Schmader, Johns, & Forbes, 2008, for a review).

Although stereotype threat is considered an important source of achievement gaps between social groups, negative stereotypes may also cause inequities by improving the performance of members of the social category that is not directly addressed by the negative stereotype. This phenomenon, called stereotype lift, is defined as "the performance boost caused by the awareness that an out-group is negatively stereotyped" (Walton & Cohen, 2003, p. 456; see also Croizet et al., 2004; Spencer, Steele, & Quinn, 1999; Steele & Aronson, 1995). Importantly, this phenomenon differs from the effects that are observed when positive in-group stereotypes are activated. Positive stereotypes may increase the performance of in-group members (e.g., Ambady, Shih, Kim, & Pittinsky, 2001; Shih, Pittinsky, &

Ambady, 1999), but may also impair it through individuals choking under the pressure of high expectations (e.g., Cheryan & Bodenhausen, 2000; Smith & Johnson, 2006). In contrast, stereotype lift is a non-threatening situation for members of the non-stereotyped group as they are not directly concerned by the stereotype. This phenomenon should therefore not be confused with the effects of positive stereotypes.

This chapter will focus predominantly on stereotype lift. The reason is that this phenomenon has been far less thoroughly investigated compared to stereotype threat. For example, a search on Psycinfo (in May 2012) yielded 376 results for articles with "stereotype threat" in the title, and only six results for stereotype-lift articles. As a consequence, we know a great deal about the processes through which stereotype threat occurs, as well as its moderators, and the different outcomes it affects (Aronson et al., 1999; Inzlicht & Kang, 2010; Rydell, R., Rydell, M., & Boucher, 2010; Schmader, 2002; Schmader et al., 2008; Spencer et al., 1999), while at the same time little is known about how and when stereotype lift occurs.

Why is research on stereotype lift so scarce? A possible reason is that this effect is small. In their meta-analysis, Walton and Cohen (2003) showed that stereotype lift represented half the magnitude of stereotype-threat effects, and nearly 30% of the studies included in the meta-analysis did not show performance boosts in the non-stereotyped group. This small effect size may be explained by the fact that stereotype lift may occur only when the social identity of the non-stereotyped group is activated. When negative out-group stereotypes are made salient, non-stereotyped individuals need more "pushing" to activate their social identity than stereotyped individuals, because the non-stereotyped individuals are not threatened by the stereotype (Marx & Stapel, 2006). However, stereotype-lift effects may be as strong as stereotype-threat effects when the social identity of the non-stereotyped group is activated, or when individuals possess certain characteristics (e.g., strong endorsement of stereotypes or prejudice, adoption of a fixed theory of ability) (e.g., Chatard, Selimbegovic, Konan, & Mugny, 2008; Danso & Esses, 2001; Josephs, Newman, Brown, & Beer, 2003; Mendoza-Denton, Khan, & Chan, 2008).Therefore, more studies are needed for a better understanding of stereotype lift.

What do we know about the mechanisms that drive stereotype lift? According to Walton and Cohen (2003), stereotype lift is caused by the awareness that an out-group is negatively stereotyped. This awareness encourages downward social comparisons with a denigrated out-group, which may elevate self-efficacy and motivation, which in turn increases performance. When we started investigating stereotype lift, no study had empirically tested these assumed processes. A possible reason for this may have been that the use of self-report measures of motivation is problematic because people may not have full conscious access to their motivational processes (Blascovich & Mendes, 2000), as well as social desirability and self-presentational reasons. For example, in the current context, men may be reluctant to acknowledge that they are "inspired by female weakness" during their workout.

Our goal was thus to assess more directly the role of motivation and self-perceptions in stereotype-lift processes. More particularly, in the two studies that will be described below (Study 1: Chalabaev, Stone, Sarrazin, & Croizet, 2008; Study 2: Chalabaev, Brisswalter, & Maazoun, 2013), we examined the role of motivational processes (measured by physiological indices) and subjective feelings and thoughts (measured by self-reports) in the relation between out-group stereotypes and motor performance (as indexed by balance in Study 1, and force output in Study 2).

Heart Rate and Heart Rate Variability as Indices of Motivation: Methodological Considerations

The use of physiological measurement in intergroup relations studies may be driven by different objectives. A first goal is to explore the health-related consequences of being the target of stereotypes or prejudice. In this case, physiological measures are used as indicators of health-related outcomes. Common measures include blood pressure (e.g., Blascovich et al., 2001) and heart rate variability (HRV) (e.g., Utsey & Hook, 2007). A second approach consists of using physiological measures to index psychological variables that may not be easily assessed via self-reports, notably when they are subject to self-presentational biases. In this case, the main concern of the researcher was to find an index that is closely related to the psychological construct of interest. In line with this second approach, our goal was to examine whether gender stereotypes may affect physiological responses that are related to motivation.

Motivation was defined as effort, which refers to the quantity of energetic and attentional resources used by the individual. We focused on effort because it may be associated with physiological modifications—notably CV measures. Indeed, several psychophysiological theorists consider, in line with Obrist (1981), that the autonomic nervous system (ANS) modulation influence on the CV system is proportional to effort, and specifically heart rate (HR) and heart rate variability (HRV), the latter corresponding to the inter-beat interval fluctuations (e.g., Blascovich & Mendes, 2000; Blascovich & Tomaka, 1996; Thayer & Lane, 2009; Wright & Gendolla, 2012; Wright & Kirby, 2001; see also Sarrazin, Roberts, Cury, Biddle, & Famose, 2002, for an illustration in the motor area).

While HR reflects general arousal, HRV indexes more specifically the mobilization of attentional resources. Indeed, there is evidence that HRV is sensitive to cognitive processing demands (e.g., Jorna, 1992; Luft, Takase, & Darby, 2009; Mulder, G., Mulder, L., Meijman, Veldman, & van Roon, 2000), which is due to the fact that during mental effort, brain activity generates increased metabolic demands (e.g., Kramer & Weber, 2000). Specifically, studies have found that an increase in cognitive workload is associated with a reduction in HRV (e.g., Capa,

Audiffren, & Ragot, 2008; Jorna, 1992; Mulder, G. & Mulder, L., 1981; Mulder et al., 2000; Scerbo et al., 2001).

More particularly, different parameters of HRV may be calculated, notably through time-domain and frequency-domain analyses. Time-domain analysis involves calculating the variability of successive inter-beat intervals. The root mean square of differences between successive inter-beat intervals (RMSSD) is considered a reliable time-domain parameter of HRV (Task Force of the European Society of Cardiology [TFESC] & North American Society of Pacing and Electrophysiology [NASPE], 1996). Whereas analyzing HRV in the time domain provides information about the total variability of inter-beat intervals, performing frequency-domain analysis allows for decomposition of this variability into different components. This is done through spectral analysis (generally Fast Fourier Transformation), which assesses HRV in different frequency bands. This analysis notably distinguishes the low frequency (0.04–0.15 Hz) and high frequency (HF) (0.15–0.40 Hz) components of HRV; in other words, the different frequencies of oscillations in the intervals between consecutive heartbeats. The HF component of HRV is the recommended parameter in the frequency domain (TFESC & NASPE, 1996). It is highly correlated to RMSSD (Berntson et al., 1997). We therefore examined the effects of the stereotype manipulation on HF and RMSSD reactivity scores.

A further methodological aspect that is specific to cardiac measurement is that examining CV responses is meaningful when values recorded after the experimental manipulation are compared to the resting values. This is done by calculating a difference score (i.e., a reactivity score) between the two values. It is thus necessary to include in the protocol an initial phase during which participants are instructed to wait quietly without speaking or moving, in order to record their physiological values during rest. A related constraint imposed by CV measurement concerns the fact that associating CV responses with psychological constructs is possible only when the individual stays still. Indeed, when the individual moves, CV responses may reflect the physical exertion required by moving, adding error variance to the interpretation of the data. As a consequence, whereas CV responses may be recorded continuously during completion of cognitive tasks, this is not the case for motor tasks. Instead, we assessed the effects of the stereotype manipulation on CV responses measured just before the beginning of the task. Finally, given that the respiration rate may influence the HF band of HRV (e.g., Althaus, Mulder, L., Mulder, G., van Roon, & Ruud, 1998), it was controlled by asking participants to breathe normally when HRV was recorded.

What Are the Mechanisms Behind Stereotype Lift? Two Investigations in the Motor Domain

In the present research, we examined whether activating gender stereotypes would affect two aspects of effort: general arousal, which was measured by HR

(Study 1), and mental effort, which was assessed by HRV (Study 2). In order to examine further the relationships between effort and CV modulation, self-reports were measured along with these physiological indices. Indeed, our work is based on the idea that the mind and the body are closely interconnected; in other words, that interoceptions are associated with subjective feelings, thoughts, and motivation. In Study 1, HR measurement was combined with self-reported measures of anxiety and self-confidence. We reasoned that an elevation in physiological arousal may sometimes be associated with positive feelings, and sometimes with negative feelings (e.g., Blascovich & Tomaka, 1996; see also Blascovich, Chapter 12 in this volume). Given that stereotype lift is supposed to encourage downward social comparisons with a denigrated out-group, we expected that non-stereotyped individuals would experience this situation positively. This would result in an increased physiological arousal associated with increased self-confidence and no change in anxiety. In Study 2, HRV measurement was combined with an open-ended question assessing the thoughts that occurred to the individual during the task. This measure aimed to help us identify whether HRV modulation reflected mental effort or self-related worries. If stereotype lift is a positively experienced situation, as we propose, then self-related worries should be lower, and mental effort higher, under conditions of stereotype lift.

The procedure used in both studies was similar to the one commonly employed in stereotype-threat and stereotype-lift research. Specifically, stereotypes were experimentally manipulated by instructing participants that the test they were about to take had either been shown to produce gender differences or not. In the two studies, electrocardiograms were obtained non-invasively using an ambulatory device (Polar cardio frequency meter: Polar Electro—model s610i in Study 1, which assesses HR; and model RS800 in Study 2, which allows measurement of HRV). Cardio frequency meters are lightweight instruments that are strapped on a belt linked to the participant by three electrodes placed on the chest. The meter is equipped with a microprocessor that measures the time lapse between two heartbeats with millisecond precision.

Study 1

We investigated how females and males respond when confronted with negative in-group and out-group stereotypes, by assigning participants to a 3 (female stereotype, male stereotype, control) × 2 (females, males) between-subject factorial design. All participants were identified with the athletic domain. In order to induce stereotypes about either females or males, we chose a task that was not associated with gender stereotypes; namely, a balance task. We confirmed in a pre-test that this task was gender-neutral.[2] In the main study, stereotypes were artificially activated by telling participants that either females or males perform poorly on the task (individuals in the control condition were told that there were no gender differences). We reasoned that these instructions would easily activate well-known stereotypes: Females' poor performance could indeed be attributed to the stereotype that women have less athletic ability than men; and males' poor performance could be attributed to the stereotype that men have poor finesse ability. The instructions put emphasis on the poor performance of a group without mentioning the other group, in order to ensure that negative stereotypes were activated (not positive stereotypes). This emphasis was necessary because there are only two sex categories. Therefore, a negative stereotype about the out-group may imply a positive stereotype for the in-group.

We examined whether the stereotype manipulation would affect performance on the balance task. Participants had to maintain their balance in a standing position on a stabilometer (i.e., platform rotating around a central pivoting shaft) for as long as possible. Performance was defined as the time during which the platform did not touch the floor. Time in balance was automatically measured with a stopwatch linked to the stabilometer by an electrical system. The test period lasted 4 minutes and the performance score corresponded to the mean time of the three best trials during this period. We also studied potential mediators of stereotype effects on balance performance: effort (indexed by HR reactivity), somatic and cognitive anxiety, and self-confidence (assessed with the Revised Competitive State Anxiety Inventory-2; Cox, Martens, & Russell, 2003). These variables were measured just before participants performed the task (for more details about the method, see Chalabaev, Stone et al., 2008).

Analyses of variance showed that both females and males performed better after negative out-group stereotypes were explicitly linked to performance compared to the control condition. However, activating negative in-group stereotypes did not decrease balance performance either for females or for males. In other words, the results showed significant stereotype-lift effects and no stereotype-threat effects. In addition, mediation analyses (Baron & Kenny, 1986) revealed that among males, stereotype lift was independently mediated by enhanced self-confidence and HR reactivity. This suggests that the extra self-confidence and energy mobilization have enabled males to be immediately efficient while performing the task (e.g., Wright, Contrada, & Patane, 1986). Although this study supports the main assumptions of Walton and Cohen (2003)— i.e., that effort is the main mechanism behind stereotype lift—the gender difference in mediation was not expected. It is possible that the stereotypes about males and females that were induced varied in their accessibility. Given that the general domain of motor performance is male-oriented (e.g., Koivula, 1999), instructions provided in the female stereotype condition may have activated the stereotype that women have less athletic ability than men more strongly than instructions in the male stereotype condition activated the stereotype about the poor movement finesse of men. The difficulty for females to rely on a strong stereotype could explain why they were not more self-confident and involved in the task when males were said to be inferior.

Study 2

In order to investigate stereotype-lift mechanisms further, we examined in Study 2 the potential mediating role of attention. This study differed from Study 1 in two additional ways: Only negative stereotypes about females were induced, and participants had to perform a task that was associated with existing cultural stereotypes. A strength task was chosen because this ability is associated with pervasive gender stereotypes, in which females are believed to be less strong than males. Stereotypes were manipulated in a 2 (female stereotype, control) × 2 (females, males) design, by telling participants either that females had been shown to underperform on the test relative to males, or that the test had not been shown to produce gender differences. Strength performance was measured twice, before and after the stereotype manipulation. Specifically, female and male participants performed three 5-second isometric maximal voluntary contractions (MVC) of the quadriceps. An isokinetic dynamometer (Biodex^(r))[3] measured the maximal voluntary isometric torque of the knee extensor muscles, the axis of the dynamometer being aligned with the anatomical knee flexion–extension axis. The subjects warmed up on the isokinetic machine by repeating 20 2-second, sub-maximal (50% MVC) isometric contractions. For the three 5-second MVC measures (with a 3-minute rest between attempts), subjects were instructed to extend the leg as hard and fast as possible. The highest peak torques were defined as MVC (Nm) values.

The effects of the stereotype manipulation were examined on strength performance and attention-related variables, which have been shown to predict performance on a strength task (e.g., Coombes et al., 2009). Specifically, we studied whether inducing negative stereotypes about females would enhance males' mobilization of attentional resources and whether it would affect their attentional focus. The quantitative component of attention was indexed by HRV modulation, which corresponded to the difference between HRV measured after the stereotype manipulation (but before the beginning of the task) and baseline HRV, each measure being recorded for a duration of 4 minutes. This allowed us to examine whether the increase in mobilization of attentional resources starts before the task begins.

The qualitative component of attention was assessed using self-reports. We reasoned that the performance boost that characterizes stereotype lift could be due to the fact that individuals focus their attention more on task-relevant thoughts and less on task-irrelevant thoughts. Indeed, task-irrelevant thoughts (i.e., whether one is good enough to meet stereotypic expectancies) seem more indicative for the operation of positive in-group stereotypes, rather than the more subtle operation of the stereotype-lift phenomenon. To test this hypothesis, we used the thought-listing technique that measures participants' spontaneous thoughts during the task, by asking them to write anything that comes to their mind (Cadinu, Maass, Rosabianca, & Kiesner, 2005; Krohne, Pieper, Knoll, & Breimer, 2002; Petty & Cacioppo, 1977, 1979). The thought-listing sentences were coded using different categories, notably including task-related thoughts (e.g., "I focused on pushing as hard as possible") and evaluation-related thoughts (e.g., "I was wondering what my level would be compared to others").

As in Study 1, the results showed that inducing negative stereotypes about females increased males' maximal voluntary contraction compared to the control condition, suggesting stereotype lift. However, activation of this stereotype did not significantly decrease the performance of females, indicating that no stereotype-threat effect occurred. To understand better the performance boost observed among males, we examined whether the stereotype manipulation affected HRV modulation and the thoughts during the task. The stereotype activation decreased both HF and RMSSD reactivity among males, reflecting a decreased parasympathetic influence. This shows that inducing downward social comparisons with a denigrated out-group increased males' mental workload in the minutes preceding the beginning of the task. The stereotype manipulation did not affect females' HRV modulation. Finally, analyses of the thought-listing sentences showed that the stereotype manipulation did not affect task-related thoughts. In other words, males did not direct their attention more onto meeting task goals when negative out-group stereotypes were induced compared to the control condition. However, the stereotype manipulation significantly affected evaluation-related thoughts. When stereotypes were induced, males reported fewer thoughts related to evaluation of the self than did females. In other words, males directed their attention less onto task-irrelevant thoughts than females did when negative stereotypes about females were induced. This suggests that the HRV reduction observed under stereotype lift did not reflect the extra cost of self-related worries, but instead an increased mental effort. However, care should be taken in interpreting these results, as mediation analyses showed that neither HRV nor thoughts significantly contributed to the performance boost observed among males. These non-significant results may be due to the difficulty of relating variables that are assessed by different instruments (i.e., self-reported and physiological measures). Another possible reason is that inducing stereotypes may affect several variables (motivation, cognitions, and performance), but these variables are affected independently of one another.

Summary

To summarize, results of these two studies revealed stereotype-lift effects among males and females when negative out-group stereotypes were made salient in motor test-taking situations. Among males, this performance boost was mediated by increased self-confidence and energy mobilization. Inducing negative stereotypes about females also increased males' mobilization of attentional resources and decreased task-irrelevant thoughts. Taken together, these results suggest that stereotype lift enhances both the energetic (HR) and attentional (HRV) aspects of effort during preparation of the task. Interestingly, the two studies showed significant stereotype-lift effects, but no stereotype-threat effects. This suggests that there may be particular conditions in which only stereotype lift occurs, reinforcing the need to investigate this phenomenon more thoroughly.

Added Value and General Comments with Regard to the Use of Physiological Measures

Several methodological and theoretical comments need to be made about the use of physiological measures in psychology. First, there are a number of methodological constraints and precautions that are specific to this type of research, and that require expertise in physiology. Collaborating with physiologists is therefore important to ensure the reliability of the physiological measures. Besides the monetary cost of physiological equipment—compared to other physiological equipment, cardio frequency meters are not very expensive (a few hundred euros), but still cost more than paper-and-pencil studies—a practical constraint of using physiological measures is that this methodology is time-consuming. Experimental sessions are often long because CV responses need to be recorded during several minutes (at baseline and following the experimental manipulation) so that the measure can be considered as reliable.

Data analysis is also time-consuming, notably because of the need to conduct artifact correction. For example, we had to manually check each participant's CV data in order to detect occasional irregular heartbeats, and to replace them with interpolated adjacent values. Other constraints imposed by physiological measures include conditions for participation. In our studies, participation was limited to individuals who did not have any CV problems and who had not ingested caffeine or nicotine for at least 3 hours prior to the experiment. They were also required to avoid any kind of exercise and not to consume any alcohol in the 24 hours before the experimental session. Moreover, the moment at which the session took place (morning or afternoon) was controlled across conditions. Finally, there is a specific constraint that may pose validity issues. Participants have to stay still in a predetermined position while physiological measures are recorded (in our studies, they had to sit with their hands on their thighs). Based on the literature on embodied cognition, one may wonder to what extent constraining participants' posture may affect their psychological functioning (e.g., Harmon-Jones & Peterson, 2009). For example, before performing the strength task, perhaps some participants wanted to stand up in order to prepare better for action.

Despite these practical constraints, we believe that using physiological measures may help to move forward the field of psychology in general, and intergroup relations in particular. Specifically, these measures may provide information that completes knowledge issuing from more traditional psychological measures. Indeed, contrary to self-reports, physiological measures are not sensitive to social desirability or defensiveness influences. This is especially useful when examining psychological variables such as effort or anxiety. For example, there is evidence that individuals voluntarily inhibit their expression of anxiety when they are under stereotype threat (Johns, Inzlicht, & Schmader, 2008). Using only self-reports would therefore limit our understanding of the mechanisms through which stereotypes affect individuals. In our studies, using CV measures allowed us to identify more effectively how activation of negative out-group stereotypes affects effort mobilization. Relying on such measures appeared to be necessary given that stereotype lift may be particularly subject to self-presentation biases. Indeed, non-stereotyped individuals (and especially men in the motor domain!) may be reluctant to acknowledge that comparing themselves with a denigrated out-group "boosts" their behavior. In addition, assessing effort with CV measures allowed individuals to prepare for action without being bothered by questionnaires.

However, although physiological measures may provide information that may not be obtained with self-reports, researchers need to be very cautious when using such measures. While physiological measures may address the limits of self-reports (notably self-presentational biases), they raise other issues, notably with regard to their validity. The main one concerns the psychological interpretation of physiological responses. Finding that the activation of stereotypes is related to physiological modifications may appear fascinating, but this observation is especially useful when these modifications may be associated with psychological variables. The problem is that associating physiological responses with motivational, affective, or cognitive states is not straightforward. For example, Blascovich and his colleagues (e.g., Blascovich, Chapter 12 in this volume; Blascovich & Katkin, 1982; Blascovich & Mendes, 2000; Blascovich & Tomaka, 1996) have shown that an increase in HR may be experienced very differently depending on whether the situation is appraised as a threat or as a challenge. Whereas the adaptive pattern of physiological arousal is reflected by pure sympathetic-adrenomedullary (SAM) activation, which results in enhanced cardiac performance and vasodilation, the maladaptive pattern is reflected by a dual activation of the SAM and pituitary–adrenocortical axes, resulting in enhanced cardiac performance and inhibition of cardiac output and vasodilation. Therefore, although there is agreement that HR is proportional to effort, this measure is not sufficient to determine whether this arousal is associated with a threat or a challenge state. Similarly, although HRV has been consistently associated with mental workload, this may reflect either an adaptive response (mental effort) or a maladaptive response (self-related worries) (e.g., Thayer & Lane, 2009). This illustrates the need to rely on well-developed biopsychosocial models when conducting psychophysiological research.

Conclusion

This chapter aimed at demonstrating the added value provided by physiological measures in social psychological research, and more particularly, how these measures may allow a better understanding of the processes involved in stereotype-lift effects on males' motor performance. However, beyond the benefits described in this chapter, the limits raised by the difficulty of interpreting physiological measures in psychological terms suggest that using such measures may be particularly informative when they are associated with more traditional psychological ones (i.e., self-reports). In other words, we believe that physiological indices should not be used in place of psychological measures, but as a complement: Using different methodologies may address the limits of each one and bring complementary insights about the phenomena under investigation. Indeed, interoceptions and subjective feelings are closely interconnected: Individuals' behaviors are determined by complex interactions between physiological modifications; subjective experience of these modifications; and individuals' cognitions, motivations, and affects. Using a multidisciplinary approach thus provides a more accurate picture of these complex relationships—see also Major & Kunstman (Chapter 17) and Scheepers (Chapter 13) in this volume. For example, results of Study 1 showed that it is the association between an increase in HR and an increase in self-confidence that resulted in higher performance on the balance task. Measuring only physiological arousal or self-confidence would have limited our understanding of the mechanisms underlying the effects of negative stereotypes on non-stereotyped individuals. This is why we believe that incorporating new methodologies may help to move the field of intergroup relations forward when they are used in a multidisciplinary perspective.

Notes

1 Gender differences in the motor area are domain-specific: Activities that require physical abilities (e.g., strength) are dominated by males, whereas activities that require expressivity are dominated by females (e.g., Riemer & Visio, 2003). The present chapter focuses on male-dominated activities.

2 In the pilot, participants (52 kinesiology students; 25 females) rated the gender typicality of the balance task on a 7-point scale ranging from (1) masculine to (7) feminine. Results showed that the task was perceived as gender-neutral ($M = 3.98$; $SD = 0.94$). Pilot participants did not take part in the main study.

3 Biodex Medical Systems, Inc., Shirley, NY, USA.

Suggestions for Further Reading

Blascovich, J., & Tomaka, J. (1996). The biopsychosocial model of arousal regulation. In M. P. Zanna (Ed.), *Advances in experimental social psychology* (Vol. 28, pp. 1–51). San Diego, CA: Academic Press.

This article presents a biopsychosocial model which identifies CV measures that differentiate challenge and threat states in motivated situations. An interesting point of this model is that it allows us to distinguish positive and negative experiences of stressful situations.

Schmader, T., Johns, M., & Forbes, C. (2008). An integrated process model of stereotype threat effects on performance. *Psychological Review, 115*, 336–356.

This article presents a model of stereotype threat that integrates cognitive, motivational, and affective, as well as physiological processes of this phenomenon.

Thayer, J. F., & Lane, R. D. (2009). Claude Bernard and the heart-brain connection: Further elaboration of a model of neurovisceral integration. *Neuroscience and Biobehavioral Reviews*, *33*, 81–88.

This article describes the role of HRV in the regulation of physiological, affective, and cognitive processes, based on evidence provided by neuroanatomical studies.

Wright, R. A., & Kirby, L. D. (2001). Effort determination of cardiovascular response: An integrative analysis with applications in social psychology. In M. P. Zanna (Ed.), *Advances in experimental social psychology* (Vol. 33, pp. 255–307). New York: Academic Press.

This chapter presents evidence of the relationships between effort and cardiovascular responses, based on Jack Brehm's theories of motivation intensity (Brehm, Wright, Solomon, Silka, & Greenberg, 1983) and emotion intensity (Brehm, 1999).

References

Aarts, H., Custers, R., & Marien, H. (2008). Preparing and motivating behaviour outside of awareness. *Science*, *319*, 1639.

Althaus, M., Mulder, L. J. M., Mulder, G., van Roon, A. M., & Ruud, B. M. (1998). Influence of respiratory activity on the cardiac response pattern to mental effort. *Psychophysiology*, *35*, 420–430.

Ambady, N., Shih, M., Kim, A., & Pittinsky, T. L. (2001). Stereotype susceptibility in children: Effects of identity activation on quantitative performance. *Psychological Science*, *12*, 385–390.

Aronson, J., Lustina, M. J., Good, C., Keough, K., Steele, C. M., & Brown, J. (1999). When White men can't do math: Necessary and sufficient factors in stereotype threat. *Journal of Experimental Social Psychology*, *35*, 29–46.

Baron, R. M., & Kenny, D. A. (1986). The moderator-mediator variable distinction in social psychological research: Conceptual, strategic, and statistical considerations. *Journal of Personality and Social Psychology*, *51*, 1173–1182.

Berntson, G. G., Bigger, J. T. Jr., Eckberg, D. L., Grossman, P., Kaufmann, P. G., Malik, M. . . . van der Molen, M. W. (1997). Heart rate variability: Origins, methods, and interpretive caveats. *Psychophysiology*, *34*, 623–648.

Blascovich, J., & Katkin, E. S. (1982). Arousal-based social behaviors as a function of individual differences in visceral perception. *Review of Personality and Social Psychology*, *3*, 73–96.

Blascovich, J., & Mendes, W. B. (2000). Challenge and threat appraisals: The role of affective cues. In J. P. Forgas (Ed.), *Feeling and thinking: The role of affect in social cognition* (pp. 59–82). New York: Cambridge University Press.

Blascovich, J., Spencer, S. J., Quinn, D., & Steele, C. (2001). African Americans and high blood pressure: The role of stereotype threat. *Psychological Science*, *12*, 225–229.

Blascovich, J., & Tomaka, J. (1996). The biopsychosocial model of arousal regulation. In M. P. Zanna (Ed.), *Advances in experimental social psychology* (Vol. 28, pp. 1–51). San Diego, CA: Academic Press.

Bosson, J. K., Haymovitz, E. L., & Pinel, E. C. (2004). When saying and doing diverge: The effects of stereotype threat on self-reported versus non-verbal anxiety. *Journal of Experimental Social Psychology*, *40*, 247–255.

Brehm, J. W. (1999). The intensity of emotion. *Personality and Social Psychology Review*, *3*, 2–22.

Brehm, J. W., Wright, R. A., Solomon, S., Silka, L., & Greenberg, J. (1983). Perceived difficulty, energization, and the magnitude of goal valence. *Journal of Experimental Social Psychology, 19*, 21–48.

Cacioppo, J. T., Priester, J. R., & Berntson, G. G. (1993). Rudimentary determinants of attitudes: Arm flexion and extension have differential effects on attitudes. *Journal of Personality and Social Psychology, 65*, 5–17.

Cadinu, M., Maass, A., Rosabianca, A., & Kiesner, J. (2005). Why do women underperform under stereotype threat? Evidence for the role of negative thinking. *Psychological Science, 16*, 572–578.

Capa, R. L., Audiffren, M., & Ragot, S. (2008). The effects of achievement motivation, task difficulty, and goal difficulty on physiological, behavioral, and subjective effort. *Psychophysiology, 45*, 859–868.

Chalabaev, A., Brisswalter, J., & Maazoun, M. (2013). Effects of stereotype lift on heart rate variability before the beginning of a strength task. Manuscript in preparation.

Chalabaev, A., Sarrazin, P., & Fontayne, P. (2009). Stereotype endorsement and perceived ability as mediators of the girls' gender orientation–soccer performance relationship. *Psychology of Sport and Exercise, 10*, 297–299.

Chalabaev, A., Sarrazin, P., Stone, J., & Cury, F. (2008). Do achievement goals mediate stereotype threat? An investigation on females' soccer performance. *Journal of Sport and Exercise Psychology, 30*, 143–158.

Chalabaev, A., Sarrazin, P., Trouilloud, D., & Jussim, L. (2009). Can sex-undifferentiated teacher expectations mask an influence of sex stereotypes? Alternative forms of sex bias in teacher expectations. *Journal of Applied Social Psychology, 39*, 2469–2498.

Chalabaev, A., Stone, J., Sarrazin, P., & Croizet, J.-C. (2008). Investigating physiological and self-reported mediators of stereotype lift effects on a motor task. *Basic and Applied Social Psychology, 30*, 18–26.

Chatard, A., Selimbegovic, L., Konan, P., & Mugny, G. (2008). Performance boosts in the classroom: Stereotype endorsement and prejudice moderate stereotype lift. *Journal of Experimental Social Psychology, 44*, 1421–1424.

Cheryan, S., & Bodenhausen, G. V. (2000). When positive stereotypes threaten intellectual performance: The psychological hazards of "model minority" status. *Psychological Science, 11*, 399–402.

Coombes, S. A., Higgins, T., Gamble, K. M., Cauraugh, J. H., & Janelle, C. M. (2009). Attentional control theory: Anxiety, emotion, and motor planning. *Journal of Anxiety Disorders, 23*, 1072–1079.

Cox, R. H., Martens, M. P., & Russell, W. D. (2003). Measuring anxiety in athletics: The revised Competitive State Anxiety Inventory-2. *Journal of Sport and Exercise Psychology, 25*, 519–533.

Cretenet, J., & Dru, V. (2004). The influence of unilateral and bilateral armflexion vs. extension on judgments: An exploratory case of a motor congruence hypothesis. *Emotion, 4*, 282–294.

Croizet, J.-C., Després, G., Gauzins, M.-E., Huguet, P., Leyens, J.-P., & Méot, A. (2004). Stereotype threat undermines intellectual performance by triggering a disruptive mental load. *Personality and Social Psychology Bulletin, 30*, 721–731.

Danso, H. A., & Esses, V. M. (2001). Black experimenters and the intellectual test performance of White participants: The tables are turned. *Journal of Experimental Social Psychology, 37*, 158–165.

DeCaro, M. S., Thomas, R. D., Albert, N. B., & Beilock, S. L. (2011). Choking under pressure: Multiple routes to skill failure. *Journal of Experimental Psychology: General, 140*, 390–406.

Eagly, A. H. (1987). *Sex differences in social behavior: A social-role interpretation.* Hillsdale, NJ: Erlbaum.

Farmer, S. F., Harrison, L. M., Mayston, M. J., Parekh, A., James, L. M., & Stephens, J. A. (2004). Abnormal cortex-muscle interactions in subjects with X-linked Kallmann's syndrome and mirror movements. *Brain, 127*, 385–397.

Filaire, E., Alix, D., Ferrand, C., & Verger, M. (2009). Psychophysiological stress in tennis players during the first single match of a tournament. *Psychoneuroendocrinology, 34*, 150–157.

Fredricks, J. A., & Eccles, J. S. (2005). Family socialization, gender, and sport motivation and involvement. *Journal of Sport and Exercise Psychology, 27*, 3–31.

Goff, P. A., Steele, C. M., & Davies, P. G. (2008). The space between us. *Journal of Personality and Social Psychology, 94*, 91–107.

Gray, R. (2004). Attending to the execution of a complex sensorimotor skill: Expertise differences, choking and slumps. *Journal of Experimental Psychology: Applied, 10*, 42–54.

Harmon-Jones, E., & Peterson, C. K. (2009). Supine body position reduces neural response to anger evocation. *Psychological Science, 20*, 1209–1210.

Inzlicht, M., & Kang, S. K. (2010). Stereotype threat spillover: How coping with threats to social identity affects aggression, eating, decision making, and attention. *Journal of Personality and Social Psychology, 99*, 467–481.

Johns, M., Inzlicht, M., & Schmader, T. (2008). Stereotype threat and executive resource depletion: The influence of emotion regulation. *Journal of Experimental Psychology: General, 137*, 691–705.

Jorna, P. (1992). Spectral analysis of heart rate and psychological state: A review of its validity as workload index. *Biological Psychology, 34*, 237–257.

Josephs, R. A., Newman, M. L., Brown, R. P., & Beer, J. M. (2003). Status, testosterone and human intellectual performance: Stereotype threat as status concern. *Psychological Science, 14*, 164–168.

Knisel, E., Opitz, S., Wossmann, M., & Keteihuf, K. (2009). Sport motivation and physical activity of students in three European schools. *International Journal of Physical Education, 46*, 40–53.

Koivula, N. (1999). Sport participation: Differences in motivation and actual participation due to gender typing. *Journal of Sport Behavior, 22*, 360–380.

Kramer, A. F., & Weber, T. (2000). Applications of psychophysiology to human factors. In J. T. Cacioppo, L. G. Tassinary & G. G. Berntson (Eds.), *Handbook of psychophysiology* (2nd ed., pp. 794–814). Cambridge, UK: Cambridge University Press.

Krohne, H. W., Pieper, M., Knoll, N., & Breimer, N. (2002). The cognitive regulation of emotions: The role of success vs. failure experience and coping dispositions. *Cognition and Emotion, 16*, 217–243.

Luft, C., Takase, E., & Darby, D. (2009). Heart rate variability and cognitive function: Effects of physical effort. *Biological Psychology, 82*, 186–191.

Marx, D. M., & Stapel, D. A. (2006). Understanding stereotype lift: On the role of the social self. *Social Cognition, 24*, 776–791.

Mendoza-Denton, R., Kahn, K., & Chan, W. Y. (2008). Can fixed views of ability boost performance in the context of favorable stereotypes? *Journal of Experimental Social Psychology, 44*, 1187–1193.

Mulder, G., & Mulder, L. J. M. (1981). Information processing and cardiovascular control. *Psychophysiology, 18*, 392–402.

Mulder, G., Mulder, L. J. M., Meijman, T. F., Veldman, J. B. P., & van Roon, A. M. (2000). A psychophysiological approach of working conditions. In R. W. Backs &

W. Boucsein (Eds.), *Engineering psychophysiology* (pp. 139–159). Mahwah, NJ: Lawrence Erlbaum.

Obrist, P. A. (1981). *Cardiovascular psychophysiology: A perspective*. New York: Plenum.

Olds, T., Tomkinson, G., Léger, L., & Cazorla, G. (2006). Worldwide variation in the performance of children and adolescents: An analysis of 109 studies of the 20-m shuttle run test in 37 countries. *Journal of Sports Sciences, 24*, 1025–1038.

Omlor, W., Patino, L., Hepp-Reymond, M.-C., & Kristeva, R. (2007). Gamma-range corticomuscular coherence during dynamic force output. *NeuroImage, 34*, 1191–1198.

Payen, V., Elliot, A. J., Coombes, S. A., Chalabaev, A., Brisswalter, J., & Cury, F. (2011). Viewing red prior to a strength test inhibits motor output. *Neuroscience Letters, 495*, 44–48.

Petty, R. E., & Cacioppo, J. T. (1977). Forewarning, cognitive responding, and resistance to persuasion. *Journal of Personality and Social Psychology, 35*, 645–655.

Petty, R. E., & Cacioppo, J. T. (1979). Effects of forewarning of persuasive intent and involvement on cognitive responses and persuasion. *Journal of Personality and Social Psychology, 5*, 173–176.

Riemer, B. A., & Visio, M. E. (2003). Gender typing of sports: An investigation of Metheny's classification. *Research Quarterly for Exercise and Sport, 74*, 193–204.

Rydell, R. J., Rydell, M. T., & Boucher, K. L. (2010). The effect of negative performance stereotypes on learning. *Journal of Personality and Social Psychology, 99*, 883–896.

Sarrazin, P., Roberts, G., Cury, F., Biddle, S., & Famose, J.-P. (2002). Exerted effort and performance in climbing among boys: The influence of achievement goals, perceived ability, and task difficulty. *Research Quarterly for Exercise and Sport, 73*, 425–436.

Scerbo, M. W., Freeman, F. G., Mikulka, P. J., Parasuraman, R., Di Nocero, F., & Prinzel, L. J. III (2001). *The efficacy of psychophysiological measures for implementing adaptive technology* (NASA/TP-2001-211018). Washington, DC: National Aeronautics and Space Administration.

Schmader, T. (2002). Gender identification moderates stereotype threat effects on women's math performance. *Journal of Experimental Social Psychology, 38*, 194–201.

Schmader, T., & Johns, M. (2003). Converging evidence that stereotype threat reduces working memory capacity. *Journal of Personality and Social Psychology, 85*, 440–452.

Schmader, T., Johns, M., & Forbes, C. (2008). An integrated process model of stereotype threat effects on performance. *Psychological Review, 115*, 336–356.

Shih, M., Pittinsky, T. L., & Ambady, N. (1999). Stereotype susceptibility: Identity salience and shifts in quantitative performance. *Psychological Science, 10*, 80–83.

Smith, J. L., & Johnson, C. S. (2006). A stereotype boost or choking under pressure? Positive gender stereotypes and men who are low in domain identification. *Basic and Applied Social Psychology, 28*, 51–63.

Spencer, S. J., Steele, C. M., & Quinn, D. M. (1999). Stereotype threat and women's math performance. *Journal of Experimental Social Psychology, 35*, 4–28.

Steele, C. M. (1997). A threat in the air: How stereotypes shape intellectual identity and performance. *American Psychologist, 52*, 613–629.

Steele, C. M., & Aronson, J. (1995). Stereotype threat and the intellectual test performance of African Americans. *Journal of Personality and Social Psychology, 69*, 797–811.

Task Force of the European Society of Cardiology (TFESC) & North American Society of Pacing and Electrophysiology (NASPE). (1996). Heart rate variability: Standards of measurement, physiological interpretation and clinical use. *Circulation, 93*, 1043–1065.

Thayer, J. F., & Lane, R. D. (2009). Claude Bernard and the heart–brain connection: further elaboration of a model of neurovisceral integration. *Neuroscience and Biobehavioral Reviews, 33*, 81–88.

Thomas, J. R., & French, K. E. (1985). Gender differences across age in motor performance: A meta-analysis. *Psychological Bulletin, 98*, 260–282.

Utsey, S. O., & Hook, J. N. (2007). Heart rate variability as a physiological moderator of the relationship between race-related stress and psychological distress. *Cultural Diversity and Ethnic Minority Psychology, 13*, 250–253.

Walton, G. M., & Cohen, G. L. (2003). Stereotype lift. *Journal of Experimental Social Psychology, 39*, 456–467.

Wright, R. A., Contrada, R. J., & Patane, M. J. (1986). Task difficulty, cardiovascular response, and the magnitude of goal valence. *Journal of Personality and Social Psychology, 51*, 837–843.

Wright, R. A., & Gendolla, G. H. E. (Eds.). (2012). *How motivation affects cardiovascular response: Mechanisms and applications*. Washington, DC: APA Press.

Wright, R. A., & Kirby, L. D. (2001). Effort determination of cardiovascular response: An integrative analysis with applications in social psychology. In M. P. Zanna (Ed.), *Advances in experimental social psychology* (Vol. 33, pp. 255–307). New York: Academic Press.

PART III

Overcoming Implicit Prejudice

9

IMPLICIT PREJUDICE AND THE REGULATION OF INTERGROUP RESPONSES

Theoretical Contributions of the Social Neuroscience Approach

David M. Amodio

Prejudice researchers are typically interested in two things, neither of which directly concerns the brain: the situational causes and consequences of prejudice for individuals in society, and the psychological mechanisms that drive biased perceptions and discriminatory behavior. Perhaps surprisingly, the field of intergroup relations is the domain in which social neuroscience research has been arguably the most active. This is partly due to a coincidence—many of the first social psychologists with neuroscience training happened to be interested in intergroup questions (Amodio, Harmon-Jones, & Devine, 2003; Bartholow, Pearson, Gratton, & Fabiani, 2003; Ito & Urland, 2003; Phelps et al., 2000; Vanman, Paul, Ito, & Miller, 1997). A more substantive reason, however, is that the field of intergroup relations provides one of the most fertile, multifaceted areas of research for studying the interplay of multiple psychological processes at multiple levels of analysis. And, at its core, the social neuroscience approach is interested in connections across levels of analysis. By studying intergroup processes, social neuroscientists can examine the rich interplay of neural function with cognition, emotion, and behavior as they operate within the individual, dyads, groups, and societies. Prejudice researchers who take this approach hope to probe the psychological mechanisms of intergroup bias while also elucidating brain function.

Social neuroscientists have studied intergroup bias from many different angles (see reviews by Amodio, 2008; Ito & Bartholow, 2009; Quadflieg & Macrae, 2011; Van Bavel & Cunningham, 2011). One approach has been to explore the physiological correlates of intergroup responses, with an emphasis on the patterns of affect (i.e., emotion) and arousal that often accompany prejudiced attitudes. This approach characterizes much of the early psychophysiological research on intergroup relations, in which measures of the autonomic nervous system, such as the galvanic skin response (GSR) or cardiovascular changes, were used as surreptitious indices of affect and, by extension, attitudes (e.g., Rankin & Campbell, 1955; see Guglielmi, 1999, for a review). These methods provided a way to assess responses to race while circumventing the participants' concerns about self-presentation. The goal of the psychophysiological approach is typically to assess the physiological correlates of psychological responses, but not necessarily to probe neural mechanisms that may underlie these responses.

The contemporary social neuroscience approach to prejudice is increasingly characterized by an emphasis on measures of brain activity and a focus on psychological mechanism. Whereas social psychophysiology approaches provide useful indicators of psychological responses, the cognitive neuroscience approach offers theoretical models of neural function that can be used to generate new and increasingly refined models of socio-cognitive processes relevant to intergroup bias. It is this mechanistic approach that has the greatest potential to inspire new theoretical insights about the psychology of prejudice, stereotyping, and discrimination because it brings ideas from neuroscience to bear on social psychology. Given the strong influence of cognitive neuroscience, this approach is often referred to as *social cognitive neuroscience* (but in this chapter, *social neuroscience* is used to describe the overarching field).

In this chapter, I describe some of the ways in which the social neuroscience approach has begun to shed light on two major questions in the psychology of intergroup bias: How do implicit forms of bias operate in the mind and behavior? And how are intergroup responses regulated? In doing so, I include some personal notes on the circumstances that led me—a researcher trained initially in social psychology—to use neuroscience approaches to study mechanisms of intergroup bias.

Implicit Prejudice and Stereotyping

Human society is a collective of social categories, and people are often perceived in terms of visual characteristics associated with race or ethnicity. In America, racial stereotypes about African Americans and other minority groups are pervasive. The perception of a racial out-group member can automatically trigger feelings, evaluations, and stereotypic thoughts which can then unintentionally influence one's behavior (Devine, 1989; Fazio, Jackson, Dunton, & Williams, 1995). Although the effects of automatic and implicit forms of bias have been studied extensively in behavioral research (for reviews, see Amodio & Mendoza, 2010; Fazio & Olson, 2003), the mechanisms that drive these effects have been difficult to identify. At the time I began to study implicit bias, as a doctoral student working with Patricia Devine at the University of Wisconsin–Madison in the late 1990s, some of the first social neuroscience studies were seeking to use neuroscience techniques to shed light on the processes driving automatic and implicit forms of racial bias.

Implicit Prejudice

The term "prejudice" has complex connotations, but in social psychology, it typically refers to a negative attitude toward a person or group. The discovery that prejudices toward minority group members, such as African Americans, could operate without one's awareness or intention galvanized the field of social cognition (Devine, 1989; Fazio et al., 1995; Gaertner & McLaughlin, 1983). As a graduate student, I was especially interested in the cognitive mechanisms underlying implicit forms of racial bias. I thought that it was necessary to understand these mechanisms in order to develop effective strategies for mitigating their effects. However, the way that implicit bias was measured at the time concerned me. Prejudice is traditionally thought to reflect an affect-laden attitude toward outgroup members (Fiske, 1998), but implicit prejudice was typically measured using reaction-time assessments of word associations. I was interested in emotional aspects of implicit bias, but it seemed like the patterns of bias assessed by popular implicit tasks, such as the sequential evaluative priming task used by Fazio et al. (1995), could be fully explained by *cognitive* associations between a social group category and positive or negative words. I wanted to determine whether implicit racial biases could also be rooted in emotion. Furthermore, I thought that this distinction was important because cognitive and affective forms of bias might require different strategies for their control or reduction.

My advisor, Patricia Devine, began discussing similar issues with her colleague Richard Davidson, an expert in affective neuroscience. He suggested that we read some emerging research on the role of the amygdala in emotion. At the time, the body of research on the amygdala and emotion included lesion studies in the animal literature, and startle–eyeblink studies in the human literature (the first human neuroimaging studies of the amygdala did not appear until the mid-1990s). In particular, emerging findings from the animal neuroscience literature on the neural structures underlying fear and classical conditioning seemed very relevant to our questions about implicit prejudice. This neuroscience work focused on the amygdala as a key neural substrate for the learning and expression of threat cues, in the context of its broader role in attention and motivation (see Figure 9.1; Davis, 2006; Fendt & Fanselow, 1999; LeDoux, Iwata, Cicchetti, & Reis, 1988; Whalen, 1998). Given other research suggesting that fear-related affect might underlie observations of implicit racial bias (Fazio et al., 1995; Greenwald, McGhee, & Schwartz, 1998), the amygdala seemed like a strong candidate for a neural substrate of implicit prejudice. Also at this time, Eddie Harmon-Jones—an expert in psychophysiological approaches to social psychology—joined the faculty at Wisconsin. Eddie and I started a weekly journal club, and in our conversations at these meetings we began to talk about ways to measure amygdala activity associated with prejudice; before long we were collaborating on this line of research.

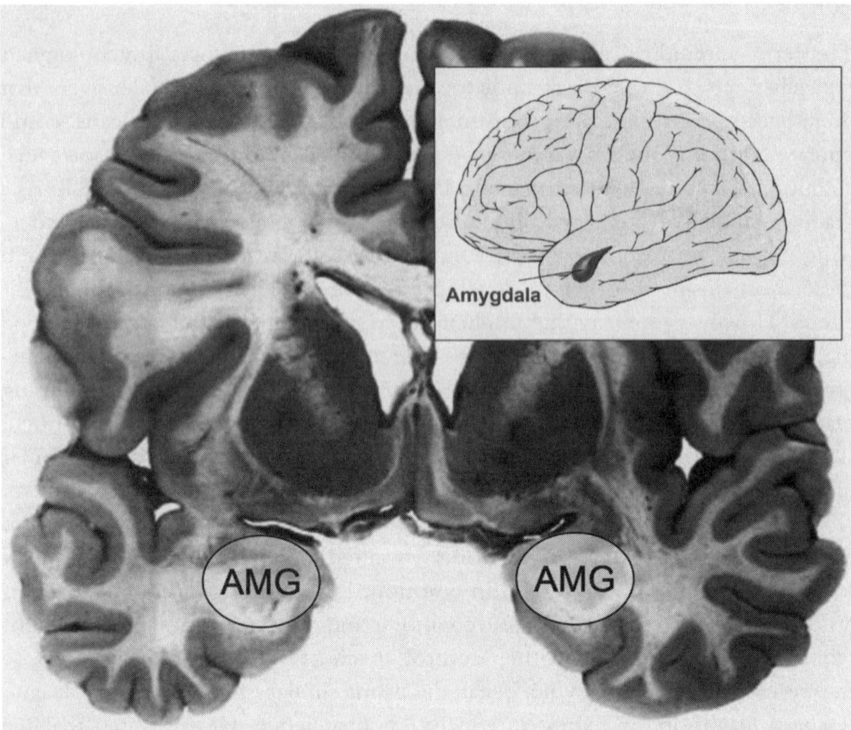

FIGURE 9.1 View of Coronal Slice Through Brain Illustrating the Amygdala (AMG)

While conducting our first experiment, we learned that other researchers had similar ideas and were using neuroimaging methods to examine amygdala activity involved in responses to race. A pair of early studies, by Hart et al. (2000) and Phelps et al. (2000), used functional magnetic resonance imaging (fMRI) to measure participants' brain activity while they viewed faces of in-group and out-group members. The goal was to see whether the amygdala might be a neural correlate of implicit prejudice. Both studies used blocked designs, such that White American participants would view a series of all White faces in one block, and a series of all Black faces in a different block. Faces were viewed for several seconds, and measures of brain activity were averaged across the entire block, as a function of race. Thus, these studies focused on identifying the location of brain activity while viewing Black compared with White faces, but were unable to discern the timing of these responses.

Interestingly, neither study found a significant difference in amygdala activity between the viewing of Black and White faces. The effects were a bit more complex. Hart et al. (2000) observed that although passive viewing of in-group and out-group faces both elicited a similar degree of amygdala activity, amygdala responses to in-group faces habituated more quickly than amygdala responses to out-group faces. That is, in early blocks of trials, amygdala activity was high to both White and Black faces; but in later blocks, activity to White faces subsided while activity to Black faces remained higher. In Phelps et al. (2000), subjects viewed faces in the context of a memory task, indicating whether each face was the same as or different from the one presented on the previous trial. Although the authors did not observe a difference in amygdala activity to Black compared with White faces on average, individual differences in amygdala response to Black versus White faces from the fMRI measure were correlated with subjects' scores on a behavioral measure of implicit racial bias, the Implicit Association Test (IAT), and also with scores on a startle–eyeblink measure of amygdala activity. None of these measures were associated with explicit self-report measures of prejudiced beliefs.

These neuroimaging findings suggested that amygdala activity might be correlated with implicit responses to race. This made sense, given that the amygdala was already known to respond quickly to threat. Although these studies were not necessarily intended to address a hypothesis regarding the social psychological process of implicit bias, by identifying a potential neural substrate of implicit racial bias, one could draw on knowledge about this brain structure from the neuroscience literature to shed new light on psychological models of implicit bias, such as how implicit associations are learned, activated, expressed in behavior, and potentially extinguished (Amodio, 2008; Amodio & Ratner, 2011b). For example, certain nuclei of the amygdala respond very rapidly to learned threats and engage physiological responses associated with inhibition (i.e., freezing) and active avoidance. In a human social interaction, this would correspond to inhibited and awkward social behavior. The amygdala-based fear response also operates without conscious deliberation and, once a fear association is acquired, it is extremely difficult to extinguish. Furthermore, whereas the amygdala response is associated with emotion, motivation, and behavior, it is not associated with language and high-level cognition. This suggests that knowledge about amygdala function might inform intergroup emotion and motivation, but probably not stereotypes.

In Amodio et al. (2003), we were interested in the amygdala's role in implicit bias because it could shed light on social psychological questions about prejudice. Because we were interested in very rapid, automatic responses to race, we needed a method that could detect millisecond changes in amygdala activity. At the time we began this research, in the late 1990s, most fMRI experiments used blocked designs, in which brain activity was measured continuously in response to a series of stimuli. This method was not appropriate for measuring automatic responses to race. Furthermore, even when event-related fMRI methods are used, changes in brain activity are measured on the order of seconds rather than milliseconds (the temporal resolution at which automatic processes are thought to unfold).

Given these issues, and given our new collaboration with Eddie Harmon-Jones, we decided to use the startle–eyeblink method to assess amygdala activity in response to faces of White and Black people. The eyeblink is part of the whole-body startle reflex, which can be measured using electromyography (EMG) by placing bioelectrodes over the muscle surrounding the eye (*orbicularis oculi*). The eyeblink is caused by contraction of this muscle. The startle reflex, and the corresponding eyeblink, are reliably amplified when a startling event (e.g., a sudden loud noise) occurs when one is already in a threatened state (Lang, Bradley, & Cuthbert, 1990). By contrast, the reflex is reduced, or inhibited, when one is already in an appetitive state. Therefore, the relative size of one's blink reflex indicates one's emotional state just prior to the onset of the startle probe. Importantly, this "affective modulation" of the startle–eyeblink response is driven by the amygdala.

The advantage of the startle–eyeblink measure is that it is able to index changes in amygdala activity within milliseconds of a stimulus being presented. Furthermore, the startle–eyeblink measure is sensitive to activity in the central nucleus of the amygdala—the specific region of the amygdala associated with fear conditioning and responses to threat. Both of these measurement attributes were critical for establishing the implicit nature of participants' response to race. By comparison, fMRI's slow temporal resolution, on the order of seconds, allows for both awareness and control. Furthermore, the spatial resolution of fMRI is too coarse to discern the different nuclei of the amygdala such as the central nucleus, which is associated with threat processing, from other nuclei, such as the basal nucleus, which is associated with appetitive processing.

Amodio et al. (2003) found that startle–eyeblink amplitudes were, on average, larger to Black than White faces, showing an effect of race on amygdala activity for the first time. Furthermore, by associating implicit racial bias with activity in the amygdala's central nucleus, this finding linked implicit bias to a uniquely affective response that is associated with a classical fear-conditioning learning mechanism. Through this link, we could draw on the fear-conditioning literature to develop a model of how implicit affective associations may be learned, expressed, and potentially unlearned. For example, and as noted above, fear-conditioned responses are learned rapidly, often after a single exposure to the stimulus in a threatening context. Once formed, such associations are very difficult and perhaps impossible to extinguish (Bouton, 1994; but see Schiller, Freeman, Mitchell, Uleman, & Phelps, 2009). Instead of being unlearned, fear-conditioned responses can be overridden by newer learned responses. This suggests that strategies to override implicit biases (e.g., through positive intergroup contact, approach training, or the use of goal strategies; see Crisp & Turner, 2009; Kawakami, Phills, Steele, & Dovidio, 2007; Mendoza, Gollwitzer, & Amodio, 2010), rather than to undo them directly, might be more effective for reducing prejudice. This idea is consistent with "replacement" models of prejudice control (Devine, 1989) and with research showing that behavioral strategies for responding without prejudice can be very effective (e.g., Mendoza et al., 2010; Monteith, Ashburn-Nardo, Voils, & Czopp, 2002).

Interestingly, most theories of implicit social cognition suggest that implicit associations are learned slowly, only after repeated exposure, and that they can be unlearned through the same mechanism (e.g., Smith & DeCoster, 2000). Although such theories are consistent with models of semantic (i.e., conceptual) associations (Sloman, 1996), they are inconsistent with research on the fear-conditioning mechanism underlying affective associations (Bouton, 1994). Furthermore, fear-conditioned responses are expressed primarily in autonomic responses and non-verbal behaviors (such as freezing and avoidance)—the same types of responses exhibited by White participants in intergroup situations, particularly among those scoring higher on behavioral tests of implicit evaluative bias (Amodio & Devine, 2006; Dovidio, Kawakami, Johnson, C., Johnson, B., & Howard, 1997; Fazio et al., 1995; McConnell & Leibold, 2001).

A second goal of Amodio et al. (2003) was to examine individual differences in amygdala responses to Black versus White faces as a function of internal and external motivations to respond without prejudice. Previous research showed that some self-avowed egalitarians are more prone to expressions of implicit prejudice than others (Devine, Plant, Amodio, Harmon-Jones, & Vance, 2002). Devine et al. showed that, among participants who reported strong internal motivation to respond without prejudice, those who also reported being highly concerned about social pressures to respond without prejudice exhibited more prejudice on implicit measures than participants who were not worried about such pressures. Although "high internal/high external" individuals hold egalitarian beliefs, their concern about appearing prejudiced in front of others is a major source of anxiety. Thus, though on the surface it might seem like this additional source of motivation should lead these individuals to be more effective in responding without prejudice, the anxiety it entails may actually undermine their efforts (Plant & Devine, 1998). Other research suggests that high internal/high external individuals have not fully internalized their egalitarian beliefs, and therefore are less skilled at responding without prejudice (Devine et al., 2002).

Amodio et al. (2003) tested whether the different behavioral profiles of high internal/low external individuals and high internal/high external individuals might relate to differences in the amygdala response to race. Indeed, we found that eye-blink responses at 4,000 milliseconds (ms) to Black and White faces did not differ among high internal/low external participants (i.e., those not worried about social pressure), suggesting no bias. By comparison, high internal/high external participants (i.e., those who were worried about social pressure) showed a significantly larger startle response to Black than to White faces, suggesting greater amygdala activity in response to Black faces. This effect was similar in magnitude to the difference observed among highly prejudiced participants (i.e., those reporting low internal motivation). This finding suggested that high internal/high external subjects, who typically report egalitarian attitudes, but show evidence of implicit prejudice (Devine et al., 2002), expressed implicit bias because of deeply-rooted fear-conditioned associations with Black people.

It is notable that in Amodio et al. (2003), we initially assumed that the startle–eyeblink responses measured 4,000 ms after the presentation of a face could reflect a controlled emotional response (whereas eyeblink responses occurring only 400 ms following face presentation could not). However, our results showed no evidence of control at 4,000 ms. This result was puzzling at first because we assumed that controlled processes should operate directly on the source of bias, much like the common dualistic notion of reason controlling one's "passion." However, Study 2 of Amodio et al. (2003) revealed that control was only expressed in behavior, in the form of self-reported emotional responses to Black faces compared with White faces. This pattern of results was consistent with the replacement model of control, in which an intended response replaces the biased tendency (e.g., Devine, 1989), and not with the notion of direct down-regulation, in which control directly inhibits the source of bias. In later research (e.g., Amodio & Ratner, 2011a), it became clear that this pattern also comported with the functional neuroanatomy of the amygdala, which is not directly connected with regions of the frontal cortex typically associated with control (described below).

Since these initial studies, several event-related fMRI studies have observed that amygdala activity is greater to out-group than to in-group faces under some conditions, but not others, and that these effects are typically subtle (Cunningham et al., 2004; Lieberman, Hariri, Jarcho, Eisenberger, & Bookheimer, 2005; Ronquillo et al., 2007; Van Bavel, Packer, & Cunningham, 2008; Wheeler & Fiske, 2005). These studies reveal that differences in amygdala activity to Black versus White faces tend to emerge most strongly in the context of a minimally demanding task, such as during the passive viewing of faces. As tasks become more demanding, such as when searching for a "dot" on the image (Wheeler & Fiske, 2005), imagining whether the target likes a particular vegetable (Wheeler & Fiske, 2005), or attempting to match the face to written group labels (Lieberman et al., 2005), amygdala activations tend to diminish. Other research suggests that amygdala effects are lessened when the out-group face is not looking directly at the subject (i.e., has averted gaze; Richeson, Todd, Trawalter, & Baird, 2008). These studies have helped to demarcate the boundary conditions for the amygdala's role in implicit bias.

An important broader implication of research connecting implicit racial affect and attitudes with amygdala activity is that, although implicit attitudes are conceived in terms of positive versus negative valence in many social psychological models, the mechanisms underlying implicit prejudice do not conform to a simple valence-based organization. That is, no brain region corresponds to "good" or "bad," which are conceptual abstractions in high-level cognition (cf., Cunningham, Raye, & Johnson, 2004). Rather, neural and biological processes are organized around more basic processes, such as perceptual and behavioral processes to coordinate approach, inhibition, and avoidance (Harmon-Jones, 2003; Harmon-Jones, E., Harmon-Jones, C., Amodio, & Gable, 2011). The amygdala in particular reflects this organization, such that it coordinates freezing, approach, and active avoidance functions in a way that cannot be adequately characterized in terms of valence (Amodio, 2008; Lang et al., 1990). To the extent that behavior-based measures of implicit prejudice assess activity of the amygdala and other neural structures, these measures should be interpreted in terms of an individual's goals, motivations, and behaviors rather than abstract (i.e., cognitive) appraisals of whether an out-group member is "good" or "bad" (e.g., Amodio & Devine, 2006; Dovidio et al., 1997; Fazio et al., 1995).

Implicit Stereotyping

Whereas much research has examined the neural correlates of race-related affect and evaluation, very little has investigated the neural correlates of social stereotypes. Stereotypes are cognitive structures stored in memory that represent attributes associated with a social group (Stangor & Lange, 1994). Building on our research on the affective roots of implicit prejudice, Amodio and Devine (2006; see also Amodio, 2008; Amodio & Mendoza, 2010) proposed that stereotypes are rooted in mechanisms of semantic memory and selection, which are associated with neural activity in the temporal lobe (Figure 9.2) and lateral posterior PFC (e.g., Brodmann areas 45 and 47). Behavioral and neuroscience research on semantic learning systems has uncovered the dynamics of how such associations are learned and expressed in behavior. Neuroscience research on the connectivity of these brain regions provides further clues about other psychological processes that are most directly influenced by these semantic associations, such as judgments and impressions (Gabrieli, 1998). By linking stereotypes to this literature, researchers can apply findings from the memory literature to understand stereotyping processes (Amodio, 2008; Amodio & Hamilton, 2012; Amodio & Ratner, 2011b). For example, whereas affective associations are learned quickly and are relatively indelible, semantic associations may be learned and unlearned through a process of repeated exposure to pairings and non-pairings. Hence, social psychological models of implicit associations (e.g., Smith & DeCoster, 2000) are consistent with semantic forms of learning and memory. As suggested by studies of neural connectivity, semantic associations are more likely to be expressed in trait impressions, goal representations, and instrumental behaviors in comparison to affective associations, and they are also more likely to emerge in verbal responses (Amodio & Devine, 2006) and even perceptual biases (Ofan, Rubin, & Amodio, 2011).

Some fMRI studies have examined neural activity associated with the completion of stereotyping tasks (Knutson, Mah, Manly, & Grafman, 2007; Mitchell, Ames, Jenkins, & Banaji, 2009; Quadflieg et al., 2009), but this work has not yet addressed the mechanisms of stereotypes per se. Rather, these studies have focused on brain activity associated with more general aspects of task completion, such as response conflict and inhibition or face perception. In a related brain lesion study,

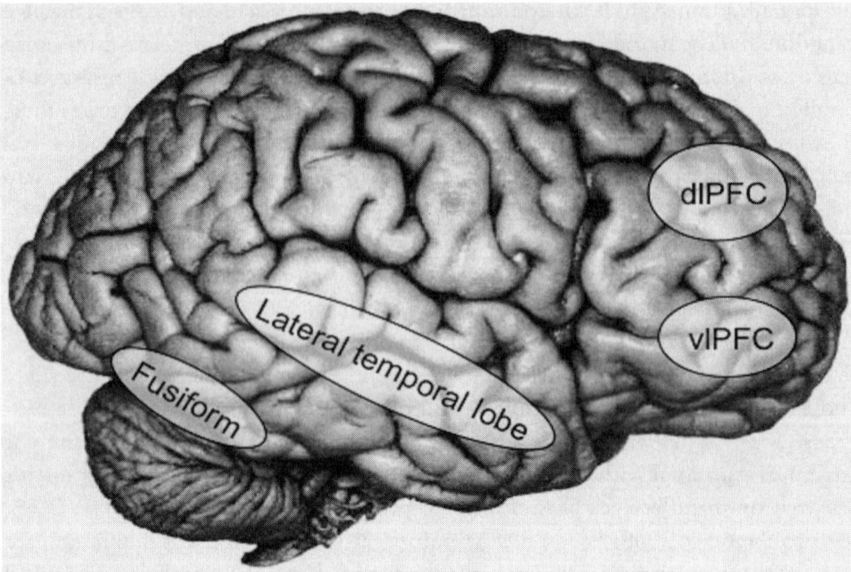

FIGURE 9.2 Lateral Aspect of the Right Hemisphere of the Brain Illustrating the Dorsolateral Prefrontal Cortex (dlPFC), Ventrolateral Prefrontal Cortex (vlPFC), Lateral Temporal Lobe, and Fusiform Regions

patients with damage to their medial prefrontal cortex (mPFC; see Figure 9.3) did not show bias on a male versus female IAT (Milne & Grafman, 2001). However, it is likely that this finding reflects the important role of the mPFC in the cognitive control processes that give rise to the IAT effect (i.e., response slowing on high-conflict trials), rather than suggesting that the mPFC contains a representation of stereotype knowledge (a function typically ascribed to lateral regions of the PFC and the temporal lobes). Understanding the specific representation of semantic associations related to stereotyping, in relation to attitudes and affective responses, is an important emerging topic in this literature (e.g., Gilbert, Swencionis, & Amodio, 2012).

Control and the Regulation of Intergroup Bias

Given the pervasiveness of implicit racial attitudes and stereotypes, how do people with egalitarian beliefs manage to respond without prejudice? Social psychological research on intergroup relations suggests that egalitarians must often exert control over their race-biased tendencies (Crandall & Eshleman, 2003; Devine, 1989). However, the process of control is complex. How is the need for control detected? Does control rely on conscious awareness? And what aspects of a prejudiced response are affected by controlled processes? Neuroscience theories and methods have proved useful for addressing questions such as these (Amodio et al., 2004;

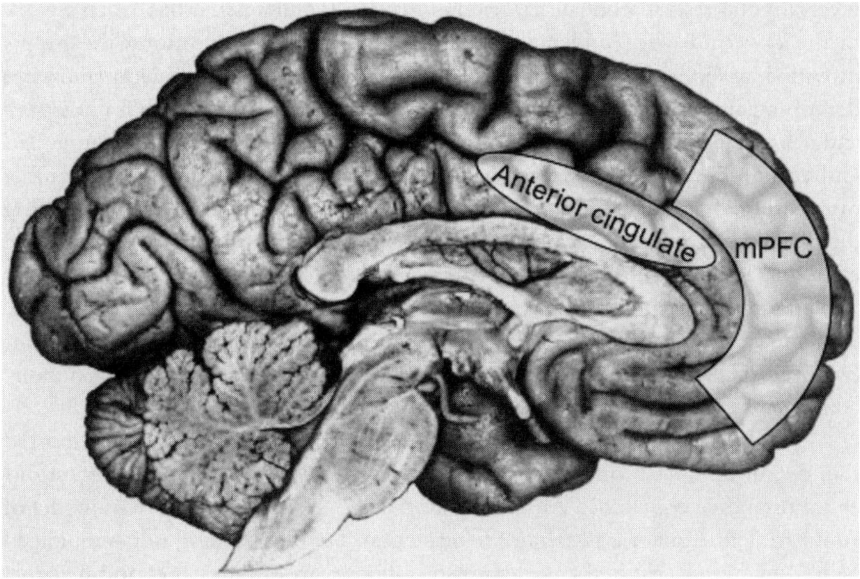

FIGURE 9.3 Medial Aspect of the Left Hemisphere of the Brain Showing the Anterior Cingulate and Medial Frontal Regions

Notes: mPFC = medial prefrontal cortex.

Bartholow, Dickter, & Sestir, 2006; for reviews, see Amodio & Devine, 2010; Amodio, Devine, & Harmon-Jones, 2007; see also Bartholow, Chapter 10 in this volume). In this section, I describe some of the ways in which neuroscience research has helped to advance theories regarding the regulation of intergroup responses.

Detecting the Need for Control

When I began conducting research on the control of intergroup bias, influential theories of control focused on the prerequisites for control, such as motivation and ability (Devine, 1989; Fazio, 1990), and on the specific steps involved in implementing a controlled response (e.g., Monteith, 1993; Wegener & Petty, 1997). These models assumed that the deployment of controlled processing was deliberative and conscious. But they did not address how the need for control is detected in the first place.

At the time I was pondering these issues, I was enrolled in a graduate course focusing on new research by prominent neuroscientists who studied issues of emotion and health. This course culminated in a symposium at which these neuroscientists spoke (the Wisconsin Symposium on Emotion), and students in the course were assigned to moderate the talks and had the opportunity to interact with speakers afterward. One of the speakers that year was Jonathan Cohen and, given my interests in control, I signed up to lead the discussion of Cohen's work in the class and then moderate his talk at the symposium. Among the papers provided by Cohen for our class preparations was a then unpublished manuscript describing a new cognitive neuroscience perspective on how control is initiated and deployed. This paper, later published as Botvinick, Braver, Barch, Carter, and Cohen (2001), proposed that control involves two major components: conflict monitoring and regulative control. *Conflict monitoring* is the process for detecting the need for control, and it specifically refers to conflict between higher-level goals and lower-level motor responses. *Regulative control* refers to the process of implementing an intended response after the need for control has been detected. This model addressed the issue that I sought to understand in social cognition: How is the need for control detected? And how can it be deployed non-consciously?

Working again with Eddie Harmon-Jones and Patricia Devine, we imported the idea of conflict monitoring into social psychology to address our questions about prejudice control. In Amodio et al. (2004), we proposed that this model of control could explain a puzzling phenomenon: Despite holding non-prejudiced beliefs, many self-avowed egalitarians still often express stereotypes and negative affect in their behavior. We wondered whether such "slips" might reflect a failure to detect the need for control by the conflict monitoring system, or rather a failure to implement control once its need has been detected. To test these competing hypotheses, we examined participants' attempts to control stereotype-biased responses, while measuring patterns of brain activity associated with conflict monitoring in the anterior cingulate cortex (ACC; see Figure 9.3). The behavioral task in this study was Payne's (2001) weapons identification task, in which participants must identify a target object as being either a gun or a tool, following the presentation of Black or White face primes. Participants are typically more accurate in identifying guns following Black faces, presumably due to stereotypes of Black people as dangerous and being involved in violent gangs. Thus, this task requires enhanced control to override an automatic stereotype on some trials (e.g., when identifying a "tool" following a Black face), but not others (e.g., when identifying a "gun" following a Black face). Importantly, this task provided the necessary conditions to examine responses when control is needed, compared with when it is not needed. It also yielded a behavioral index of control that could be parsed from the participants' degree of automatic stereotyping (Payne, 2001).

To measure activity of the ACC, we again turned to the psychophysiology lab to use event-related brain potential (ERP) methods—see chapters by Bartholow (10), Derks (3), and Ito (2) in this volume for methodological details. Because we were interested, in part, in detecting a non-deliberative form of control, it was important to use a method that could measure rapid changes in neural activity. Event-related brain potential methods are well-suited to this purpose, whereas the slow temporal resolution of fMRI would not allow us to differentiate more or less deliberative forms of processing. In this study, we focused on two ERP components known to reflect ACC activity—the *error-related negativity* (ERN) component and the *N2* component (also called the *correct-related negativity*). The amplitude of the ERN wave is associated with the detection of response conflict associated with an erroneous response, whereas the amplitude of the N2 wave reflects the detection of conflict that results in a successfully controlled response. We found that both were more strongly activated on trials where control over stereotypes was needed (Amodio et al., 2004; see Figure 9.4). That is, heightened ACC activity was observed both when control succeeded and when it failed. This finding suggested that participants were generally good at detecting the need for control, and when unintentional expressions of bias occurred, it reflected a failure to implement the intended response. This new perspective on control helped to explain why racial biases sometimes slip through into behavior despite a person's egalitarian intentions (see also Bartholow et al., 2006; Correll, Urland, & Ito, 2006; Bartholow, Chapter 10, and Ito, Chapter 2 in this volume).

In a follow-up study, we used this neuroscience model of control to address why some low-prejudice people—those who have strong internal motivation to respond without prejudice, but are also very concerned about external social pressures—are especially prone to unintended expressions of bias (Amodio, Devine, & Harmon-Jones, 2008). We proposed that these individuals might have trouble controlling their expressions of bias because their neural systems are less sensitive to conflict between an activated stereotype and their task-level goal of responding without prejudice (despite being hyper-sensitive to external social cues; Plant & Devine, 1998). Consistent with this hypothesis, these high-internal/high-external-motivation subjects showed low levels of ACC activity on trials that required control over automatic stereotypes—a pattern matching that of participants with highly prejudiced beliefs. By contrast, participants who were highly internally motivated and unconcerned about external pressures showed strong ACC responses when stereotype control was needed, and they were better at exerting control over their behavior. In another study, we asked whether the conflict monitoring function is also involved when the conflict arises from social pressure to respond without prejudice. We found that the engagement of control in response to external social cues involved activity in more anterior regions of the ACC and in the mPFC (Amodio, Kubota, Harmon-Jones, & Devine, 2006)—regions shown in other work to be important for inferring the perspective of another person and coordinating behavior with that person (Amodio & Frith, 2006; Harris, Todorov, & Fiske, 2005; Mitchell, 2009).

Together, these social neuroscience studies have revealed an important component of control that was missing in prior social psychological models: the process whereby conflict between an automatic tendency and one's intended response is detected and the need for control is registered. We showed that this aspect of control occurred rapidly and without deliberation, such that control could be deployed implicitly. This work has also distinguished the processes of detecting conflict involving one's own goals from conflict involving the goals of other people

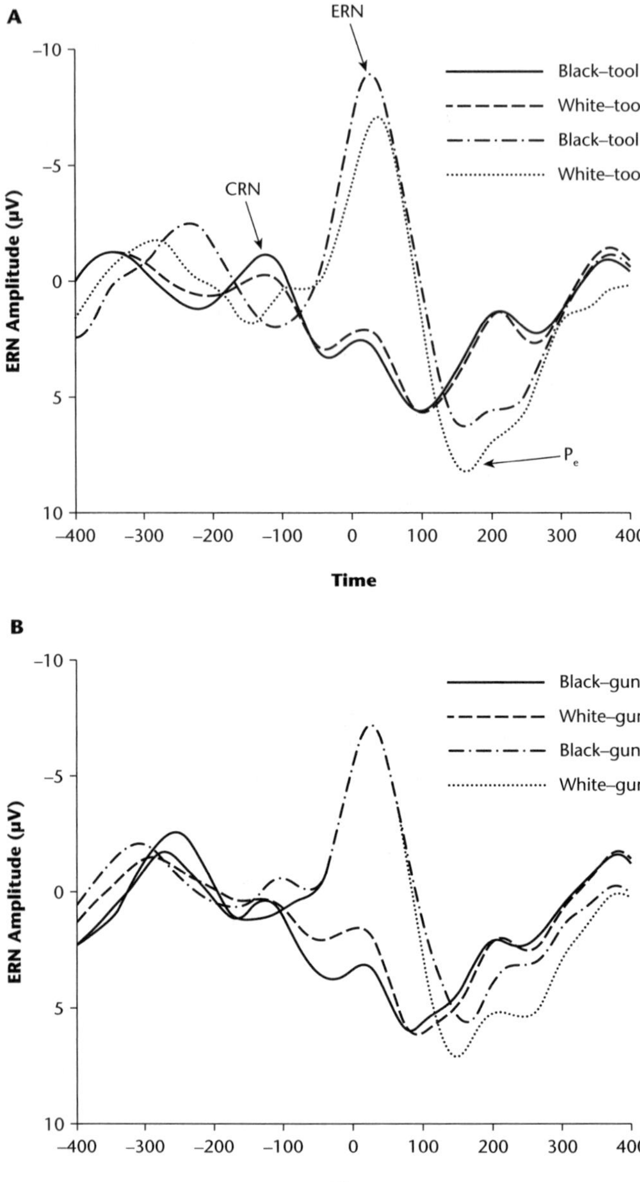

FIGURE 9.4 The Successful Regulation of a Race-biased Response Involves the Detection of Conflict Between a Stereotype-biased Tendency and One's Task Goal

Notes: Conflict detection is associated with the rapid activation of the anterior cingulate cortex and indexed by the error-related negativity (ERN) and the N2 (or correct-related negativity) components of the event-related potential.

Adapted from Amodio et al. (2004)

(e.g., peers or authority figures). These findings, made possible by the integration of theories and methods from neuroscience, have helped to advance mainstream social psychological theories of self-regulation (e.g., Harmon-Jones, E., Amodio, & Harmon-Jones, C., 2009; Monteith, Lybarger, & Woodcock, 2009; Sherman et al., 2008).

Implementing Control

Many theories in social and cognitive psychology posit that once the need for control is detected, other mechanisms are engaged to implement a controlled response (Devine, 1989; Shiffrin & Schneider, 1977). Neuroscience research has implicated the prefrontal cortex (PFC) in this function (Figure 9.2, above; Badre & Wagner, 2007; Botvinick et al., 2001; Kerns et al., 2004). However, the specific target of control is not always clear—it could be the emotional response to an out-group member, the stereotype itself, the behavior tendency, one's perception of the out-group member, or some other process. Although social psychological theories typically consider each of these targets of control to be equally plausible, an analysis of the neural circuitry of the PFC suggests that some of these processes are more likely to be modulated by control mechanisms than others. In particular, this circuitry suggests that the PFC modulates goal-directed action processes as well as sensory input and perceptual processing (Miller & Cohen, 2001). By contrast, the PFC has few connections to subcortical regions associated with emotion, such as the amygdala (Barbas & Zikopaulous, 2007). Given that the amygdala is often identified as a neural correlate of implicit racial affect and evaluation, this analysis suggests that the notion of direct down-regulation of implicit bias is unlikely.

Several studies have observed patterns of PFC activity while White participants viewed pictures of White and Black people. However, the role of these PFC activations in controlled processing is not always clear. In the earliest example, Richeson et al. (2003) had White American subjects view Black and White faces while their brain activity was measured using fMRI. Later, these participants had an interaction with a Black experimenter, after which they completed a behavioral measure of cognitive control. The authors found that participants who showed stronger PFC activity to Black than White faces in the first session performed more poorly on the cognitive control task following an interracial interaction. The authors reasoned that subjects who showed greater PFC activity when viewing faces had engaged in spontaneous control and were thus also more likely to engage control during the interracial interaction, which in turn produced a more "depleted" pattern of performance on the Stroop task. However, given the task design, it was difficult to know how exactly the PFC activity related to some form of control, since participants simply viewed pictures of faces with neutral expressions without being observed or having to make any race-relevant responses. It is possible that when faces of Black people appeared, participants felt that they needed to control some aspect of their response that was not measured or manipulated in the experiment. But without a task designed to create a response conflict in which control is needed, such as the weapons identification task described above, the extent to which PFC activity in response to a Black face was associated with control is ambiguous.

A study by Cunningham et al. (2004) also used fMRI to measure brain activity in response to faces of Black and White people. In the task, participants indicated whether each face appeared on the right or left side of the visual field. Like Richeson et al. (2003), Cunningham et al. (2004) assumed that people spontaneously engage some form of control when simply viewing faces of out-group members. Similarly, a study by Lieberman et al. (2005) found an increase in some PFC regions (e.g., the right ventrolateral PFC; see Figure 9.2, above) when participants viewed Black versus White faces and were instructed to match the faces to written labels. But because these tasks did not involve a conflict of some sort in which control was needed, it was difficult to determine exactly what was being controlled in these studies. Thus, the PFC activity observed in these studies may have reflected the control of prejudice, but it could also have reflected other psychological processes, such as the enhanced attention to racial features or an increase in task focus.

To characterize a pattern of brain activity as relating to control, brain activity needs to be measured in the context of an experimental task that manipulates the need for control and, if possible, yields a quantitative behavioral index of successful control. To this end, I have used experimental tasks to manipulate the control of racially biased tendencies. That is, instead of simply presenting out-group faces to a participant and assuming that the participant will engage in some form of control, we have used tasks that require participants to override or correct their racially biased tendencies. For example, in Amodio et al. (2007), participants were given bogus feedback that they had responded with prejudice on a picture-viewing task, which made them feel guilty. However, when given the opportunity to reduce their degree of prejudice, participants who reported more guilt expressed greater desire to engage in prejudice-reduction behaviors, and these intentions were associated with stronger left PFC activity.

In other research, we have used priming tasks, such as Payne's (2001) weapons identification task, in which the accurate categorization of a target (e.g., as a gun or a tool) requires participants to override race-biased response tendencies elicited by the primed face of a Black, in comparison with a White, individual. In one study using the weapons identification task, we found that better response control on the task was predicted by stronger proactive engagement of the left PFC throughout the task (Amodio, 2010). Furthermore, we found that the effect of left PFC activity on behavior control was mediated through rapid attentional orienting to Black face primes, compared with White face primes. That is, it appeared that greater task motivation, evidenced by greater left PFC activity, was associated with stronger orienting to cues that control is needed (i.e., Black faces), which in turn enabled participants to override the influence of racial stereotypes on their task responses. Studies such as these have provided direct evidence that the PFC is involved in the control of intergroup responses.

Although the PFC is often hypothesized to "down-regulate" the activity of racial bias (e.g., Cunningham et al., 2004), my research suggests that this is not the case (Amodio et al., 2003). Rather, the PFC is involved in coordinating attention and action in the service of a task goal (Amodio, 2010; Amodio et al., 2007), as described previously. These findings comport with Monteith's (1993) self-regulation model, which posits that once the goal to control intergroup responses is formed, an individual becomes vigilant to cues that control is needed, and these cues facilitate behavioral forms of regulation (see also Richeson & Trawalter, 2008). Furthermore, research on neural function in the animal literature suggests that the PFC primarily regulates attention, perception, and action, but does not directly modulate activity in the amygdala associated with negative affect and evaluation (as reviewed in Amodio & Ratner, 2011a; see also Barbas & Zikopoulos, 2007). Thus, although down-regulation models of control remain popular, social neuroscience research has revealed that control operates primarily by proactively coordinating attention, perception, and action, and that changes in the sources of bias—race-biased affect, attitudes, and stereotypes—occur indirectly, as a result of these proactive processes.

Conclusions

Although the social neuroscience approach to intergroup bias is still relatively new, it has already made significant contributions to the psychological literature on prejudice, stereotyping, and discrimination. Early social psychophysiological approaches provided researchers with invaluable methods for assessing prejudiced responses which were otherwise concealed in overt behaviors and self-reports. More recently, social neuroscientists have begun to invoke models of brain structure and function to understand the socio-cognitive and emotional mechanisms that underlie implicit racial bias and the regulation of intergroup responses. This research has shed new light on how intergroup biases are acquired, expressed, and potentially changed, and on how intergroup behaviors are regulated. In doing so, this research has already begun to provide important theoretical advances that would not have been anticipated by traditional behavioral approaches to the study of intergroup bias.

With the neurosciences expanding with increasing pace, and with increasing cross-talk between neuroscientists and social psychologists, the biggest breakthroughs in the social neuroscience of prejudice are surely yet to come. As this field develops, social neuroscientists' success in advancing psychological theory will depend on their ability to harness the power of neuroscience for probing mechanisms of the social mind.

Suggestions for Further Reading

Amodio, D. M. (2008). The social neuroscience of intergroup relations. *European Review of Social Psychology, 19*, 1–54.

A broad review of neuroscience approaches to the study of intergroup prejudice and stereotyping.

Amodio, D. M. (2010). Coordinated roles of motivation and perception in the regulation of intergroup responses: Frontal cortical asymmetry effects on the P2 event-related potential and behavior. *Journal of Cognitive Neuroscience, 22*, 2609–2617.

Presents a test of the proactive control model, whereby the PFC coordinates the control of intergroup bias by tuning attention to racial cues.

Amodio, D. M., & Hamilton, H. K. (2012). Intergroup anxiety effects on implicit racial evaluation and stereotyping. *Emotion, 12*, 1273–1280.

Presents support for the memory systems model of implicit bias, whereby implicit racial attitudes and stereotypes are rooted in different neural substrates (e.g., the amygdala and left temporal lobe, respectively). Here, intergroup anxiety, associated with amygdala activity, selectively amplified implicit racial attitudes without affecting the activation of implicit stereotypes.

Amodio, D. M., & Harmon-Jones, E. (2012). Neuroscience approaches in social and personality psychology. In M. Snyder & K. Deaux (Eds.), *Handbook of social and personality psychology* (pp. 11–150). New York: Oxford University Press.

A comprehensive review of social neuroscience approaches to core areas of social and personality psychology.

Amodio, D. M., Harmon-Jones, E., Devine, P. G., Curtin, J. J., Hartley, S. L., & Covert, A. E. (2004). Neural signals for the detection of unintentional race bias. *Psychological Science, 15*, 88–93.

An initial demonstration of conflict monitoring as a mechanism involved in the control of racial stereotypes.

Amodio, D. M., & Ratner, K. G. (2011). A memory systems model of implicit social cognition. *Current Directions in Psychological Science, 20*, 143–148.

Presents a theoretical model of implicit social cognition derived from neuroscience research on learning and memory, with an emphasis on implications for implicit stereotyping and prejudice.

References

Amodio, D. M. (2008). The social neuroscience of intergroup relations. *European Review of Social Psychology, 19*, 1–54.

Amodio, D. M. (2010). Coordinated roles of motivation and perception in the regulation of intergroup responses: Frontal cortical asymmetry effects on the P2 event-related potential and behavior. *Journal of Cognitive Neuroscience, 22*, 2609–2617.

Amodio, D. M., & Devine, P. G. (2006). Stereotyping and evaluation in implicit race bias: Evidence for independent constructs and unique effects on behavior. *Journal of Personality and Social Psychology, 91*, 652–661.

Amodio, D. M., & Devine, P. G. (2010). Regulating behavior in the social world: Control in the context of intergroup bias. In R. R. Hassin, K. N. Ochsner & Y. Trope (Eds.), *Self control in society, mind and brain* (pp. 49–75). New York: Oxford University Press.

Amodio, D. M., Devine, P. G., & Harmon-Jones, E. (2007). A dynamic model of guilt: Implications for motivation and self-regulation in the context of prejudice. *Psychological Science, 18*, 524–530.

Amodio, D. M., Devine, P. G., & Harmon-Jones, E. (2008). Individual differences in the regulation of intergroup bias: The role of conflict monitoring and neural signals for control. *Journal of Personality and Social Psychology, 94*, 60–74.

Amodio, D. M., & Frith, C. D. (2006). Meeting of minds: The medial frontal cortex and social cognition. *Nature Reviews Neuroscience*, 7, 268–277.

Amodio, D. M., & Hamilton, H. K. (2012). Intergroup anxiety effects on implicit racial evaluation and stereotyping. *Emotion, 12*, 1273–1280.

Amodio, D. M., Harmon-Jones, E., & Devine, P. G. (2003). Individual differences in the activation and control of affective race bias as assessed by startle eyeblink responses and self-report. *Journal of Personality and Social Psychology, 84*, 738–753.

Amodio, D. M., Harmon-Jones, E., Devine, P. G., Curtin, J. J., Hartley, S. L., & Covert, A. E. (2004). Neural signals for the detection of unintentional race bias. *Psychological Science, 15*, 88–93.

Amodio, D. M., Kubota, J. T., Harmon-Jones, E., & Devine, P. G. (2006). Alternative mechanisms for regulating racial responses according to internal vs. external cues. *Social Cognitive and Affective Neuroscience, 1*, 26–36.

Amodio, D. M., & Mendoza, S. A. (2010). Implicit intergroup bias: Cognitive, affective, and motivational underpinnings. In B. Gawronski and B. K. Payne (Eds.), *Handbook of implicit social cognition* (pp. 353–374). New York: Guilford Press.

Amodio, D. M., & Ratner, K. (2011a). Mechanisms for the regulation of intergroup responses: A social neuroscience analysis. In J. Decety & J. T. Cacioppo (Eds.), *Handbook of social neuroscience* (pp. 729–741). New York: Oxford University Press.

Amodio, D. M., & Ratner, K. G. (2011b). A memory systems model of implicit social cognition. *Current Directions in Psychological Science, 20*, 143–148.

Badre, D., & Wagner, A. D. (2007). Left ventrolateral prefrontal cortex and the control of memory. *Neuropsychologia, 45*, 2883–2901.

Barbas, H., & Zikopoulos, B. (2007). The prefrontal cortex and flexible behavior. *The Neuroscientist, 13*, 532–545.

Bartholow, B. D., Dickter, C. L., & Sestir, M. A. (2006). Stereotype activation and control of race bias: Cognitive control of inhibition and its impairment by alcohol. *Journal of Personality and Social Psychology, 90*, 272–287.

Bartholow, B. D., Pearson, M. A., Gratton, G., & Fabiani, M. (2003). Effects of alcohol on person perception: A social cognitive neuroscience approach. *Journal of Personality and Social Psychology, 85*, 627–638.

Botvinick, M. M., Braver, T. S., Barch, D. M., Carter, C. S., & Cohen, J. D. (2001). Conflict monitoring and cognitive control. *Psychological Review, 108*, 624–652.

Bouton, M. E. (1994). Conditioning, remembering, and forgetting. *Journal of Experimental Psychology: Animal Behavior Processes, 20*, 219–231.

Correll, J., Urland, G. R., & Ito, T. A. (2006). Event-related potentials and the decision to shoot: The role of threat perception and cognitive control. *Journal of Experimental Social Psychology, 42*, 120–128.

Crandall, C. S., & Eshleman, A. (2003). A justification-suppression model of the expression and experience of prejudice. *Psychological Bulletin, 129*, 414–446.

Crisp, R. J., & Turner, R. N. (2009). Can imagined interactions produce positive perceptions? Reducing prejudice through simulated social contact. *American Psychologist, 64*, 231–240.

Cunningham, W. A., Johnson, M. K., Raye, C. L., Gatenby, J. C., Gore, J. C., & Banaji, M. R. (2004). Separable neural components in the processing of Black and White Faces. *Psychological Science, 15*, 806–813.

Cunningham, W. A., Raye, C. L., & Johnson, M. K. (2004). Implicit and explicit evaluation: fMRI correlates of valence, emotional intensity, and control in the processing of attitudes. *Journal of Cognitive Neuroscience, 16*, 1717–1729.

Davis, M. (2006). Neural systems involved in fear and anxiety measured with fear-potentiated startle. *American Psychologist, 61*, 741–756.

Devine, P. G. (1989). Stereotypes and prejudice: Their automatic and controlled components. *Journal of Personality and Social Psychology, 56*, 5–18.

Devine, P. G., Plant, E. A., Amodio, D. M., Harmon-Jones, E., & Vance, S. L. (2002). The regulation of explicit and implicit race bias: The role of motivations to respond without prejudice. *Journal of Personality and Social Psychology, 82*, 835–848.

Dovidio, J. F., Kawakami, K., Johnson, C., Johnson, B., & Howard, A. (1997). On the nature of prejudice: Automatic and controlled processes. *Journal of Experimental Social Psychology, 33*, 510–540.

Fazio, R. H. (1990). Multiple processes by which attitudes guide behaviour: The MODE model as an integrative framework. In M. P. Zanna (Ed.), *Advances in experimental social psychology* (Vol. 23, pp. 75–109). New York: Academic Press.

Fazio, R. H., Jackson, J. R., Dunton, B. C., & Williams, C. J. (1995). Variability in automatic activation as an unobtrusive measure of racial attitudes: A bona fide pipeline? *Journal of Personality and Social Psychology, 69*, 1013–1027.

Fazio, R. H., & Olson, M. A. (2003). Implicit measures in social cognition: Their meaning and use. *Annual Review of Psychology, 54*, 297–327.

Fendt, M., & Fanselow, M. S. (1999). The neuroanatomical and neurochemical basis of conditioned fear. *Neuroscience & Biobehavioral Reviews, 23*, 743–760.

Fiske, S. T. (1998). Stereotyping, prejudice, and discrimination. In D. T. Gilbert, S. T. Fiske & G. Lindzey (Eds.), *The handbook of social psychology* (Vol. 2, pp. 357–411). New York: McGraw-Hill.

Gabrieli, J. D. (1998). Cognitive neuroscience of human memory. *Annual Review of Psychology, 49*, 87–115.

Gaertner, S. L., & McLaughlin, J. P. (1983). Racial stereotypes: Associations and ascriptions of positive and negative characteristics. *Social Psychology Quarterly, 46*, 23–30.

Gilbert, S. J., Swencionis, J. K., & Amodio, D. M. (2012). Evaluative vs. trait representation in intergroup social judgments: Distinct roles of anterior temporal lobe and prefrontal cortex. *Neuropsychologia, 50*, 3600–3611.

Greenwald, A. G., McGhee, D. E., & Schwartz, J. K. L. (1998). Measuring individual differences in implicit cognition: The Implicit Association Test. *Journal of Personality and Social Psychology, 74*, 1464–1480.

Guglielmi, R. S. (1999). Psychophysiological assessment of prejudice: Past research, current status, and future directions. *Personality and Social Psychology Review, 3*, 123–157.

Harmon-Jones, E. (2003). Clarifying the emotive functions of asymmetrical frontal cortical activity. *Psychophysiology, 40*, 838–848.

Harmon-Jones, E., Amodio, D. M., & Harmon-Jones, C. (2009). Action-based model of dissonance. *Advances in Experimental Social Psychology, 41*, 119–166.

Harmon-Jones, E., Harmon-Jones, C., Amodio, D. M., & Gable, P. A. (2011). Attitudes toward emotions: Conceptualization and measurement of evaluations of specific emotions. *Journal of Personality and Social Psychology, 101*, 1332–1350.

Harris, L. T., Todorov, A., & Fiske, S. T. (2005). Attributions on the brain: Neuro-imaging dispositional inferences, beyond Theory of Mind. *NeuroImage, 28*, 763–769.

Hart, A. J., Whalen, P. J., Shin, L. M., McInerney, S. C., Fischer, H., & Rauch, S. L. (2000). Differential response in the human amygdala to racial outgroup vs. ingroup face stimuli. *NeuroReport, 11*, 2351–2355.

Ito, T. A., & Bartholow, B. D. (2009). The neural correlates of race. *Trends in Cognitive Sciences, 13*, 524–531.

Ito, T. A., & Urland, G. R. (2003). Race and gender on the brain: Electrocortical measures of attention to race and gender of multiply categorizable individuals. *Journal of Personality and Social Psychology, 85*, 616–626.

Kawakami, K., Phills, C. E., Steele, J. R., & Dovidio, J. F. (2007). (Close) distance makes the heart grow fonder: Improving implicit racial attitudes and interracial interactions through approach behaviors. *Journal of Personality and Social Psychology, 92*, 957–971.

Kerns, J. G., Cohen, J. D., MacDonald, A. W., Cho, R. Y., Stenger, V. A., & Carter, C. S. (2004). Anterior cingulate conflict monitoring and adjustments in control. *Science, 303*, 1023–1026.

Knutson, K. M., Mah, L., Manly, C. F., and Grafman, J. (2007). Neural correlates of automatic beliefs about gender and race. *Human Brain Mapping, 28*, 915–930.

Lang, P. J., Bradley, M. M., & Cuthbert, B. N. (1990). Emotion, attention, and the startle reflex. *Psychological Review, 97*, 377–395.

LeDoux, J. E., Iwata, J., Cicchetti, P., & Reis, D. J. (1988). Different projections of the central amygdaloid nucleus mediate autonomic and behavioral correlates of conditioned fear. *Journal of Neuroscience, 8*, 2517–2529.

Lieberman, M. D., Hariri, A., Jarcho, J. M., Eisenberger, N. I., & Bookheimer, S. Y. (2005). An fMRI investigation of race-related amygdala activity in African-American and Caucasian-American individuals. *Nature Neuroscience, 8*, 720–722.

McConnell, A. R., & Leibold, J. M. (2001). Relations among the Implicit Association Test, discriminatory behavior, and explicit measures of racial attitudes. *Journal of Experimental Social Psychology, 37*, 435–442.

Mendoza, S. A., Gollwitzer, P. M., & Amodio, D. M. (2010). Reducing the expression of implicit stereotypes: Reflexive control through implementation intentions. *Personality and Social Psychology Bulletin, 36*, 512–523.

Miller, E. K., & Cohen, J. D. (2001). An integrative theory of prefrontal cortex function. *Annual Review of Neuroscience, 24*, 167–202.

Milne, E., & Grafman, J. (2001). Ventromedial prefrontal cortex lesions in humans eliminate implicit gender stereotyping. *The Journal of Neuroscience, 21*, 1–6.

Mitchell, J. P. (2009). Social psychology as a natural kind. *Trends in Cognitive Sciences, 13*, 246–251.

Mitchell, J. P., Ames, D. L., Jenkins, A. C., & Banaji, M. R. (2009). Neural correlates of stereotype application. *Journal of Cognitive Neuroscience, 21*, 594–604.

Monteith, M. J. (1993). Self-regulation of prejudiced responses: Implications for progress in prejudice-reduction efforts. *Journal of Personality and Social Psychology, 65*, 469–485.

Monteith, M. J., Ashburn-Nardo, L., Voils, C. I., & Czopp, A. M. (2002). Putting the brakes on prejudice: On the development and operation of cues for control. *Journal of Personality and Social Psychology, 83*, 1029–1050.

Monteith, M. J., Lybarger, J. E., & Woodcock, A. (2009). Schooling the cognitive monster: The role of motivation in the regulation and control of prejudice. *Social and Personality Compass, 3*, 211–226.

Ofan, R. H., Rubin, N., & Amodio, D. M. (2011). Seeing race: N170 responses to race and their relation to automatic racial attitudes and controlled processing. *Journal of Cognitive Neuroscience, 23*, 3152–3161.

Payne, B. K. (2001). Prejudice and perception: The role of automatic and controlled processes in misperceiving a weapon. *Journal of Personality and Social Psychology, 81*, 181–192.

Phelps, E. A., O'Connor, K. J., Cunningham, W. A., Funayama, E. S., Gatenby, J. C., Gore, J. C., & Banaji, M. R. (2000). Performance on indirect measures of race evaluation predicts amygdala activation. *Journal of Cognitive Neuroscience, 12*, 729–738.

Plant, E. A., & Devine, P. G. (1998). Internal and external motivation to respond without prejudice. *Journal of Personality and Social Psychology, 75*, 811–832.

Quadflieg, S., & Macrae, C. N. (2011). *European Review of Social Psychology, 22*, 215–273.

Quadflieg, S., Turk, D. J., Waiter, G. D., Mitchell, J. P., Jenkins, A. C., & Macrae, C. N. (2009). Exploring the neural correlates of social stereotyping. *Journal of Cognitive Neuroscience, 21*, 1560–1570.

Rankin, R. E., & Campbell, D. T. (1955). Galvanic skin response to negro and white experimenters. *Journal of Abnormal and Social Psychology, 51*, 30–33.

Richeson, J. A., Baird, A. A., Gordon, H. L., Heatherton, T. F., Wyland, C. L., Trawalter, S., & Shelton, J. N. (2003). An fMRI investigation of the impact of interracial contact on executive function. *Nature Neuroscience, 6*, 1323–1328.

Richeson, J. A., Todd, A. R., Trawalter, S., & Baird, A. A. (2008). Eye-gaze direction modulates race-related amygdala activity. *Group Processes and Intergroup Relations, 11*, 233–246.

Richeson, J. A., & Trawalter, S. (2008). The threat of appearing prejudiced and race-based attentional biases. *Psychological Science, 19*, 98–102.

Ronquillo, J., Denson, T. F., Lickel, B., Lu, Z., Nandy, A, & Maddox, K. B. (2007). The effects of skin tone on race-related amygdala activity: An fMRI investigation. *Social Cognitive and Affective Neuroscience, 2*, 39–44.

Schiller, D., Freeman, J. B., Mitchell, J. P., Uleman, J. S., & Phelps, E. A. (2009). A neural mechanism of first impressions. *Nature Neuroscience, 12*, 508–514.

Sherman, J. W., Gawronski, B., Gonsalkorale, K., Hugenberg, K., Allen, T. J., & Groom, C. J. (2008). The self-regulation of automatic associations and behavioral impulses. *Psychological Review, 115*, 314–335.

Shiffrin, R., & Schneider, W. (1977). Controlled and automatic human information processing: II. Perceptual learning, automatic attending, and a general theory. *Psychological Review, 84*, 127–190.

Sloman, S. A. (1996). The empirical case for two systems of reasoning. *Psychological Bulletin, 119*, 3–22.

Smith, E. R., & DeCoster, J. (2000). Dual process models in social and cognitive psychology: Conceptual integration and links to underlying memory systems. *Personality and Social Psychology Review, 4*, 108–131.

Stangor, C., & Lange, J. (1994). Mental representations of social groups: Advances in understanding stereotypes and stereotyping. *Advances in Experimental Social Psychology, 26*, 357–416.

Van Bavel, J. J., & Cunningham, W. A. (2011). A social neuroscience approach to self and social categorisation: A new look at an old issue. *European Review of Social Psychology, 21*, 237–284.

Van Bavel, J. J., Packer, D. J., & Cunningham, W. A. (2008). The neural substrates of in-group bias: A functional magnetic resonance imaging investigation. *Psychological Science, 19*, 1131–1139.

Vanman, E. J., Paul, B. Y., Ito, T. A., & Miller, N. (1997). The modern face of prejudice and structural features that moderate the effect of cooperation on affect. *Journal of Personality and Social Psychology*, *73*, 941–959.

Wegener, D. T., & Petty, R. E. (1997). The flexible correction model: The role of naïve theories of bias in bias correction. In M. P. Zanna (Ed.), *Advances in experimental social psychology* (Vol. 29, pp. 141–208). Mahwah, NJ: Erlbaum.

Whalen, P. J. (1998). Fear, vigilance, and ambiguity: Initial neuroimaging studies of the human amygdala. *Current Directions in Psychological Science*, 7, 177–188.

Wheeler, M. E., & Fiske, S. T. (2005). Controlling racial prejudice and stereotyping: Social cognitive goals affect amygdala and stereotype activation. *Psychological Science*, *16*, 56–63.

10

EVENT-RELATED BRAIN POTENTIALS AND THE ROLE OF COGNITIVE CONTROL IN IMPLICIT RACE BIAS

Bruce D. Bartholow

Long Day's Journey into Social Neuroscience

It seems strangely difficult to imagine now, but it was not very long ago when only a tiny minority of social psychologists used any kind of psychophysiological or neuroimaging measures in their research programs. Even those who did so probably had no idea that they were at the forefront of what would become a major force within the field. If we can consider the social neuroscience movement to be a "new dawn" of inquiry, in which social and biological approaches to understanding social behavior have been integrated in a theoretically rich and meaningful way (and one that has been recognized as important by a majority of the field), then it was in the hazy early morning hours of this new dawn that I took my first tentative steps onto the trail blazed by pioneers such as John Cacioppo, Jim Blascovich, and their collaborators. In this chapter, I will present a brief overview of my experience as a social psychologist utilizing psychophysiological measures to study the mental events that give rise to social behaviors, focusing on racial categorization and bias. The main focus of this chapter will be on recent work in which my colleagues and I have investigated the extent to which performance on tasks measuring implicit bias is driven by cognitive conflict and control processes. However, I begin with a story of how I first came to use psychophysiological measures to address a longstanding question in person perception, an experience that formed the foundation of my excitement about the possibilities of social cognitive neuroscience.

Although my interest in using psychophysiological measures actually dates a bit further back than this, I will begin with my experiences as a graduate student struggling to find ways to measure cognitive and affective responses to expectancy violations—that is, how people react when they encounter information about another person that is somehow unexpected (e.g., based on the person's social group) or atypical of that person (e.g., based on what is already known or assumed about the person's traits). Although numerous theoretical and empirical reports had documented the notion that unexpected information triggers more elaborative cognitive processing than expected information (see Srull & Wyer, 1989; Stangor & McMillan, 1992), I was especially captivated by the possibility that expectancy violations might elicit an affective or emotional response in perceivers (e.g., Mandler, 1990; Olson, Roese, & Zanna, 1996). In particular, I was struck by the idea that expectancy violations will always produce a negative affective response, at least initially, even if the violating information is positive in nature (Olson et al., 1996). Although my advisor and I conducted numerous experiments investigating cognitive and affect-related responses to expectancy violation (e.g., Kernahan, Bartholow, & Bettencourt, 2000), I was frustrated by the lack of precision afforded by the behavioral measures we were using for addressing the issue of potentially fleeting affective reactions.

Fortunately for me, several things coincided around the time I was pondering these issues that led me to a solution. First, Cacioppo and his colleagues had recently published several articles investigating responses to attitude inconsistency using event-related brain potentials (ERPs) (Cacioppo, Crites, Berntson, & Coles, 1993; Cacioppo, Crites, Gardner, & Berntson, 1994). This technique seemed very promising for my interests given that ERPs can register brain responses to stimuli much more rapidly than any behavioral measure can, and indeed can register responses in the absence of any behavioral measure. Second, Eric Vanman and colleagues (Vanman, Paul, Ito, & Miller, 1997) had just published their landmark paper using facial electromyography (EMG) to study prejudice-related affective responses. Their work, coupled with earlier work by Cacioppo and colleagues (Cacioppo, Petty, Losch, & Kim, 1986) showing the sensitivity of facial EMG measures to even tiny and fleeting differences in affective responding, suggested that facial EMG would be sensitive to the affective consequences of expectancy violations.

Third, and by far most important for my story, the department in which I was trained had just hired two cognitive psychophysiologists, Gabriele Gratton and Monica Fabiani. Despite not knowing me at all and not being affiliated with the social psychology training area, they were open to considering my ideas and allowing me to test them in their laboratory. Thus, with extensive assistance from Professors Gratton and Fabiani and their students (as well as my advisor, Professor Ann Bettencourt), I designed what turned out to be my dissertation project. The ERP, EMG and behavioral data that emerged from that study (see Bartholow, Fabiani, Gratton, & Bettencourt, 2001) combined to tell an important story about how unexpected information is processed; especially that:

- Expectancy violations did not universally elicit negative affective responses (rather, the valence of response depended upon the valence of the unexpected information).

- The "unexpectedness" of behavior affected responding earlier in the time course of processing than did the valence of behavior.
- The kinds of social expectancy violations we investigated were processed very differently from semantic violations, despite the fact that we used verbal stimuli.

Although for most people the dissertation experience tends to represent a culmination of theoretical and methodological training in a line of research that one has been pursuing for years, in my case, this was only partly true. My dissertation certainly represented a culmination of my theoretical training, but the methodological piece was completely new to me, which left me with the uneasy feeling that I did not know what I was doing. Therefore, I pursued a post-doctoral fellowship in cognitive neuroscience in order to bolster my ERP training. I ended up staying in that position for only about 6 months, however, because a tenure-track faculty job was offered to me that was too good to pass up. Thus, much of my hands-on ERP training has come "on the job," and through continued consultation with my very generous mentors and friends in the field. I will always be grateful for their assistance.

What Questions Do We Address Using ERPs?

Since conducting that original study (and a follow-up; see Bartholow, Pearson, Gratton, & Fabiani, 2003), my students and I have used ERP measures in a variety of different paradigms in order to test ideas pertaining to a number of questions in social cognition (and cognition more generally). One common theme that has emerged over the years is an interest in understanding the interplay between more-or-less automatic and control-related processes in determining implicit bias, and whether ERPs might provide new insights into this issue. A more general, related aim has been to discover the psychological loci or mechanisms of experimental effects using measures of brain activity. As noted by Rugg and Coles (1995, p. 27), "Given that cognitive processes are implemented by the brain, it seems to make sense to explore the possibility that measures of brain activity can provide insights into their nature."

Although the idea that social behaviors represent some combination of control-related and automatic or unintended processes is commonplace (see Chaiken & Trope, 1999), social cognition has always favored the view that automatic processes are of primary interest (see Bargh & Chartrand, 1999). Responses on laboratory tasks thought to measure implicit race bias, for example, are often characterized as "under the control of automatically activated evaluations" (Greenwald, McGhee, & Schwartz, 1998, p. 1464). Two examples of such tasks are the well-known Implicit Association Test (IAT) (Greenwald et al., 1998) and the Affective Priming Task (Fazio, Sanbonmatsu, Powell, & Kardes, 1986), thought to assess the automatic activation of attitudes or evaluations, which also has been adapted for studying race bias (Fazio, Jackson, Dunton, & Williams, 1995).

Despite the seemingly widespread view that performance on such tasks represents the influence of automatic processes exclusively (see De Houwer, Teige-Mocigemba, Spruyt, & Moors, 2009), in recent years a number of challenges to the notion of "process-pure" implicit measures have emerged (see Amodio & Mendoza, 2010; Sherman, Klauer, & Allen, 2010), based in part on recent theoretical and empirical developments in cognitive neuroscience. Many have noted, first of all, that performance indices are derived from "explicit" (i.e., observable) behavioral responses, which are always subject to the influence of controlled processes to some degree (see Jacoby, 1991).

A more fundamental critique can be derived from theoretical differences concerning the structure and function of the information-processing system. Many social-cognitive researchers appear to have adopted some version of a discrete-stage model of processing (e.g., Sanders, 1990), which assumes that earlier or more quickly occurring stages must run to completion before later or more slowly developing stages can begin. Applied to implicit race bias tasks, such models predict that early, automatic aspects of processing (e.g., activation of evaluations) occur prior to slower-developing, control-related processes (e.g., conscious response commission), and that automatic evaluation is completed before responses are made.

In contrast to discrete-stage models, so-called "continuous flow" models posit that multiple aspects of processing occur simultaneously (see Eriksen & Schultz, 1979; McClelland, 1979). Numerous psychophysiological experiments have provided evidence in favor of the continuous flow model, primarily by demonstrating that response preparation does not wait for stimulus evaluation to finish, but rather begins while evaluative information about stimuli is still accumulating (see Coles, Smid, Scheffers, & Otten, 1995). This finding means that differences in reaction time (RT) between various conditions in a laboratory task can reflect differences in stimulus evaluation, response activation, or both. Specifically, slower responses can result from early, cursory stimulus evaluation leading to initial activation of an incorrect response, which can lead to conflict if continuing stimulus evaluation eventually indicates that the opposite response should be made (see Gratton, Coles, Sirevaag, Eriksen, & Donchin, 1988).

This discovery has important implications for interpreting reaction-time differences on implicit race bias tasks, most of which tend to involve an element of response conflict. Evaluative priming tasks provide a good example (but see Bartholow, 2010, for a more general discussion and extension to other tasks). In such tasks, participants are presented with a prime (e.g., a valenced word, or a face representing a racial group), followed quickly by a word which must be categorized as positive or negative as quickly as possible (see Fazio et al., 1986, 1995). The classic finding is that trials in which the prime and target are evaluatively or stereotypically congruent produce faster responses than trials in which the prime and target are incongruent (i.e., the *affective congruency effect*). This was initially interpreted in terms of spreading of activation (Fazio et al., 1986), but a number of researchers later argued that the effect could instead be the result of response conflict occurring on incongruent trials if the primes elicit preparation of responses that conflict with the response ultimately required by the targets (De Houwer, Hermans, Rothermund, & Wentura, 2002; Klinger, Burton, & Pitts, 2000; Wentura & Rothermund, 2003). Our lab provided the first neurophysiological evidence that primes activate response preparation in the motor cortex, which produces conflict in response output if the response required by the target differs from the one prepared following prime onset (Bartholow, Riordan, Saults, & Lust, 2009).

What Are ERPs, Anyway?

The ERP is an index of brain activity derived from the electroencephalogram (EEG), which reflects electrical activity generated by the firing of cortical neurons (for an extensive discussion, see Fabiani, Gratton, & Federmeier, 2007). The continuous recording of EEG (e.g., during a psychological task) indexes changes in patterns of brain voltage over time, the amplitude of which normally ranges from approximately –100 to +100 microvolts. When measured in the context of an experimental task involving specific events (e.g., stimuli or responses), it becomes possible to examine epochs of the EEG that reflect neural processes uniquely associated with those events. This event-related EEG response is called the ERP. Physiologically, ERPs represent the summation of activity created by post-synaptic potentials from populations of synchronously active, primarily cortical neurons whose activity summates to produce a signal recordable at the scalp (see Allison, Wood, & McCarthy, 1986). Psychologically, ERPs represent neural manifestations of specific information-processing activities associated with a stimulus or response event. The ERP waveform is typically comprised of a series of positive and negative voltage deflections, often referred to as components (but see Luck, 2005, for problems with this nomenclature). Specific deflections/components often are associated with a particular information-processing operation or set of operations (see Fabiani et al., 2007), though it is quite likely that any given component represents numerous simultaneously occurring processes (see Coles & Rugg, 1995). In general, the amplitude of a given ERP component represents the extent to which those operations are engaged by a stimulus or response event, and the latency at which the component peaks is thought to index an upper limit for the time required to complete those operations (see Fabiani et al., 2007).

ERPs can be measured non-invasively from the surface of the scalp using electrodes (generally made of tin or silver coated with silver chloride), often embedded in a stretch-nylon cap that can be worn by the participant. These electrodes are connected to amplifiers that magnify the very weak electrical signals emitted by the neurons by a factor of 10,000 to 50,000 so they can be measured accurately. These analog signals are digitally sampled at a frequency ranging from

100 to around 1,000 Hz (samples per second) and stored to a computer hard drive for later cleaning and analyses.

In addition to the typical experimental design issues with which one has to contend in crafting social psychological experiments, measuring ERPs requires some additional considerations (for more detailed discussion of these issues, see Bartholow & Amodio, 2009). First and foremost, measurement of ERPs requires that discrete stimulus and response events occur within the experimental task, and that it be possible precisely to specify and denote the time at which each event of interest occurs. The "cleanest" ERP studies are those in which each trial contains a single stimulus event and a single response (in cases where overt responses are desired), and in which trials are separated by at least 1 or 2 seconds. This is because every "event" (i.e., stimulus presentation or participant response) will be associated with changes in EEG voltage, and the occurrence of multiple events in close temporal proximity can lead to difficulties separating neural responses to different events. However, there are various techniques (e.g., subtraction of one kind of response from another) that permit separation of multiple responses under some conditions, and these are often used in paradigms involving multiple stimulus events per trial (e.g., priming studies). Another key factor in ERP studies is that participants must be able to sit still, because muscle movements represent large sources of electrical artifact in EEG data. Thus, it is not really feasible to record ERPs while subjects interact face-to-face or move around.

Related to this issue, and perhaps the most dramatic difference between ERP experiments and typical social cognition experiments, is the fact that only a single participant can be run at a given time in an ERP study. This is primarily a function of most labs having only one or two EEG amplifier set-ups and data recording rooms (which ideally should be isolated and quiet, free from other people and distractions). Thus, an experiment involving 100 participants that might take a week to complete by running multiple sessions per day, each involving 8–10 participants, in a large room equipped with separate computer workstations will easily take several months to complete when recording ERPs. In addition to running fewer people at once, ERP studies generally require considerably more trials in each condition of interest than tend to be used in behavioral studies. As a general rule of thumb (varying considerably on the basis of anticipated effect sizes, numbers of participants, etc.), at least 50 trials representing each condition of interest should be used to ensure clean and reliable ERP signals (see Luck, 2005). Moreover, ERP experiments require considerably more time for set-up and clean-up compared with typical behavioral experiments (in general, you need to add 40–80 minutes to the time it actually takes participants to complete the study tasks).

What Have We Discovered Using ERPs?

Despite the challenges posed by recording ERPs, one of the key advantages they provide is the ability to examine, in a temporally sensitive way, the extent to which stimuli engage different, at least partially separable information-processing operations, and to estimate the influence of those different operations on behavioral outcomes. By way of example, consider the use of RT differences between two experimental conditions to infer differences in cognitive processing between those conditions. The time taken by a participant to make a behavioral response in an experimental task represents an *outcome* of a number of underlying neurophysiological, psychological, and skeletomotor responses, all working synchronously (and often simultaneously) to produce a button press, but it does not represent a direct and unambiguous measure of any of those processes. However, augmenting RT with a temporally sensitive measure of neural activity such as ERPs can help to reveal the psychological locus of experimental effects that often are expressed in RT differences (see Kutas, McCarthy, & Donchin, 1977).

One of the areas in which my colleagues and I have adopted this approach is in understanding RT effects commonly found in priming-type tasks used to assess the influence of racial stereotypes. In particular, my colleagues and I have focused on testing the role of response conflict and control-related processes that many social-cognitive theorists have assumed play very little role in such tasks (e.g., Fazio et al., 1986, 1995; Greenwald & Banaji, 1995). Although the response conflict hypothesis and more traditional models (e.g., spreading of activation) make similar predictions regarding behavioral effects (i.e., faster RT to stereotype-congruent than to stereotype-incongruent trials), the psychological processes ascribed to these differing hypotheses are associated with very different patterns of neural activity, which can be used to specify the locus of experimental effects in ways that behavioral measures cannot (see Bartholow, 2010; Luck, 2005).

For instance, research has suggested that categorizing individuals by race is more difficult in the presence of stereotype-incongruent information relative to stereotype-congruent information (e.g., see Livingston & Brewer, 2002; Richeson & Trawalter, 2005). This effect could result from differences in spreading activation, whereby activation of stereotypes facilitates recognition of stereotype-congruent categories, because activation of stereotypic content spreads more quickly to those categories than to categories not associated with that content. This mechanism suggests a locus of experimental effects in the evaluative categorization process (i.e., evaluative categorization of race should be faster/more efficient in the presence of stereotype-congruent information). A particular component of the ERP, the P3 or P300, has long been associated with evaluative categorization. Specifically, the latency at which the P3 peaks following stimulus onset is thought to reflect the speed with which the stimulus is categorized and its evaluative implications understood, as indicated by the finding that P3 latency increases as categorization becomes more difficult (e.g., Kutas et al., 1977). In addition, the P3 is believed to be independent of processes responsible for response output (e.g., Magliero, Bashore, Coles, & Donchin, 1984), and therefore represents a relatively "pure" or direct reflection of stimulus categorization.

In contrast, the response conflict hypothesis predicts that stereotype-incongruent stimulus pairings lead to slower RTs because of conflict in response preparation and execution. In other words, if two stimuli within the same experimental trial are associated, either implicitly or explicitly, with opposing responses (e.g., a Black face and a positive or stereotype-incongruent word), attending to both stimuli can result in both response channels (e.g., left-hand and right-hand responses) becoming activated, either simultaneously or consecutively, within that trial. Conflict between these opposing responses results in slowing of ultimate response output, relative to situations in which all stimuli within a given trial are associated with only a single response option.

Two components of the ERP provide information concerning the occurrence of response conflict. First, the N2 component, a negative-going deflection in the stimulus-locked ERP generally peaking around 200–300 milliseconds (ms) post-stimulus and maximal at fronto-central scalp locations, has consistently been associated with conflict monitoring processes in the medial prefrontal cortex (see van Veen & Carter, 2002). Second, the lateralized readiness potential (LRP) directly reflects activation of the motor cortex as participants prepare to execute behavioral responses, and is specifically sensitive to hemispheric differences associated with activation of responses made with the left versus right hand (see Coles et al., 1995). Lateralized readiness potential data from several studies have shown that conflict between competing response representations can lead to initial, sub-threshold preparation and activation of the incorrect response, followed by (delayed) activation and execution of the correct response (e.g., Gratton, Coles, & Donchin, 1992; Gratton et al., 1988).

With this prior ERP work as a foundation, Bartholow and Dickter (2008) tested the role of response conflict in producing RT differences in a racial categorization task by recording ERPs while participants completed a modified flanker task, in which centrally presented target faces of Black men and White men were flanked on four sides by words reflecting American cultural stereotypes for Blacks and Whites. That is, some trials involved stereotype-congruent target–flanker pairings (e.g., Black faces flanked by Black-stereotypic words), whereas others involved stereotype-incongruent pairings (e.g., Black faces flanked by White-stereotypic words). Participants' task was simply to categorize each target face according to race as quickly as possible by pressing buttons with their left and right index fingers.

Bartholow and Dickter's (2008) behavioral findings revealed a typical stereotype-congruency effect in RT, such that categorization responses were facilitated on trials in which faces were flanked by stereotype-congruent relative to stereotype-incongruent words. Importantly, however, the latency of the P3 was not affected by stereotype congruency of the flanker words, a pattern generally inconsistent with the idea that evaluative categorization of race is more difficult in the context of stereotype-incongruent information. However, the N2 component was significantly increased on stereotype-incongruent trials, and the LRP revealed significant response conflict on such trials. Specifically, participants tended initially to activate the incorrect response on stereotype-incongruent trials prior to "reversing," and activating and executing the correct response. This effect was especially pronounced in blocks where stereotype-congruent trials were highly probable, suggesting that participants developed a response strategy in which their initial responses were driven by the flanker words, despite the fact that those words were irrelevant to their task. Critically, the amplitude of this LRP response conflict effect was significantly associated with the size of the stereotype-congruency effect in RT, providing evidence that the slower RTs seen on stereotype-incongruent trials was determined by response conflict.

Although Bartholow and Dickter (2008) provided evidence that response conflict *can* occur and influence responses in a race categorization task, the task used in that study arguably does not typify many race bias tasks in at least two ways. First, the participants' task was to categorize the race of faces while attempting to ignore stereotype-related words. In most paradigms, faces or other category cues tend to be used as task-irrelevant stimuli (i.e., primes) along with words or other objects that must be classified. Second, the targets (faces) and task-irrelevant words were presented simultaneously, whereas in most paradigms a prime appears prior to a task-relevant target to which participants must respond.

These issues were addressed in a different study (Bartholow et al., 2009), in which participants completed a standard evaluative priming task (e.g., Fazio et al., 1986) while ERPs were recorded. As in the Bartholow and Dickter (2008) study, the primary aim of Bartholow et al. (2009) was to determine the extent to which typical congruency effects might be caused by response conflict and strategic control. Indeed, the ERP measures in that study showed, for the first time, that primes activate response preparation in the motor cortex in a way that leads to response conflict on incongruent trials, and that these differences in conflict predict affective congruency effects behaviorally. Moreover, by manipulating the proportion of congruent and incongruent trials across trial blocks (see also Klauer, Rossnagel, & Musch, 1997), Bartholow et al. (2009) were able to show that conflict effects can be reversed when congruent trials are relatively infrequent. In other words, participants are able strategically to control response activation on the basis of whether prime valence predicts target valence, regardless of whether prime valence and target valence are congruent.

In addition to using ERPs to investigate the locus of experimental effects observed in behavioral responses, we (and others) have used ERPs to test basic questions about the role of control-related processes in performance on implicit bias tasks. Studies conducted by our friend and colleague David Amodio provided considerable inspiration for our use of this approach, and thus I provide a brief description of some of his work here. Whereas some of the work just described (Bartholow & Dickter, 2008; Bartholow et al., 2009) investigated ways in which control-related processes influence correct responses in implicit tasks, Amodio and his colleagues have examined the role of conflict monitoring when control fails, and how such failures ultimately contribute to self-regulation of race bias. For example, Amodio et al. (2004; see also Amodio, Chapter 9 in this volume) addressed a longstanding question about the control of stereotyping: Do people sometimes fail to override automatic stereotypes because (a) they are unable to detect the unwanted influence of the stereotype; or (b) because they are unable to implement control, even though the unwanted influence of the stereotype was detected? Building on research in cognitive neuroscience (e.g., Botvinick, Braver, Barch, Carter, & Cohen, 2001), Amodio et al. (2004) suggested that the self-regulation of responses to stereotyped targets involves the coordination of two complementary processes: an initial, evaluative conflict monitoring mechanism sub-served by activity in the anterior cingulate cortex (ACC); and a regulative mechanism, associated with activity in the lateral prefrontal cortex (see Kerns et al., 2004), which strengthens the influence of intentional responses to override an unwanted response tendency.

Participants in Amodio et al.'s study (2004) completed the Weapons Identification Task (WIT; see Payne, 2001), in which participants must rapidly classify visual targets as either guns or tools following the presentation of White or Black faces (primes), while ERPs were recorded. Amodio et al. focused on a response-locked component of the ERP, known as the error-related negativity (ERN), known to reflect ACC activity (Dehaene, Posner, & Tucker, 1994; van Veen & Carter, 2002); this has been associated with an evaluative, conflict monitoring function important for self-regulation (see Yeung, Botvinick, & Cohen, 2004). Amodio et al. (2004) found that ERN amplitudes were largest on trials indicative of race bias (i.e., pressing the "gun" key following a Black face prime when the target was actually a tool); and that the amplitude of this "race-bias ERN" predicted performance in the task overall, such that participants who showed the largest ERNs on race-bias errors (relative to other errors) made fewer such errors overall, and showed a stronger influence of control-related processes (using the process dissociation procedure [PDP]; see Jacoby, 1991) on their performance.

Recent research by Bartholow, Henry, Lust, Saults, and Wood (2012) extended the work of Amodio et al. (2004) by investigating the effects of alcohol on the evaluative control component of race bias regulation. Bartholow et al. randomly assigned participants to consume either a moderate dose of alcohol (0.80 g/kg ethanol), a placebo beverage that participants were told contained alcohol, or a control beverage (alcohol-free, and participants knew it) prior to completing the WIT. Consistent with predictions, alcohol had no effect on RTs, indicating no effects of alcohol on automatic activation of stereotypic associations (see also Bartholow, Dickter, & Sestir, 2006), but those who consumed alcohol displayed more bias (i.e., more Black–tool errors) than participants in the other beverage conditions. Also, replicating Amodio et al. (2004), Bartholow et al. (2012) found that the amplitude of the bias-related ERN significantly correlated with PDP-control estimates computed from task performance. However, this pattern held only among participants in the placebo and control conditions. For participants who consumed alcohol, the ERN was significantly reduced overall (see also

Ridderinkhof et al., 2002), and was unrelated to PDP-control. Moreover, whereas placebo participants (and, to a lesser extent, control participants) made control-related adjustments in their responses following errors, alcohol participants did not. These findings suggest that alcohol increases expression of race bias, at least in part, by significantly impairing the evaluative conflict-monitoring component of cognitive control. More generally, these findings suggest that differences in responses on implicit bias tasks do not necessarily reflect differences in the strength of automatic associations or their activation.

Around the same time that Amodio et al. (2004) were investigating the role of evaluative conflict-monitoring processes in the regulation of bias, Bartholow et al. (2006) were focusing on the role of the regulative control component in regulating bias. Bartholow et al. hypothesized that alcohol intoxication interferes with the regulative function of control, such that even if people who are drunk realize the need to implement control they simply cannot do so. As indicated previously, regulative control has been linked to activity in lateral prefrontal cortical areas (see Kerns et al., 2004). Such activity appears to be reflected in a component of the ERP known variously as the negative slow wave (NSW)/frontal slow wave (see West & Alain, 2000) or the conflict sustained potential (conflict SP) (see West, Jakubek, Wymbs, Perry, & Moore, 2005), a relatively slow/sustained negativity appearing primarily over frontal scalp locations that tends to be larger on high-conflict trials (e.g., incongruent Stroop trials) in which conflict is successfully resolved. Thus, Bartholow et al. (2006) predicted that the NSW would be enhanced in a bias regulation task on trials requiring inhibition of a prepotent, stereotype-congruent response, but that this effect would be reduced following alcohol consumption.

Participants in Bartholow et al.'s (2006, Experiment 2) study completed a version of the stop signal task adapted to involve stereotype-congruent and stereotype-incongruent responses. Specifically, participants were asked to indicate as quickly as possible whether trait words (some were stereotypic and some were counter-stereotypic) could be used to describe Black and White faces (primes), but on 25% of the trials (i.e., "stop" trials) a red "X" appeared shortly after the onset of the trait word, signaling participants to halt the response they should have initiated. As predicted, intoxicated participants made more inhibition errors indicative of race bias (i.e., failing to stop stereotype-congruent responses). Moreover, the NSW was larger for sober versus intoxicated subjects, and this difference was most pronounced for the more difficult stereotype-congruent stop trials, indicating that alcohol impairs the ability to implement the enhanced regulative control needed to inhibit race bias.

More recently, some colleagues and I have undertaken a collaborative effort to understand the role of individual differences in cognitive control abilities—often referred to as executive functions, or EFs—in determining performance on tasks intended to measure implicit race bias. That is, whereas performance on such measures typically has been ascribed to individual differences in the strength of underlying racial attitudes (see Fazio et al., 1995; Greenwald et al., 1998), we hypothesize that such associations tell only part of the story. Numerous models of the development of prejudiced attitudes (e.g., Arkes & Tetlock, 2004; Karpinski & Hilton, 2001) suggest that the seeds of bias are planted through early exposure to a culture in which some groups are stereotypically associated with negative traits, and that this occurs in spite of one's personal egalitarian values and motives. Thus, most, if not all, people in such cultures are likely to develop negative automatic associations for stigmatized groups (see Devine, 1989). Therefore, differences in the expression of those negative associations (e.g., on laboratory measures of race bias) probably stem not from differences in the types or strength of those associations, but rather from variation in the ability and/or desire to control their expression.

Although my colleagues and I certainly were not the first scholars to propose this general idea (see Klauer, Schmitz, Teige-Mocigemba, & Voss, 2010; Sherman et al., 2010; Stewart, von Hippel, & Radvansky, 2009), our project promises to be the most comprehensive examination of this issue to date. Unlike most previous studies, in which only a single EF ability and/or a single bias task have been examined (e.g., Payne, 2005; von Hippel, Silver, & Lynch, 2000), and extending the recent approach used by Klauer et al. (2010), our project involves comprehensive assessment of the three core components of EF (task switching, working memory updating, and response inhibition), each assessed with three representative tasks, as well as a "comprehensive" assessment of race bias using multiple tasks—the IAT, the WIT, and the First Person Shooter Task (FPS) (see Correll, Park, Judd, & Wittenbrink, 2002), in which participants must quickly decide whether to "shoot" or "not shoot" (by pressing a key) White and Black target individuals on the basis of whether they are holding a gun or an innocuous object, such as a cell phone. Moreover, we used a much larger sample than has been used in most previous studies ($N = 485$), a portion of whom ($N = 180$) completed the bias tasks while ERPs were recorded, permitting examination of a number of hypotheses concerning the neurophysiological correlates of individual differences in specific EF abilities as manifest in specific bias tasks.

Some of the key findings to emerge from this project (Ito et al., 2013) to date include, first, that the implicit bias measures show very poor correlations with one another. Performance on the WIT and FPS correlates very modestly ($r = .21$, $p < .05$), and performance on the IAT does not correlate with either of the other two measures ($rs < .06$, $ps > .20$). Thus, one initial conclusion to be drawn from this study is that tasks that are assumed to all measure the same basic "bias" construct in fact appear to be measuring rather different aspects of bias. One possible reason for these poor patterns of association is that different EF component abilities underlie performance on the three bias tasks. Indeed, our data suggest that, whereas shifting ability is a significant (negative) predictor of bias as measured by the IAT (see also Klauer & Mierke, 2005; Klauer et al., 2010), but not the other bias measures, a broader, "Common EF" latent factor (representing common variance among all three latent EF factors) negatively predicts bias as measured by the WIT and FPS, but not the IAT. Perhaps most interesting of all are moderator models showing that, whereas estimates of automatic association (i.e., PDP-automatic) strongly predict bias on all three bias measures, this association is significantly moderated by the Common EF factor. In other words, automatic associations predict bias less strongly for individuals with better overall EF ability.

Finally, the ERP data have also proven quite useful in understanding how EF abilities relate to bias expression. For example, ERN amplitudes measured when participants committed bias-related errors in both the WIT (mistakenly pressing "gun" to tools following Black primes) and FPS ("shooting" an unarmed Black target) were highly similar, and thus were collapsed to create a composite race-bias ERN measure. This measure was then tested as a mediator of the relationship between Common EF ability and a composite control estimate (PDP-control across FPS and WIT) in a structural equation model. This model indicated that, as predicted, the Common EF factor predicted race-bias ERN, which also predicted the composite control estimate. Importantly, bootstrapping indicated a significant indirect effect of Common EF on the control estimate via the race-bias ERN factor, indicating significant mediation. The large, positive association between Common EF and the composite control estimate was also significantly reduced by including the race-bias ERN variable in the model. This analysis indicates that better overall EF ability predicts stronger engagement of evaluative conflict- monitoring processes in the ACC on bias-related errors, which ultimately predicts better self-regulation of bias.

The Question of Added Value

Based on the challenges inherent in recording ERPs pointed out above (see "What Are ERPs, Anyway?"), at this point, you might well be thinking, "Why would I ever want to bother using ERPs in my research?" That is an excellent question, and one that requires careful consideration. Not all research programs are amenable to ERPs, and just as with any other measure that you might consider using, ERPs should only be used if they help to address a theoretical question that would be difficult to address using simpler means. With this framework in mind, the approach used in my lab has been to use behavioral measures whenever possible, including to pilot-test new paradigms, and to augment those measures with ERPs when we want to know more than behavioral measures alone can tell us.

As I hope I have made clear, ERPs have proved very useful to my colleagues and me in addressing questions related to understanding implicit bias, focusing especially on the influence of control-related processes. After many years of models in which bias expression in the lab was explained primarily (or entirely) in terms of the strength of underlying associations between categories and evaluations (e.g., Fazio et al., 1995; Greenwald et al., 1998), and a general focus on the power of automaticity to drive social behaviors more generally (Bargh & Chartrand, 1999), the work reviewed here (and much other work besides) has forced a re-examination of these issues, which ultimately makes for much richer understanding of the social mind and how it gives rise to social behavior.

Use of ERPs to measure neural responses as participants complete tasks intended to measure implicit social-cognitive processes provides one example of the ways in which a social neuroscience approach can illuminate previously misunderstood or ignored elements of the information-processing machinery that mediates the relationship between task goals, stimuli, and responses. Studies in which the theory and methods of cognitive neuroscience are brought to bear in understanding fundamental aspects of social behavior represent some of the best of what social neuroscience can and should be doing for the field, in terms of integrating theory and methods in ways that produce significant advancement of social psychological theory. In contrast to this approach, researchers should strive to avoid application of the tools of cognitive neuroscience simply because it is possible to do so, or because this kind of work is deemed "hot" or trendy at the moment; in other words, one should avoid measuring the brain merely—to paraphrase famed British mountaineer George Mallory—because it's there.

Related to this point, researchers should also carefully consider which psychophysiological techniques are most appropriate for their research questions (see also Mendes, Chapter 16 in this volume). For example, functional magnetic resonance imaging (fMRI) has risen in prominence quite rapidly since its introduction as a research tool in the mid-1990s, primarily due to its promise in allowing researchers to discover the specific neural structures involved in various mental processes. However, although revealing the neural structures that become active under various conditions can be important for some theoretical questions, for other questions it might be less important. Moreover, what one gains in structural specificity with fMRI is offset by very poor temporal resolution. For instance, whereas ERPs are sensitive to changes in neural activity with a millisecond level of temporal resolution, the fMRI signal is several orders of magnitude slower, on the order of 4–6 seconds. This is because ERPs directly sample voltage changes associated with neuronal firing, which travel through the brain at essentially the speed of light (see Luck, 2005); fMRI measures a more sluggish blood-flow response that occurs some seconds after the neural activity of interest has occurred. For many theoretical questions in social cognition, the timing of mental events is of greater importance than their specific neural sources, and therefore ERPs would be a better choice of method in those cases.

Suggestions for Further Reading

Amodio, D. M., & Bartholow, B. D. (2011). Event-related potential methods in social cognition. In A. Voss, C. Stahl & C. Klauer (Eds.), *Cognitive methods in social psychology* (pp. 303–339). New York: Guilford Press.

This chapter provides much more detail on a number of the conceptual issues touched upon in the current chapter, and discusses use of the ERP approach in other domains.

Cacioppo, J. T., Berntson, G. G., Lorig, T. S., Norris, C. J., Rickett, E., & Nusbaum, H. (2003). Just because you're imaging the brain doesn't mean you can stop using your head: A primer and set of first principles. *Journal of Personality and Social Psychology, 85*, 650–661.

As the title implies, this paper provides basic conceptual grounding for anyone considering the use of neuroscience methods.

Cacioppo, J. T., Berntson, G. G., Sheridan, J. F., & McClintock, M. K. (2000). Multi-level integrative analyses of human behavior: Social neuroscience and the complementing nature of social and biological approaches. *Psychological Bulletin, 126*, 829–843.

Ochsner, K. N., & Lieberman, M. D. (2001). The emergence of social cognitive neuroscience. *American Psychologist, 56*, 717–734.

These articles provide excellent conceptual and theoretical overviews of the social cognitive neuroscience approach of investigating phenomena at the social, cognitive and neural levels of analysis.

Ito, T. A., & Bartholow, B. D. (2009). The neural correlates of race. *Trends in Cognitive Sciences, 13*, 524–531.

This paper presents a concise overview of the neural activity involved in processing race and racial bias, and provides citations for other relevant readings.

References

Allison, T., Wood, C. C., & McCarthy, G. M. (1986). The central nervous system. In M. G. H. Coles, E. Donchin & S. W. Porges (Eds.), *Psychophysiology: Systems, processes, and applications* (pp. 5–25). New York: Guilford Press.

Amodio, D. M., Harmon-Jones, E., Devine, P. G., Curtin, J. J., Hartley, S. L., & Covert, A. E. (2004). Neural signals for the detection of unintentional race bias. *Psychological Science, 15*, 88–93.

Amodio, D. M., & Mendoza, S. A. (2010). Implicit intergroup bias: Cognitive, affective and motivational underpinnings. In B. Gawronski & B. K. Payne (Eds.), *Handbook of implicit social cognition* (pp. 353–374). New York: Guilford.

Arkes, H. R., & Tetlock, P. E. (2004). Attributions of implicit prejudice, or "Would Jesse Jackson 'fail' the Implicit Association Test?" *Psychological Inquiry, 15*, 257–279.

Bargh, J. A., & Chartrand, T. L. (1999). The unbearable automaticity of being. *American Psychologist, 54*, 462–479.

Bartholow, B. D. (2010). On the role of conflict and control in social cognition: Event-related brain potential investigations. *Psychophysiology, 47*, 201–212.

Bartholow, B. D., & Amodio, D. M. (2009). Using event-related brain potentials in social psychological research: A brief review and tutorial. In E. Harmon-Jones & J. S. Beer (Eds.), *Methods in social neuroscience* (pp. 198–232). New York: Guilford Press.

Bartholow, B. D., & Dickter, C. L. (2008). A response conflict account of the effects of stereotypes on racial categorization. *Social Cognition*, 26, 273–291.

Bartholow, B. D., Dickter, C. L., & Sestir, M. A. (2006). Stereotype activation and control of race bias: Cognitive control of inhibition and its impairment by alcohol. *Journal of Personality and Social Psychology, 90*, 272–287.

Bartholow, B. D., Fabiani, M., Gratton, G., & Bettencourt, B. A. (2001). A psychophysiological analysis of cognitive processing of and affective responses to social expectancy violations. *Psychological Science, 12*, 197–204.

Bartholow, B. D., Henry, E. A., Lust, S. A., Saults, J. S., & Wood, P. K. (2012). Alcohol effects on performance monitoring and adjustment: Affect modulation and impairment of evaluative cognitive control. *Journal of Abnormal Psychology, 121*, 173–186.

Bartholow, B. D., Pearson, M. A., Gratton, G., & Fabiani, M. (2003). Effects of alcohol on person perception: A social cognitive neuroscience approach. *Journal of Personality and Social Psychology, 85*, 627–638.

Bartholow, B. D., Riordan, M. A., Saults, J. S., & Lust, S. A. (2009). Psychophysiological evidence of response conflict and strategic control of responses in affective priming. *Journal of Experimental Social Psychology, 45*, 655–666.

Botvinick, M. M., Braver, T. S., Barch, D. M., Carter, C. S., & Cohen, J. D. (2001). Conflict monitoring and cognitive control. *Psychological Review, 108*, 624–652.

Cacioppo, J. T., Crites, S. L. Jr., Berntson, G. G., & Coles, M. G. H. (1993). If attitudes affect how stimuli are processed, should they not affect the event-related brain potential? *Psychological Science, 4*, 108–112.

Cacioppo, J. T., Crites, S. L. Jr., Gardner, W. L., & Berntson, G. G. (1994). Bioelectrical echoes from evaluative categorizations: I. A late positive brain potential that varies as a function of trait negativity and extremity. *Journal of Personality and Social Psychology, 67*, 115–125.

Cacioppo, J. T., Petty, R. E., Losch, M. E., & Kim, H. S. (1986). Electromyographic activity over facial muscle regions can differentiate the valence and intensity of affective reactions. *Journal of Personality and Social Psychology, 50*, 260–268.

Chaiken, S., & Trope, Y. (1999). *Dual-process theories in social psychology*. New York: Guilford Press.

Coles, M. G. H., & Rugg, M. D. (1995). Event-related brain potentials: An introduction. In M. D. Rugg & M. G. H. Coles (Eds.), *Electrophysiology of mind: Event-related brain potentials and cognition* (pp. 1–26). New York: Oxford University Press.

Coles, M. G. H., Smid, H. G. O. M., Scheffers, M. K., & Otten, L. J. (1995). Mental chronometry and the study of human information processing. In M. D. Rugg & M. G. H. Coles (Eds.), *Electrophysiology of mind: Event-related brain potentials and cognition* (pp. 86–131). New York: Oxford University Press.

Correll, J., Park, B., Judd, C. M., & Wittenbrink, B. (2002). The police officer's dilemma: Using ethnicity to disambiguate potentially threatening individuals. *Journal of Personality & Social Psychology, 83*, 1314–1329.

Dehaene, S., Posner, M. I., & Tucker, D. M. (1994). Localization of a neural system for error detection and compensation. *Psychological Science, 5*, 303–305.

De Houwer, J., Hermans, D., Rothermund, K., & Wentura, D. (2002). Affective priming of semantic categorisation responses. *Cognition and Emotion, 16*, 643–666.

De Houwer, J., Teige-Mocigemba, S., Spruyt, A., & Moors, A. (2009). Implicit measures: A normative analysis and review. *Psychological Bulletin, 135*, 347–368.

Devine, P. G. (1989). Stereotypes and prejudice: Their automatic and controlled components. *Journal of Personality and Social Psychology, 56*, 5–18.

Eriksen, C. W., & Schultz, D. W. (1979). Information processing in visual search: A continuous flow conception and experimental results. *Perception and Psychophysics, 25*, 249–263.

Fabiani, M., Gratton, G., & Federmeier, K. (2007). Event related brain potentials. In J. T. Cacioppo, L. G. Tassinary & G. G. Berntson (Eds.), *Handbook of psychophysiology* (3rd ed., pp. 85–119). New York: Cambridge University Press.

Fazio, R. H., Jackson, J. R., Dunton, B. C., & Williams, C. J. (1995). Variability in automatic activation as an unobtrusive measure of racial attitudes: A bona fide pipeline? *Journal of Personality and Social Psychology, 69*, 1013–1027.

Fazio, R. H., Sanbonmatsu, D. M., Powell, M. C., & Kardes, F. R. (1986). On the automatic activation of attitudes. *Journal of Personality and Social Psychology, 50*, 229–238.

Gratton, G., Coles, M. G. H., & Donchin, E. (1992). Optimizing the use of information: Strategic control of activation of responses. *Journal of Experimental Psychology: General, 121*, 480–506.

Gratton, G., Coles, M. G., Sirevaag, E. J., Eriksen, C. W., & Donchin, E. (1988). Pre- and poststimulus activation of response channels: A psychophysiological analysis. *Journal of Experimental Psychology: Human Perception and Performance, 14*, 331–334.

Greenwald, A. G., & Banaji, M. R. (1995). Implicit social cognition: Attitudes, self-esteem, and stereotypes. *Psychological Review, 102*, 4–27. doi:10.1037/0033-295X.102.1.4

Greenwald, A. G., McGhee, D. E., & Schwartz, J. L. K. (1998). Measuring individual differences in implicit cognition: The implicit association test. *Journal of Personality and Social Psychology, 74*, 1464–1480.

Ito, T. A., Bartholow, B. D., Miyake, A., Correll, J., Friedman, N. A., Loersch, C., & Altamirano, L. (2013). Toward a comprehensive account of the role of executive control in implicit race bias. Manuscript in preparation.

Jacoby, L. L. (1991). A process dissociation framework: Separating automatic from intentional uses of memory. *Journal of Memory and Language, 30*, 513–541.

Karpinski, A., & Hilton, J. L. (2001). Attitudes and the Implicit Association Test. *Journal of Personality and Social Psychology, 81*, 774–788.

Kernahan, C., Bartholow, B. D., & Bettencourt, B. A. (2000). Effects of category-based expectancy violation on affect-related evaluations: Toward a comprehensive model. *Basic and Applied Social Psychology, 22*, 85–100.

Kerns, J. G., Cohen, J. D., MacDonald, A. W., Cho, R. Y., Stenger, A., & Carter, C. S. (2004). Anterior cingulate conflict monitoring and adjustments in control. *Science, 303*, 1023–1026.

Klauer, K. C., & Mierke, J. (2005). Task-set inertia, attitude accessibility, and compatibility-order effects: New evidence for a task-set switching account of the Implicit Attitude Test. *Personality and Social Psychology Bulletin, 31*, 208–217.

Klauer, K. C., Rossnagel, C., & Musch, J. (1997). List-context effects in evaluative priming. *Journal of Experimental Psychology: Learning, Memory, and Cognition, 23*, 246–255.

Klauer, K. C., Schmitz, F., Teige-Mocigemba, S., & Voss, A. (2010). Understanding the role of executive control in the Implicit Association Test: Why flexible people have small IAT effects. *Quarterly Journal of Experimental Psychology, 63*, 595–619.

Klinger, M. R., Burton, P. C., & Pitts, G. S. (2000). Mechanisms of unconscious priming: I. Response competition, not spreading activation. *Journal of Experimental Psychology: Learning, Memory, and Cognition, 26*, 441–455.

Kutas, M., McCarthy, G., & Donchin, E. (1977, August 19). Augmenting mental chronometry: The P300 as a measure of stimulus evaluation time. *Science, 197*, 792–795.

Livingston, R. W., & Brewer, M. B. (2002). What are we really priming? Cue-based versus category-based processing of facial stimuli. *Journal of Personality and Social Psychology, 82*, 5–18.

Luck, S. J. (2005). *An introduction to the event-related potential technique*. Cambridge, MA: The MIT Press.

McClelland, J. L. (1979). On the time relations of mental processes: An examination of systems of processes in cascade. *Psychological Review, 86*, 287–330.

Magliero, A., Bashore, T. R., Coles, M. G. H., & Donchin, E. (1984). On the dependence of P300 latency on stimulus evaluation processes. *Psychophysiology, 21*, 171–186.

Mandler, G. (1990). A constructivist theory of emotion. In N. S. Stein, B. L. Leventhal & T. Trabasso (Eds.), *Psychological and biological approaches to emotion* (pp. 21–43). Hillsdale, NJ: Erlbaum.

Olson, J. M., Roese, N. J., & Zanna, M. P. (1996). Expectancies. In E. T. Higgins & A.W. Kruglanski (Eds.), *Social psychology: Handbook of basic principles* (pp. 211–238). New York: Guilford Press.

Payne, B. K. (2001). Prejudice and perception: The role of automatic and controlled processes in misperceiving a weapon. *Journal of Personality and Social Psychology, 81*, 181–192.

Payne, B. K. (2005). Conceptualizing control in social cognition: How executive functioning modulates the expression of automatic stereotyping. *Journal of Personality and Social Psychology, 89*, 488–503.

Richeson, J. A., & Trawalter, S. (2005). On the categorization of admired and disliked exemplars of admired and disliked racial groups. *Journal of Personality and Social Psychology, 89*, 517–530.

Ridderinkhof, K. R., de Vlugt, Y., Bramlage, A., Spaan, M., Elton, M., Snel, J., & Band, G. P. H. (2002). Alcohol consumption impairs detection of performance errors in mediofrontal cortex. *Science, 298*, 2209–2211.

Rugg, M. D., & Coles, M. G. H. (1995). The ERP and cognitive psychology: Conceptual issues. In M. D. Rugg & M. G. H. Coles (Eds.), *Electrophysiology of mind: Event-related brain potentials and cognition* (pp. 27–39). New York: Oxford University Press.

Sanders, A. F. (1990). Issues and trends in the debate on discrete versus continuous processing of information. *Acta Psychologica, 74*, 123–167.

Sherman, J. W., Klauer, K. C., & Allen, T. J. (2010). Mathematical modeling of implicit social cognition: The machine in the ghost. In B. Gawronski & B. K. Payne (Eds.), *Handbook of implicit social cognition* (pp. 156–175). New York: Guilford Press.

Srull, T. K., & Wyer, R. S. (1989). Person memory and judgment. *Psychological Review, 96*, 58–83.

Stangor, C., & McMillan, D. (1992). Memory for expectancy-congruent and expectancy-incongruent information: A review of the social and social developmental literatures. *Psychological Bulletin, 111*, 42–61.

Stewart, B. D., von Hippel, W., & Radvansky, G. A. (2009). Age, race, and implicit prejudice: Using process dissociation to separate the underlying components. *Psychological Science, 20*, 164–168.

Vanman, E. J., Paul, B. Y., Ito, T. A., & Miller, N. (1997). The modern face of prejudice and structural features that moderate the effect of cooperation on affect. *Journal of Personality and Social Psychology, 73*, 941–959.

van Veen, V., & Carter, C. S. (2002). The anterior cingulate as a conflict monitor: fMRI and ERP studies. *Physiology & Behavior*, 77, 477–482.

von Hippel, W., Silver, L. A., & Lynch, M. E. (2000). Stereotyping against your will: The role of inhibitory ability in stereotyping and prejudice among the elderly. *Personality and Social Psychology Bulletin, 26*, 523–532.

Wentura, D., & Rothermund, K. (2003). The "meddling in" of affective information: A general model of automatic evaluation effects. In J. Musch & K. C. Klauer (Eds.), *The psychology of evaluation: Affective processes in cognition and emotion* (pp. 50–86). Mahwah, NJ: Lawrence Erlbaum.

West, R., & Alain, C. (2000). Evidence for the transient nature of a neural system supporting goal-directed action. *Cerebral Cortex, 10*, 748–752.

West, R., Jakubek, K., Wymbs, N., Perry, M., & Moore, K. (2005). Neural correlates of conflict processing. *Experimental Brain Research*, *167*, 38–48.

Yeung, N., Botvinick, M. M., & Cohen, J. D. (2004). The neural basis of error detection: Conflict monitoring and the error-related negativity. *Psychological Review*, *111*, 931–959.

11

MORAL ACCOUNTABILITY AND PREJUDICE CONTROL

Evidence from Cardiovascular and EEG Responses

Naomi Ellemers and Félice Van Nunspeet

A few days after I (Naomi Ellemers) had given birth to my first son—who is now 15—my professor at the Free University (Vrije Universiteit) in Amsterdam (Gün Semin) called me on the phone. "I know this is bad timing," he said. "The deadline is in two weeks, but there is a special fund at the university for innovative research projects, and I think you should apply." My interest was in examining effects of the image of the self and others, the effects of group norms, and ways of coping with identity threat. These all raise self-presentational and defensive responses, and for some time I had felt I needed a different way to examine these issues, so I proposed to use psychophysiological measures to capture responses that people would not give or could not give in self-reports.

The research proposal was not funded. A major concern communicated by the granting committee was that there were so few examples of social psychological studies that had successfully used psychophysiological measures, they were not convinced that we could pull this off. So, to pilot some of the ideas proposed—hoping to prove the committee wrong—we designed an initial study in which we used galvanic skin responses (GSR) to indicate individual level physiological arousal resulting from information about the task performance of one's gender group (Van Rijswijk & Ellemers, 1998). Our working hypothesis was that participants should be more aroused by information about their group the more strongly that they identified with the group. To our surprise, we found exactly the opposite result. Women who indicated that they were reluctant to identify with their gender group showed the greatest evidence of physiological arousal (increased skin conductance, indicating sweat) when we noted the inferior task performance of women as a group. In retrospect, we argued that this reflected the lack of convergence between participants' own preference to be considered as a unique individual and their treatment as representatives of their gender group ("categorization threat," see also Barreto & Ellemers, 2003). This unexpected result, and the fact that the physiological stress measure did not simply mirror self-reported levels of identification convinced me (if not the granting committee!) that there was something there worth testing.

This experience convinced me of the added value of a multi-measure approach. It was clear that self-reports in which people can express their own wishes and intentions only tell part of the story. By the same token, information gets lost when physiological responses are seen as indicating the "best" or most "real" indicators of how people respond to the situation. Instead, these results made me realize more than ever that we need to examine the *interplay* between physiological responses, self-perceptions, self-reports, and (compensatory) behaviors. This would offer a way to understand how people's preferred self-views interact with the way they are perceived by others, to assess the extent to which they are aware this is the case, and to interpret the efforts they make to influence the image of themselves in their own eyes or in the eyes of others.

Still, several years passed before I could really put these ideas into practice, because the granting committee had got that part right: It is not so easy for a social psychologist to develop appropriate research designs and to learn enough about psychophysiological measures to be able to do this type of research. So in the meantime, I just continued my research on groups and social identities with the more traditional methods I knew so well.

The Importance of Being Moral

In our research on group processes and intergroup relations (for overviews, see Ellemers, 2012; Ellemers, Spears, & Doosje, 2002), for many years we had studied different strategies people can use to establish a positive social identity (Tajfel & Turner, 1979). One of our questions was how people selectively focus on specific comparative dimensions to maintain a positive image of their group (e.g., Ellemers & Van Rijswijk, 1997). We systematically compared the way people think of different groups (see also Fiske, Ames, Cikara, & Harris, Chapter 5 in this volume) by distinguishing between evaluations of the ability to achieve important goals (competence), the inclination to interact with others (sociability), and the tendency to be sincere about one's intentions (morality).

A series of studies on this topic yielded consistent results. If given a choice, people care first and foremost about the *morality* of the groups to which they belong, rather than the group's competence or its sociability (Leach, Ellemers, & Barreto, 2007). For instance, traits indicating morality (e.g., honesty, trustworthiness) explained the largest amount of variance in in-group ratings, and were considered most important as group-defining features. Group members were most inclined to infer the virtue of the group from its perceived morality, and the morality of the group—rather than its competence or sociability—was the best predictor of individual differences in group pride and in-group identification.

This experimental work was complemented with observations in more applied contexts. For instance, across a broad sample of employees, the perceived morality of the organization reliably predicted work satisfaction and commitment to the organization (Ellemers, Kingma, Van de Burgt, & Barreto, 2010). Likewise, in a financial service organization operating worldwide, middle managers who considered their organization to be more moral were more inclined to comply with organizational regulations and reported greater work satisfaction than managers who considered their organization less moral. In fact, perceived organizational integrity was a better predictor of these important job attitudes than the perceived performance or efficiency of the organization (Ellemers & Boezeman, 2013).

During the past 5 years, research in our own lab and elsewhere has accumulated evidence suggesting that morality is highly important for people's social identities. People seek membership in teams and organizations that are moral, they identify strongly with groups that are moral, derogate other groups in terms of their inferior morality, and derive more pride from their group's morality than from its competence or sociability (for an overview, see Ellemers, Pagliaro, & Barreto, in press).

Regulation of Moral Behavior in Groups

As the supreme importance of moral judgments in group contexts became clear, we followed up on this initial research to examine the further implications of these insights for group dynamic processes and intergroup relations. We argued that if it is important to uphold the morality of one's group, then the best way for individuals to show that they are good and worthy group members is to display moral behavior. Indeed, we suspected that moral appeals and moral imperatives tend to be used by religious or political group leaders as a highly effective way to regulate the behavior of individual group members (see also Ellemers & Van den Bos, 2012). We examined this by comparing the effects of injunctive group norms (what other group members indicate as the best way to behave), depending on whether the recommended behavior was related to moral evaluations (doing this makes you a moral group member) or competence evaluations (doing this makes you a competent group member).

In line with expectations, we found that group members were more inclined to adapt their behavioral choices (in this case, the decision to work for the group rather than themselves) after being informed that norm transgressions indicate a lack of morality rather than indicating a lack of competence (Ellemers, Pagliaro, Barreto, & Leach, 2008). Furthermore, we established that the impact of moral group norms is not based on its appeal to "higher" goals such as altruism or the desire to suppress selfishness. Instead, in a series of follow-up studies, we were able to demonstrate that group members were inclined to adapt their behavior to what the group identified as the moral course of action, regardless of whether this required them to display selfish or cooperative behavior. This was the case because they thought that acting in a way which the group considers moral is the best way to earn respect from other in-group members. Indeed, these same moral norms did not have such effects on individual behavior when they were communicated by the out-group (Pagliaro, Ellemers, & Barreto, 2011).

These findings were of great interest to us, and extended our understanding of how and why moral judgments are so important to people working and living together in groups. At the same time, they raised new challenges for us as researchers. The very fact that people find it so important to appear moral, and to prove to themselves and others that they are good and moral group members (i.e., behave in ways that the group proclaims to be moral), makes it difficult to reliably examine these issues through self-reports or by observing overt and deliberate behavioral choices. Manipulations that question the morality of the self or other members of the group are likely to elicit defensive responses or meet with outright denial, while opportunities to indicate the intention to behave morally tend to suffer from self-presentational biases. Fortunately, in the meantime, we had developed in some of our other research programs some experience with psychophysiological measures, which could be put to good use to advance our insights into the importance of morality.

Cardiovascular Indicators of Stress and Coping

After our initial attempts with GSR measures, considerable time passed before we further developed our use of psychophysiological measures. The delay stemmed from the realization that the standard manipulations and measures we used in our studies on groups and social identities could not easily be adapted for this purpose, but required that we invent completely new research paradigms. The breakthrough came when Daan Scheepers joined us in Leiden, expressing an interest in developing such methodologies, and was willing to invest in acquiring the necessary skills and expertise. Around the same time we successfully attracted grant money to buy the necessary equipment and set up our own psychophysiology lab. Initially, the equipment we used was rather primitive. We started by using blood pressure cuffs that were intended for household health monitoring, and asked participants at specific points during the study to press a button to record their blood pressure (e.g., Scheepers & Ellemers, 2005; Scheepers, Ellemers, & Sintemaartensdijk, 2009). We had taped off the display screen so that participants could not monitor their own blood-pressure values, and read off the measures after participants had completed the study. This also raised a new (for us) domain of procedural and ethical issues, such as the question of how to proceed when we recorded pathologically high blood-pressure values. Later we upgraded our lab and equipment step by step, and currently (2013) we can record different indicators of cardio-vascular reactivity online, from continuous measures of blood pressure and impedance cardiography (ICG) (Scheepers, Chapter 13 in this volume; see also Blascovich, Chapter 12 in this volume).

When we started using this methodology, we simply used it as an implicit online indicator of the degree to which individuals were personally affected ("aroused") by group-level outcomes. In our current work, we more specifically distinguish between cardiovascular arousal patterns indicating people's tendency to focus on the resources they have available to cope with the situation (a *challenge* response), and a response pattern characteristic for people who are preoccupied by situational demands, or consider these insurmountable (a *threat* response). When peripheral blood vessels constrict, the total peripheral resistance (TPR) increases. If this happens while the blood volume pumped out by the heart (cardiac output [CO]) remains the same or is decreased, this indicates a state of threat. This is considered a maladaptive stress response, which is associated with the development of cardiovascular diseases over time (Blascovich, 2008; Blascovich & Tomaka, 1996; see also Blascovich, Chapter 12 in this volume). When, however, increased cardiac output is accompanied by vascular dilation (indicated by decreased TPR), this indicates the emergence of challenge. The combination of low TPR and high CO allows for an efficient transport of oxygen-rich blood through the body, making available the energy and nutrients needed to take adequate action. Challenge is considered a more beneficial arousal response during task engagement, which indicates the perceived ability to cope successfully with difficult circumstances (Ellemers, Van Nunspeet, & Scheepers, in press; see also Scheepers, Chapter 13 in this volume).

To assess these specific response patterns, disposable electrodes are attached to participants in our studies, to assess changes in electrical resistance (through impedance cardiography [ICG]) indicating the volume and timing of blood flow in different parts of the body. In addition to ICG, measures of electrocardiography (ECG) and continuous blood pressure are taken which, in combination, eventually can yield measures of CO and TPR as well as other associated cardiovascular measures (heart rate, pre-ejection period). We typically examine changes in cardiovascular reactivity while people engage in a task, in comparison to baseline measures. Such responses are best captured while participants perform a task that does not involve any physical exertion or movement that can alter cardiovascular activity. We tend to use a "speech task" for this purpose where participants are invited orally to state their attitudes or intentions regarding a focal aspect of the study; this is recorded (e.g., with a webcam), and can also be content-analyzed. Combining the information from different cardiovascular indicators while participants engage in this task allows us to differentiate between the emergence of threat versus challenge response patterns.

Moral Accountability as a Source of Stress

The distinction between threat and challenge cardiovascular responses proved highly useful when we proceeded with our research on morality. We had already established that people want to behave in ways that are considered moral by others in their group. As a result, people should be highly motivated to communicate the importance that they attach to shared moral goals. However, this does not inform us of their estimated feasibility of achieving these goals, nor of their ability to turn goodwill into effective practice. We argued that exactly because moral goals are so self-relevant and socially consequential, reminding people of these goals, or—even worse—pointing out their moral shortcomings, might not necessarily be the best way to induce changes in their moral behavior (Young, Monin, & Owens, 2009).

We know from other literatures that increasing people's accountability for goals that they already find important is likely to backfire. For instance, people may ruminate over the thought that their behavior will be evaluated in this way, they become distracted by the awareness that their behavior is monitored by others (Manstead & Semin, 1980), or they may simply choke under the pressure of having to perform well (Steele, 1997). Attempts to control and improve individual moral behavior typically emphasize the costs of not adhering to moral goals (Skitka & Mullen, 2002). However, to the extent that people already subscribe to these goals, the enhanced awareness that they should not fail to achieve them is bound to raise stress. The type of stress experienced, however, is likely to be maladaptive. Criticizing people for failing to achieve goals they find important is not going to help them to improve—it will only make them feel miserable. Over time, such stress experiences and the threat to the self that these imply may even cause people to disengage from important goals as a way to shield themselves from such negative emotions in the future (e.g., Derks, Van Laar, & Ellemers, 2007). Thus, we argued that confronting people with moral failures is likely to raise a negative state of threat, while greater adherence to moral guidelines depends on conditions that make people feel they can cope with the situation and elicit positive challenge.

We put these ideas to the test by assessing cardiovascular threat and challenge responses of people who were confronted with past moral failures of themselves or other in-group members, or whose moral values were challenged by others who held opposing values. Across different studies we consistently found that reminding people of past moral failures raised defensive responses instead of motivating them toward change (Kouzakova, Ellemers, Harinck, & Scheepers, 2012). For instance, when research participants were criticized for their past immoral behavior, when their moral values were called into question (Kouzakova, Ellemers, Harinck, & Scheepers, 2013), or even when they were reminded of the lack of morality of another member of their group (Van der Lee, Ellemers, & Scheepers, 2013), they were more likely to display a cardiovascular threat pattern than when they were confronted with other types of shortcomings or disagreements.

Turning Threat into Challenge

Thus, our initial research on cardiovascular responses to moral evaluations suggests that simply reminding people about the moral implications of their behavior (e.g., "discrimination is morally wrong!") is not going to help them successfully regulate their behavior (e.g., control their prejudice), as it mainly raises threat and elicits defensive responses leading to self-justification instead of self-improvement. We wanted to find out how to overcome this, in the hope of developing a more effective strategy to persuade people to behave more morally. Inspired by the literature on stress and coping, we argued that instead of emphasizing situational demands (i.e., by stressing the importance of morality), it might be important to focus people's attention on their coping resources, and encourage them to take advantage of improvement opportunities.

Our initial results seemed to support this, as we observed that people's responses to others noting their immoral behavior differed, depending on whether they were preoccupied with past moral failures or considering future opportunities to behave morally. When being confronted with negative evaluations of fellow group members for their past immoral behavior, participants in our research reported increased levels of negative emotions, which diminished their perceived ability to cope with the situation. However, when they were offered a new opportunity to display moral behavior, this reduced the experience of negative emotions, and increased their self-reported coping ability (Van der Lee et al., 2013). This would suggest that moral admonitions are most likely to be effective when these invite people to consider future possibilities to elicit attempts at moral self-improvement. By contrast, reiterating past moral failures that cannot be changed only makes them feel miserable.

We built on these findings to examine how this might apply to people's desire to overcome racial prejudice and their willingness to support affirmative action and social change (Does, Derks, & Ellemers, 2011). For this line of research, we developed a paradigm in which native Dutch participants were given labor statistics to remind them of the disadvantaged position of ethnic minorities on the Dutch labor market—which was attributed to unequal employment opportunities. The experimental manipulation was contained in the way the goal of equal treatment was phrased. We compared participants' responses, depending on whether equal treatment was described as a "moral obligation" which needed to be met, or as a "moral ideal" that would be good to achieve. We then asked participants to describe in an open format what types of behaviors might contribute to equal treatment, and whether they endorsed affirmative action policies aiming to achieve this. Content-coding of participants' ideas revealed that whereas moral obligations elicited more guidelines phrased as "don'ts" (e.g., don't look down on them), moral ideals were more likely to make participants list "do's" (e.g., get to know them). In addition, participants who were invited to consider equality as a moral ideal—rather than an obligation—were more likely to endorse affirmative action measures, reported more support for racial equality, and indicated a greater willingness to incur personal costs (e.g. sacrifice career opportunities) to achieve this (Does et al., 2011).

In a follow-up study, we used similar manipulations to examine more implicit responses indicating people's motivation for and approach to the achievement of equal employment opportunities as a moral goal. In this study, we assessed cardiovascular indicators of threat and challenge, and recorded participants' speech rate while they were stating their intentions of how to achieve more equality in employment opportunities. Results of this study replicated previous findings, and again confirmed the differential effects of presenting moral goals in terms of obligations versus ideals in recruiting support for affirmative action policies. In addition, we established that these manipulations affected participants' cardiovascular threat versus challenge responses when verbalizing their ideas on how to achieve the goal of providing equal opportunities to members of different ethnic groups. That is, in the moral ideals condition, participants showed a cardiovascular response pattern indicating challenge, and displayed a higher speech rate, indicating eagerness in approaching the task. By comparison, participants in the moral obligation condition showed cardiovascular responses more indicative of threat, and displayed a lower speech rate during the task (Does, Derks, Ellemers, & Scheepers, 2012).

Taken together, the cardiovascular measures we used across different studies enabled us to draw a number of conclusions that have theoretical as well as practical value. Most importantly, these measures revealed that the motivation to be a moral person and behave morally is not enough to suppress prejudice or combat discrimination effectively, as people may also choke or walk away when they think they might fail to live up to moral expectations. The chances of people being positively engaged and working toward moral goal achievement are highest when prejudice control is described as an attractive ideal, or when the focus is on future opportunities instead of past difficulties. Only then do people respond to moral goal setting as a challenge with which they are able to cope.

Examining the Neuropsychological Implications of Morality

Up to this point, we have mainly examined how people respond to feedback from others about their moral behavior, and the cardiovascular responses they display when approaching moral goals. We still do not know, however, what the behavioral implications of these responses are. What do people actually do when trying to behave morally, what happens when they control their (implicit) prejudice? Others before us—including several authors in this book; see Amodio (Chapter 9) and Bartholow (Chapter 10) in this volume—have convincingly shown that subjective concerns relating to people's group memberships and the way they relate to members of other groups can have a real and visible impact in neural processing of relevant information (for overviews, see also Ellemers, 2012; Ellemers et al., in press). So we argued that the impact of moral accountability might also be visible at that level.

This is when Félice Van Nunspeet came on board. Her training as a developmental psychologist had prepared her to do neuroscientific research with children, but her thesis on social decision making in adolescence made her realize she wanted to do her PhD on social neuroscience. This was easier said than done, because no such projects were available at the time. But Félice volunteered to work as a research assistant in my lab for a while, to see whether we could somehow introduce brain measures into my research on morality, in the hopes of acquiring funds to start a PhD on this topic. Although this was a leap of faith for both of us, initial results were encouraging, and fortunately funds became available exactly when we needed them. We soon discovered that additional expertise was required, and we were excited to find out that our colleagues from cognitive and developmental psychology were not only willing to help us out, but even showed a genuine interest in our plans.

Our first challenge was to find an appropriate task that would require participants to behave in ways that would be (or could be made to seem) relevant to moral concerns, through responses that would allow for the reliable measurement of brain activity through electroencephalographic (EEG) or functional magnetic resonance imaging (fMRI) measures (ideally both). So far, our moral tasks had asked participants to consider complex social information before indicating their equally complex responses: Participants were requested to indicate their agreement with affirmative action policies, to speak about ways to achieve moral goals, to make decisions on moral dilemmas, or to resolve a moral disagreement with another participant. These tasks were clearly useless for our present purposes. To reliably assess relevant neurophysiological indicators of interest (specific components in EEG waveforms or a BOLD signal in fMRI), we needed a task that could contain hundreds of trials, and evidently had social implications, even though it should be performed while sitting still in isolation during EEG recording or lying in an MRI scanner.

We decided to adapt for this purpose an Implicit Association Test (IAT), which is often used as an indirect measure of prejudice (Greenwald, McGhee, & Schwartz, 1998). The IAT is a reaction-time task, assessing people's inclination to associate particular concepts (e.g., pictures of specific individuals or stereotypical ethnic names) with positive and negative attributes (evaluative words or pictures of attractive versus unattractive objects); which had already been successfully used in neuroimaging research on prejudice control (e.g., Ibáñez et al., 2010). To examine our predictions, we framed this task either as a test of morality (i.e., indicating the tendency to hold biased expectations about members of different groups) or as a test of competence (i.e., indicating the ability to quickly associate particular pictures with specific key presses). Our main prediction was that people would somehow control their implicit prejudice—to display less evidence of bias in their task performance—when the moral implications of the test were emphasized.

ERP Evidence for Prejudice Control

We set out to develop stimulus materials that would be relevant to the Dutch context, where people are mainly concerned with discriminatory treatment of

Muslim migrants. For this aim, we selected pictures of women with and without a headscarf who were seen to represent Muslim out-group and non-Muslim in-group members respectively, while being rated as equally attractive, kind, trustworthy, and competent (Van Nunspeet, Ellemers, Derks, & Nieuwenhuis, 2012). We used these pictures as stimulus materials in an IAT, which was presented to participants either as a test of their morality or as a test of their competence. As we had hoped, a first study revealed that participants indeed were more likely to adjust their performance on this task when they were reminded of the moral (rather than the competence) implications of the test. Participants who thought their moral behavior was being assessed slowed down their responses on trials that might reveal prejudicial associations to show less bias against women with a headscarf (Van Nunspeet et al., 2012).

In a follow-up study, we took EEG measures to assess specific cognitive processes online while participants performed this task, by registering event-related brain potentials (ERPs) for specific task trials. Positive (P) and negative (N) peaks in electrical activity in the brain occurring at specific points in time after stimulus presentation or during response generation are associated with particular cognitive processes. Comparing the amplitude of these peaks during different types of trials indicates the degree to which these processes play a role during task performance (see also Ito, Chapter 2 in this volume).

We argued that an increased effort to achieve prejudice control under moral task instructions might be evident from different aspects of task engagement. First, the awareness of moral concerns might change the way people attend to and process the information they receive. This should be evident from ERP components related to stimulus presentation (occurring at around 100–200 milliseconds), indicating enhanced perceptual attention and stimulus processing. Second, activation of moral goals might have an impact on the way people respond during the task. This might be visible in peaks during response generation, indicating increased concern with correct performance.

We know from prior neuroscientific research with IAT measures that perceptual processing of in- and out-group faces tends to be different (e.g., Kubota & Ito, 2007; Van Bavel, Packer, & Cunningham, 2011). Initial attention is increased when viewing out-group rather than in-group faces, although at later stages there is evidence of increased (deep) processing of in-group compared to out-group faces (Dickter & Bartholow, 2007; Ito & Urland, 2003, 2005; see also Derks, Chapter 3 in this volume and Ito, Chapter 2 in this volume). Thus, the differentiation in amplitudes when viewing in- and out-group faces indicates the degree to which people are attentive to the social category these stimuli represent. In our own study, we also found these patterns—indicating differences in perceptual attention—when participants viewed faces of women with, versus without, a headscarf. This again supports the notion that these stimuli represent social cues that are relevant to information processing in our Dutch participants. Additionally, we found ERP evidence that the differential attention to faces of women with or without a headscarf (as indicated by the N1 and P150 components) was more pronounced under moral (versus competence) task instructions. This suggests that participants approached the task differently—as they were more attentive to whether they were viewing in-group or out-group faces—when they had been made aware of the moral implications of the task (Van Nunspeet et al., 2012).

Additionally, based on the evidence we found of bias reduction in participants' behavioral responses, we argued that emphasizing the moral (rather than the competence) implications of the task might result in enhanced response monitoring, indicating participants' concern with showing a correct response. Indeed, prior research had revealed that people who indicate that they are highly internally motivated to respond without prejudice show increased amplitudes of error-related negativity (ERN) potentials when completing implicit measures of prejudice (Amodio, Devine, & Harmon-Jones, 2008); this is associated with increased behavioral control. Our study extended these prior findings, as we found that—compared to participants who were informed that the test assessed their competence in associating pictures with key presses—participants who performed the IAT under moral task instructions showed increased ERN amplitudes on incorrect trials, indicating enhanced response monitoring.

In-group Accountability and Moral Behavior

These indicators of brain activity thus suggest that participants are more attentive to the difference between in-group versus out-group stimuli and more inclined to monitor and control their responses when reminded of the moral implications of prejudice control. However, we argued (see Ellemers & Van den Bos, 2012) that people care so much about their moral behavior because of its implications for the achievement of in-group acceptance and inclusion (see also Pagliaro et al., 2011). To further test the validity of this reasoning at the level of brain activity during task performance, we went one step further. In a set of follow-up studies, we examined whether participants' concern about displaying moral behavior through prejudice control would emerge in particular when being monitored by an in-group member, rather than by someone who did not represent a shared social identity.

Our aim was to make participants think their task behavior was monitored either by an in-group or by an out-group member, without interfering with the distinction between the specific target groups that we used in our IAT. This is why we introduced a procedure to create so-called minimal groups in the lab (allegedly indicating a "P" versus "O" style in decision making and cooperation). This allowed us to present pictures of the same individual to participants, while leading them to believe that this meant their IAT performance was being monitored either by an in-group or an out-group member. At the end of each trial of the IAT, participants were shown a picture of an individual who indicated whether their response on the prior trial had been correct (smiling, thumbs up) or incorrect (frowning, thumbs down). Identical pictures were used in in-group and out-group conditions; the only cue to the alleged group membership of the individual providing visual feedback consisted of a text underneath the picture indicating "P" or "O" group membership.

In a first behavioral study in which we added visual feedback from in-group and out-group members to each IAT trial, it seemed that enhanced awareness of the moral implications of the task only reduced biased responses when these were monitored by an in-group member. In fact, when participants thought their behavior was monitored by an out-group member, they even tended to display *more* bias when the moral implications of the task had been emphasized (Van Nunspeet, Derks, Ellemers, & Nieuwenhuis, 2013). Obviously, this first evidence was encouraging and fits our reasoning as well as our prior results on in-group moral norm adherence. Nevertheless, we still wanted to see whether the differential impact of in-group versus out-group monitoring of moral behavior would also show up in participants' ERP responses.

In a follow-up study, we used the same minimal groups procedure to introduce behavioral monitoring by an in-group (versus an out-group) representative, while taking ERP measures during IAT performance. In this study, indicators of early perceptual attention and response monitoring revealed that the tendency to approach a moral test differently from a test of competence was mainly raised when being accountable to the in-group. That is, evidence of increased social categorization (on the N1) was more pronounced when one's performance on a moral (rather than a competence) task was monitored by an in-group (rather than an out-group) member. Likewise, the ERN—indicating response monitoring—showed an increased amplitude when the moral implications of the task had been emphasized and participants thought the feedback they received was provided by an in-group member. In sum, these ERP measures show the differential impact of in-group (versus out-group) accountability for moral behavior at very early stages of information processing (N1—indicating attention to social category information) and response monitoring (indicated by ERN amplitudes).

This evidence of differential brain activity complements our results on self-reports and task behaviors in important ways. First, participants show a different approach to the task immediately after stimulus presentation. This suggests that prejudice control involves more than the inhibition of inappropriate responses. Instead, it seems that people who try to avoid revealing biased associations with members of stigmatized groups literally perceive and interpret the situation differently from the outset. Second, their differential responses to in-group versus out-group monitoring remind us that the effectiveness of attempts to enhance moral behavior in others by making them accountable for their actions may be limited. That is, it seems that increasing prejudice control through behavioral accountability is not achieved because of a generic effect of social monitoring of moral behavior. Instead, our results suggest that the motivation to attend to information and control responses relevant to moral behavior can only be increased in this way when people consider the individuals who monitor their behavior as relevant to the self and their identity.

What Have We Learned?

The introduction of measures indicating cardiovascular and brain activity in our research on morality and prejudice control has yielded a number of important insights that we could not have obtained with more traditional self-reports or behavioral observations. Our initial studies on the importance of social judgments revealed that in-group accountability and awareness of the moral implications of social behavior are relevant to prejudice control and intergroup relations. However, merely knowing that people respond very strongly when being held morally accountable to themselves or to others with a relevant shared identity does not help us to change or improve their moral behavior. Indeed, our research clearly shows that attempts to make people control prejudice differ in their effectiveness, depending on what these attempts focus on and how this is communicated. Precisely because of the importance people attach to morality, being reminded of past moral failures or presenting moral behavior as an obligation they have to meet mainly elicits negative emotions and raises defensive responses. Focusing on future opportunities and moral ideals that might be achieved is less threatening, and more challenging (in a positive way).

We were able to identify specific conditions that are conducive to openness and change by taking into account cardiovascular threat/challenge indicators. We also learned that people can and do adapt their behavior when morally accountable, even if this is not indicated on self-reports. The consideration of ERP responses revealed evidence of prejudice control at very early stages of information processing and response monitoring. This brings home the message that implicit processes relevant to stereotypical associations and biased responses already emerge before deliberate judgment and decision making is possible. At the same time, our results point to concrete measures that may help people to control group-based prejudice, as they have an impact on these processes from the very start. For instance, even without self-monitoring or suppression of emerging responses, emphasizing the moral implications of their behavior may cause people simply to approach the task differently. Such interventions are most likely to be effective when being aware that one's moral behavior is monitored by someone who represents a relevant shared identity.

Remaining Puzzles and Challenges

Although we are happy about the things we have learned so far, we have only started on this new journey. Still being in the process of building a "repertoire" of appropriate designs, tasks, and materials, we are somewhat reluctant to abandon a satisfactory paradigm in search of improvement, or to examine research questions that cannot be addressed with our current procedures. In the meantime, we discovered that even little changes can have unintended effects. For instance, after we had pre-tested the IAT and found the behavioral results we were looking for, we doubled the number of trials as a way to ensure that we would have a sufficient number of incorrect responses to be able to reliably estimate the ERN. However, while optimizing task conditions for these relevant ERP indicators, we inadvertently reduced the behavioral effects, as displays of bias disappeared after a certain number of trials, probably due to a learning effect.

Another thing that has become clear is that results obtained with a specific procedure or measure cannot simply be used to draw conclusions that go beyond those findings. For instance, it seems that whereas raising people's awareness of the moral implications of their behavior helps them to control prejudicial responses, it is not self-evident that these effects are sustained over time. Indeed, further attempts to explore immediate versus more long-term implications of prejudice control suggest that this may require additional cognitive resources, which result in depletion and perhaps even rebound effects over time. For instance, we found that the motivation to behave morally on a first task (in this case a memory task) resulted in cognitive depletion on a second task, which actually undermined participants' ability to take the perspective of others on a third task (Ståhl & Ellemers, 2013). Thus, whereas moral motivation might help people to recruit the cognitive resources needed to control their responses, over time this may result in depletion and actually backfire. Clearly, these are issues that we have only begun to examine.

When we started thinking about including measures of brain activity in our research, we were aiming for triangulation, by examining the same situations with different measures. Specifically, our initial idea was to gain more insight into the timing of responses in the brain with EEG measures, and to learn more about the specific brain areas involved in these responses through localization studies using fMRI procedures. While this seemed a good idea, in the meantime, we have realized how tricky it is to study "the same" process with different measures. In fact, this even seems to be the case for research questions where we aimed to connect information about behavioral responses (e.g., coding of speech content) to explicit self-reports and implicit measures on relevant psychophysiological indicators. Here too we have come to the conclusion that—in view of the sometimes quite subtle and sophisticated effects in which we are interested—it is not always possible to connect these different types of measures in a single study. Again, we have come to the conclusion that distinct processes may be responsible for each of these responses, and optimizing aspects of the task or measures for one methodology often makes it less likely to capture another effect.

Importantly, this implies that whereas parallel responses to experimental manipulations may be observed, these different levels are not necessarily directly related (even if editors and reviewers would like to see statistical evidence of mediation). Indeed, we suspect that these different types of responses (e.g., behavior, self-reports, psychophysiology) do not necessarily constitute consecutive steps in a single causal sequence. Instead, it seems that the relation between different response types may just as well be recursive or corrective, as people become aware of their psychophysiological responses, and try to account for this on self-reports, or compensate for this in their behavior. While this sometimes makes it hard to understand what is going on, examining the interplay between different measures and when and why they are (not) related will yield more insight into the processes of self-monitoring and social accountability in which we are interested.

Conclusion

Because of the strong motivation to be moral—or at least to appear moral in the eyes of the self as well as others—research on individual and group morality cannot rely on self-report (questionnaire) data alone. This is why we have started to combine different indicators, to examine the interplay between different types of psychophysiological indicators, overt self-reports, and behavioral responses. We now use these different types of methodology to assess how the motivation to be moral affects psychophysiological responses as well as more controlled self-presentations, and how the desire to appear moral impacts on cognitive processes implicated in prejudice control.

Suggestions for Further Reading

Amodio, D. M. (2008). The social neuroscience of intergroup relations. *European Review of Social Psychology, 19*, 1–54.

This review of neuropsychological evidence very clearly explains the interplay between immediate versus deliberate processes in prejudice control.

Bartholow, B. D. (2010). On the role of conflict and control in social cognition: Event-related brain potential investigations. *Psychophysiology, 47*, 201–212.

This is a very nice overview that covers ERPs related to conflict and control during "automatic" processes in the light of social psychological issues such as race bias.

Blascovich, J., Vanman, E., Mendes, W. B., & Dickerson, S. S. (2011). *Social psychophysiology for social and personality psychology*. New York: Sage.

A very practical introduction of different measures and what they can be used for.

Kubota, J. T., Banaji, M. R., & Phelps, E. A. (2012). The neuroscience of race. *Nature Neuroscience Review, 15*, 940–948.

This overview of fMRI studies indicates the brain areas and networks implemented in racial attitudes and their control.

Woodman, G. F. (2010). A brief introduction to the use of event-related potentials in studies of perception and attention. *Attention, Perception, & Psychophysics, 72*, 2031–2046.

For researchers who want to start using EEG, this is a very useful tutorial in which issues concerning the evaluation and planning of ERP experiments (focused on perception and attention) are discussed.

References

Amodio, D. M., Devine, P. G., & Harmon-Jones, E. (2008). Individual differences in the regulation of intergroup bias: The role of conflict monitoring and neural signals for control. *Journal of Personality and Social Psychology, 94*, 60–74.

Barreto, M., & Ellemers, N. (2003). The effects of being categorised: The interplay between internal and external social identities. *European Review of Social Psychology, 14*, 139–170.

Blascovich, J. (2008). Challenge and threat. In A. J. Elliot (Ed.), *Handbook of approach and avoidance motivation* (pp. 431– 445). Mahwah, NJ: Erlbaum.

Blascovich, J., & Tomaka, J. (1996). The biopsychosocial model of arousal regulation. In M. P. Zanna (Ed.), *Advances in experimental social psychology* (Vol. 28, pp. 1–51). New York: Academic Press.

Derks, B., Van Laar, C., & Ellemers, N. (2007). Social creativity strikes back: Improving motivated performance of low status group members by valuing ingroup dimensions. *European Journal of Social Psychology, 37*, 470–493.

Dickter, C. L., & Bartholow, B. D. (2007). Racial ingroup and outgroup attention biases revealed by event-related brain potentials. *Social Cognitive and Affective Neuroscience, 2*, 189–198.

Does, S., Derks, B., & Ellemers, N. (2011). Thou shall not discriminate: How emphasizing moral ideals rather than obligations increases whites' support for social equality. *Journal of Experimental Social Psychology, 47*, 562–571.

Does, S., Derks, B., Ellemers, N., & Scheepers, D. (2012). At the heart of egalitarianism: How equality framing shapes cardiovascular challenge vs. threat in Whites. *Social Psychological and Personality Science, 3*, 747–753.

Ellemers, N. (2012). The group self. *Science, 336*, 848–852. doi: 10.1126/science.1220987

Ellemers, N., & Boezeman, E. (2013). Integrity and compliance: How perceived organizational morality and competence of a financial service organization relate to employee compliance and work satisfaction. Manuscript under editorial consideration.

Ellemers, N., Kingma, L., Van de Burgt, J., & Barreto, M. (2010). Corporate social responsibility as a source of organizational morality, employee commitment and satisfaction. *Journal of Organizational Moral Psychology, 1*, 1–28.

Ellemers, N., Pagliaro, S., & Barreto, M. (in press). Morality and behavioral regulation in groups: A social identity approach. *European Review of Social Psychology*.

Ellemers, N., Pagliaro, S., Barreto, M., & Leach, C. W. (2008). Is it better to be moral than smart? The effects of morality and competence norms on the decision to work at group status improvement. *Journal of Personality and Social Psychology, 95*, 1397–1410.

Ellemers, N., Spears, R., & Doosje, B. (2002). Self and social identity. *Annual Review of Psychology, 53*, 161–186.

Ellemers, N., & Van den Bos, K. (2012). Morality in groups: On the social-regulatory functions of right and wrong. *Social and Personality Psychology Compass, 6*, 878–889.

Ellemers, N., Van Nunspeet, F., & Scheepers, D. (in press). It's all in the mind: How social identification processes affect neurobiological responses. In M. Mikulincer & P. R. Shaver (Eds.), *Nature and formation of social connections: From brain to group*. Washington, DC: American Psychological Association.

Ellemers, N., & Van Rijswijk, W. (1997). Identity needs versus social opportunities: The use of group-level and individual-level identity management strategies. *Social Psychology Quarterly, 60*, 52–65.

Greenwald, A. G., McGhee, D. E., & Schwartz, J. L. K. (1998). Measuring individual differences in implicit cognition: The Implicit Association Test. *Journal of Personality and Social Psychology, 74*, 1464–1480.

Ibáñez, A., Gleichgerrcht, E., Hurtado, E., González, R., Haye, A., & Manes, F. F. (2010). Early neural markers of implicit attitudes: N170 modulated by intergroup and evaluative contexts in IAT. *Frontiers in Human Neuroscience, 4*, 1–14.

Ito, T. A., & Urland, G. R. (2003). Race and gender on the brain: Electrocortical measures of attention to the race and gender of multiply categorizable individuals. *Journal of Personality and Social Psychology, 85*, 616–626.

Ito, T. A., & Urland, G. R. (2005). The influence of processing objectives on the perception of faces: An ERP study of race and gender perception. *Cognitive, Affective, & Behavioral Neuroscience, 5*, 21–36.

Kouzakova, M., Ellemers, N., Harinck, F., & Scheepers, D. (2012). The implications of value conflict: How disagreement on values affects self-involvement and perceived common ground. *Personality and Social Psychology Bulletin, 38*, 798–807.

Kouzakova, M., Ellemers, N., Harinck, F., & Scheepers, D. (2013). At the heart of a conflict: Cardiovascular and self-regulation responses to value vs. resource conflicts. Manuscript submitted for publication.

Kubota, J. T., & Ito, T. A. (2007). Multiple cues in social perception: The time course of processing race and facial expression. *Journal of Experimental Social Psychology, 43*, 738–752.

Leach, C., Ellemers, N., & Barreto, M. (2007). Group virtue: The importance of morality vs. competence and sociability in the evaluation of in-groups. *Journal of Personality and Social Psychology, 93*, 234–249.

Manstead, A. S. R., & Semin, G. R. (1980). Social facilitation effects: Mere enhancement of dominant responses? *British Journal of Social and Clinical Psychology, 19*, 119–135.

Pagliaro, S., Ellemers, N., & Barreto, M. (2011). Sharing moral values: Anticipated ingroup respect as a determinant of adherence to morality-based (but not competence-based) group norms. *Personality and Social Psychology Bulletin, 37*, 1117–1129.

Scheepers, D., & Ellemers, N. (2005). When the pressure is up: The assessment of social identity threat in low and high status groups. *Journal of Experimental Social Psychology, 41*, 192–200.

Scheepers, D., Ellemers, N., & Sintemaartensdijk, N. (2009). Suffering from the possibility of status loss: Physiological responses to social identity threat in high status groups. *European Journal of Social Psychology, 39*, 1075–1092.

Skitka, L. J., & Mullen, E. (2002). The dark side of moral conviction. *Analyses of Social Issues and Public Policy, 2*, 35–41.

Ståhl, T., & Ellemers, N. (2013). Moral exhaustion: Why working towards a moral goal increases the risk of subsequent moral failures. Manuscript submitted for publication.

Steele, C. M. (1997). A threat in the air: How stereotypes shape intellectual identity and performance. *American Psychologist, 52*, 613–629.

Tajfel, H., & Turner, J. C. (1979). An integrative theory of intergroup conflict. In W. G. Austin & S. Worchel (Eds.), *The social psychology of intergroup relations* (pp. 33–47). Monterey, CA: Brooks/Cole.

Van Bavel, J. J., Packer, D. J., & Cunningham, W. A. (2011). Modulation of the fusiform face area following minimal exposure to motivationally relevant faces: Evidence of in-group enhancement (not out-group disregard). *Journal of Cognitive Neuroscience, 19*, 1131–1139.

Van der Lee, R., Ellemers, N., & Scheepers, D. (2013). The threat of a moral identity:

Cardiovascular reactivity during intragroup morality (vs. competence) evaluations. Manuscript submitted for publication.

Van Nunspeet, F., Derks, B., Ellemers, N., & Nieuwenhuis, S. (2013). Moral impression management: Evaluation by an ingroup member during a moral IAT enhances perceptual attention and response monitoring. Manuscript submitted for publication.

Van Nunspeet, F., Ellemers, N., Derks, B., & Nieuwenhuis, S. (2012). Moral concerns increase attention and response monitoring during IAT performance: ERP evidence. *Social Cognitive and Affective Neuroscience*. Doi: 10.1093/scan/nss118

Van Rijswijk, W., & Ellemers, N. (1998). Emotionele reacties op feedback over de eigen groep. [Emotional responses to feedback about one's group]. *Fundamentele Sociale Psychologie, 12*, 86–96.

Young, S. D., Monin, B., & Owens, D. (2009). Opt-out testing for stigmatized diseases: A social psychological approach to understanding the potential effect of recommendations for routine HIV testing. *Health Psychology, 28*, 675–681.

PART IV

Coping with Prejudice and Identity Threat

12

THE BIOPSYCHOSOCIAL MODEL OF CHALLENGE AND THREAT

Reflections, Theoretical Ubiquity, and New Directions

Jim Blascovich

This chapter provides a retrospective account of the evolution of the Biopsychosocial Model of Challenge and Threat (BPM-CT) (Blascovich, 2008; Blascovich & Mendes, 2000; Blascovich & Tomaka, 1996) and the development of the patterns of cardiovascular responses marking these motivational states, with a focus on threat during social interactions between members of stigmatized and non-stigmatized groups. In doing so, I hope to "reconstruct the logic," as accurately as possible, of how the theoretical model came to be; describe generally how the bases of the neurophysiological indexes of its nominal motivational states were validated and used to test important social psychological theories; and suggest a new applied direction. This chapter will outline the basic theoretical principles of the BPM-CT. As such, it provides background for several other chapters in this book that describe research in which this model is applied to prejudice and intergroup relations—see the chapters by Major and Kunstman (17), Mendes (16), Mendoza-Denton and Page-Gould (14), and Scheepers (13).

How the Cardiovascular Indexes of Challenge and Threat Came to Be

The philosopher Abraham Kaplan made it clear in *The Conduct of Inquiry* (1964) that, generally speaking, two procedures or "logics" are involved in scientific inquiry. Specifically, he contrasted "logic-in-use," how scientists actually proceed, with "reconstructed logic," or how scientists report they proceeded. In terms of experimental methods, for example, the latter is what we find necessarily but incompletely described in method sections of typical scientific psychology journal articles. However, as many have learned from unsuccessful attempts to replicate studies from published methodological details, reconstructed logic is far from the whole story. Traditionally, we have typically needed much more detailed information of the sort sometimes found in appendices or available via communication with investigators themselves (i.e., "lab lore"). Even so, the reconstructed logic story is never complete for reasons best discussed elsewhere (e.g., Kaplan, 1964; Eagleman, 2011), though, as we have argued (e.g., Blascovich et al., 2002), the use of digital immersive virtual environment technology to conduct social psychology experiments provides a complete and accurate digital record of experimental protocols underlying participants' experiences and behaviors during experiments, and therefore alleviates the need for "lab lore."

Unfortunately, the pathway to theory development appears to be more opaque. Few theorists provide a "reconstructed logic" of how their theoretical models came about, let alone provide any hints to their "logic-in-use" for creating them. Given the request of this volume's editors briefly to relate such a story, I welcome the opportunity to reconstruct here the highlights of the logic underlying the development of the BPM-CT—at least an incomplete, but short version.

As a graduate student at the University of Nevada, Reno, in the late 1960s and early 1970s, I studied risk taking and gambling both in a laboratory casino in the Psychology department and as a participant observer in actual casinos. Funding the latter work out of my own pocket, I often felt my heart pounding when I had placed a "large bet" (e.g., US$10) relative to my monthly NIH graduate fellowship stipend (i.e., US$200) at the time. Sometimes, I interpreted this visceral perception positively, at other times negatively. It was quite clear to me that I won more often in the former than in the latter evaluative state.

At about the same time, I became fascinated by Shapiro and Crider's (1969) seminal chapter in the second edition of *The Handbook of Social Psychology* on social psychophysiology. I believed that I could figure out the psychophysiology of risk taking by the time I earned my PhD by running gambling studies and simultaneously recording physiological responses; a naïve approach, as quickly became clear. Undaunted, and mesmerized by emerging but primitive digital acquisition and recording technology (shrinking the size of individual electrocardiographic [EKG] data acquisition equipment from something refrigerator-sized to something less than the size of a package of cigarettes coupled with an early Sony cassette tape recorder), I assumed that I could collect the necessary data to inform my inquiries by wiring up myself and seven interested departmental faculty members and recording EKG for each of us simultaneously during a 3-hour poker game. Unfortunately, I was more foolish than precocious, as there was no way actually to make sense out of such a massive data set with the computer technology of the time (i.e., a mainframe with a grand total of 64 kilobytes of random access memory) and even less on the basis of any coherent psychological theory.

Instead I turned to a more manageable paradigm in which participants' actions could be structured and controlled while EKG data were collected. The paradigm included an experimental game; specifically, a dyadic zero-sum competitive experimental game (i.e., what one person wins on a trial the other loses) during which the cardiovascular data were collected. Across two identical studies (Blascovich, Nash, & Ginsburg, 1978), the results indicated that eventual winners in male dyads were predicted in 19 out of 20 total dyads by their pre-game heart rates. That is, for 19 out of 20 pairs, the winners had higher heart rates than the losers. Specifically, males whose heart rates increased more from the resting baseline than their competitors—following game instructions and a couple of practice trials, but before the actual onset of the game— went on to win the game. However, such heart-rate increase disparities were not predictive of outcomes for female dyads because there was little disparity. Female winners' and losers' pre-game heart rates post instructions were both similarly elevated. We speculated at the time (mid-1970s) that women were simply less used to competitive games than men, and their heart rates had risen for some other reason yet to be determined.

It took 20 more years to figure out that risk taking is most likely a specific case of a much more ubiquitous phenomenon—motivated task performance, one that all humans face many times a day at work, at home, during play, and in social psychology experiments. During those two decades there was much learning, revision, and relearning about social psychophysiology. There were didactically enlightening and promising (but ultimately dead) ends to my quest: These included chasing down the role of neo-Jamesian inspired visceral self-perceptions such as those that I experienced as a graduate student (and still do) in the casino, and dispositional influences on cardiovascular reactivity such as "Type A" personality style.

But, there was also progress!

The Biopsychosocial Model of Challenge and Threat

Two major sources of information underlay the initial development of the BPM-CT that eventually occurred. The first was Shapiro and Crider's (1969) chapter in the second edition of *The Handbook of Social Psychology* on social psychophysiology mentioned above. The other involved a high-impact paper relevant to my interests. Dienstbier (1989) in a fascinating *Psychological Review* article reported evidence based on rodent research that supported his neuroendocrine model of physiological toughness. Basically, Dienstbier demonstrated that animals who appeared to thrive in potentially dangerous and therefore potentially threatening environments exhibited what he labeled "physiological toughness." Dienstbier empirically demonstrated that physiological toughness in rodents was associated with patterns of neuroendocrine responses driven by the sympathetic–adrenal–medullary (SAM) axis. In contrast, Dienstbier demonstrated that rodents who appeared to wither, decline, and even die in potentially threatening environments exhibited what could be labeled "physiological weakness," associated with patterns of activity driven by the SAM *and* the hypothalamic–pituitary–adrenal (HPA) axis.

Moving beyond the connection that Dienstbier made between the neuroendocrine model underlying his rodent work and Lazarus and Folkman's (1984) cognitive-appraisal theory, we (Tomaka, Blascovich, Kelsey, & Leitten, 1993) extended Dienstbier's neuroendocrine patterns of physiological toughness to include both the predictable cardiovascular responses based on it during human task performance under a rubric we labeled "challenge," and the pattern of physiological weakness under a rubric we labeled "threat."

The BPM-CT is informed by cognitive appraisal theories in the mold of, but not identical to, Lazarus and Folkman's conception (1984). Because of that subtle difference—and because we found the term "appraisal" a bit misleading because of increasing evidence of the importance of automatic or non-conscious processes in social psychology—we chose to use the term "evaluation," as it has fewer conscious and cognitivistic connotations and also allows for the inclusion of affective (emotional) influences. Hence, challenge occurs when personal resources are evaluated as equal to or exceeding situational task demands, and threat occurs when situational task demands outweigh personal resources.

Accrued validational studies of the BPM-CT are myriad, including correlational, experimental, and predictive validational ones (see Blascovich, 2008, for a review). Additionally, convergent validation of the cardiovascular markers stems from 30 or more experiments in which there was a manipulation of the resources and demand evaluations that theoretically underlie social influence phenomena such as social facilitation/inhibition, social comparison, attitude functionality, leadership, dispositional styles, etc. (for reviews, see Blascovich, 2008; Blascovich & Mendes, 2010).

Table 12.1 lists the specifics of the challenge and threat patterns such that SAM-driven increases in heart rate (HR) and ventricular contractility (VC) occur during both challenge and threat, while increases in cardiac output (CO) and decreases in total peripheral resistance (TPR) occur only in challenge. Hence, VC and/or HR increases do not discriminate the motivational states of challenge and threat. What we have argued is that they index task engagement, which we propose is necessary

TABLE 12.1 Specifics of the challenge and threat patterns in the BPM-CT

Challenge	*Threat*
– HR increases	– HR increases
– VC increases	– VC increases
– TPR decreases	– TPR stable/increases
– CO increases	– CO stable/decreases

by definition for either challenge or threat states to occur in a motivated performance situation.

In retrospect, it seems likely that the heart-rate increases in the experimental game study described above for both male and females are simply indexing task engagement. Speculatively, males with higher pre-game heart rates were more engaged, and hence, more successful than males with lower pre-game heart rates. In the case of female winners and losers having equally high pre-game heart rates, the women may have been uniformly threatened due to the competitive aspects of the novel experimental game. Hence, engagement was more universal across female participants, making the HR index of engagement uninformative in terms of predictions of success or failure.

In sum, according to the BPM-CT, the SAM[+]HPA pattern drives the distinctive differences during threat (compared to challenge). When demands and resources are in balance or the latter somewhat exceeds the former, situational approach or "challenge" motivation occurs accompanied by increases in CO and decreases in TPR. When demands somewhat outweigh resources, situational approach/avoidance or "threat" motivation occurs such that there is little change or even decreases in CO and little change or increases in TPR. Finally, when either the ratio of demands to resources or of resources to demands is quite high, situational avoidance or "task disengagement" occurs accompanied by little change in HR and VC.

Detailed descriptions of research methods, including requisite technology (electrocardiography, impedance cardiography, and continuous blood-pressure measurement systems), can be found elsewhere (Blascovich, in press, b; Blascovich, Mendes, Vanman, & Dickerson, 2011). For my purposes here, a general description is more important as several other chapters provide more specific descriptions within individual researchers' substantive contexts.

Two final notions should be made clear about the BPM-CT. First, the cardiovascular indexes of challenge–threat responses have been validated exclusively in contexts that we have labeled "active coping" (cf. Obrist, 1981) or "motivated performance situations" (Blascovich, 2008). Motivated performance situations are those that are goal- and therefore self-relevant to individuals and require some sort of instrumental cognitive or cognitive-behavioral response. In modern everyday life, such situations are ubiquitous and range from recreational ones such as game playing, to educational ones such as exam taking, to occupational ones such as teaching and problem solving, to important personal ones such as job interviews and patient encounters with medical personnel.

Second, it is also important to note that neither challenge nor threat necessarily remains in a steady state during any given motivated performance situation. Rather, individuals' resource/demand evaluations are iterative. For example, a student who is taking a college-entrance exam involving many questions may vacillate between challenge and threat as a function of the specific substance of each question and/or the ordering of the questions. Additionally, individuals may adapt to motivated performance situations that involve repetitive tasks (e.g., serial subtractions) and become less task-engaged.

Results of Challenge–Threat Studies of Stigma and Threat

It is well-known that membership in socially devalued or stigmatized groups carries many disadvantages. Interactions between stigmatized and non-stigmatized individuals have been thought to be threatening psychologically for both parties, at least since Erving Goffman (1963) proffered that hypothesis in his book, *Stigma*, a half-century ago. However, the stigma–threat relationship remained to be demonstrated unambiguously at the physiological level until the 21st century. The latter is important as social psychologists now understand better the impact of automatic, unconscious processes on all aspects of social influence and the value of implicit measures in testing theoretical hypotheses. We maintain that the physiological measures we describe here assess motivational states such as threat or challenge implicitly; that is, without conscious control on the part of individuals. Such implicit measurement is important for studying intergroup relations, particularly those involving members of stigmatized groups. Subsequently, we have conducted numerous studies on this topic.

Much of this research focused on ostensible interactions between members of non-stigmatized and stigmatized groups within North American culture. To accomplish this, we examined interactions in dyads of stigmatized and non-stigmatized individuals involving race, abominations of the body (i.e., facial birthmarks), socio-economic status, etc. Theorists (e.g., Goffman, 1963) have long hypothesized that non-stigmatized individuals interacting with stigmatized ones are threatened (Crocker, Major, & Steele, 1998). Concurrently, stigmatized individuals are susceptible to threat, including increased awareness of self-relevant negative stereotypes such as stereotype threat (Steele, 1997).

Until the turn of the millennium, little if any stigma research involved clear and physiologically sound evidence of threat. For us, the biopsychosocial model of challenge and threat provided a theoretical, conceptual, and physiological basis that could be employed to do so. Consequently, we (Blascovich, Mendes, Hunter, Lickel, & Kowai-Bell, 2001; Mendes, Blascovich, Lickel, & Hunter, 2002) hypothesized that during interactions with a stigmatized other, members of non-stigmatized groups would be threatened as indexed by the cardiovascular pattern of threat associated with the BPM-CT.

Each experiment involved conditions in which dyads of participants included a naïve participant and a confederate. The latter was experimentally manipulated to be stigmatized by virtue of some characteristic of interest (e.g., birthmark or race). The dyads performed as partners in cooperative motivated performance situations involving speech and word-finding tasks. Four experiments were conducted in which same-sex (male or female) dyads were run. In these studies, stigma was manipulated experimentally via either opaque or transparent facial birthmarks on confederates (cf., Kleck & Strenta, 1980) or by the race (Black or White) of confederates who then interacted with their non-stigmatized partner. Our results revealed that non-stigmatized partners exhibited the threat pattern of cardiovascular responses when cooperating with the stigmatized partner, and the challenge pattern when cooperating with the non-stigmatized partner (Blascovich et al., 2001, Experiments 1 and 2). Hence, both an experimentally manipulated stigma (birthmark) and a natural one (race) were tested on males and females separately. This research confirmed the long held stigma–threat hypothesis physiologically. Furthermore, using the birthmark manipulation on genuine participants (i.e., not confederates), we also demonstrated that stigmatizing individuals by virtue of an artificial birthmark created threat in these individuals during similar cooperative tasks.

Stereotype Threat

Steele and Aronson (1995) elucidated stereotype threat theory. Accordingly, in a domain in which a stereotype of one's group carries negative performance expectancies, such stigmatized individuals experience increased anxiety and uncertainty (i.e., threat) because one's performance might confirm the negative stereotype. This psychic load leads them to poorer performance that confirms the stereotype. Blascovich, Spencer, Quinn, and Steele (2001) demonstrated heightened blood-pressure responses among stereotype-threatened African Americans during performance on a verbal abilities test that are consistent with the threat pattern of cardiovascular responses via the BPM-CT. The full threat pattern was later found in a gender–mathematics stereotype study by Vick, Seery, Blascovich, and Weisbuch (2008) in which participants attempted to complete either a purportedly gender-biased mathematics test in favor of males or one that was purportedly gender-fair. Results indicated that stereotype-threatened women exhibited the cardiovascular threat pattern compared to non-stereotype-threatened women. Interestingly, male participants exhibited challenge when a male gender bias was implied, but threat when it was not.

In sum, our work has demonstrated threat effects on non-stigmatized individuals during dyadic cooperative motivated performance situations with stigmatized individuals. Furthermore, the particular type of stigma seems unlikely to matter as the threat effects have been associated with so-called visible stigmas such as race, ethnicity, and physical deformities, etc., as well as less visible ones such as socioeconomic status, religion, etc.

Importantly, we have also shown that members of stigmatized groups themselves are threatened in motivated performance situations with non-stigmatized individuals (Blascovich, Mendes, & Seery, 2002). Specifically, while task-engaged (as indexed by increased VC and HR measures), both stigmatized and non-stigmatized individuals interacting in a dyadic cooperative motivated performance situation exhibit cardiovascular threat patterns including increased

TPR and decreased CO. This research has also demonstrated that individuals are not typically consciously aware that they are threatened during such interactions.

Furthermore, attesting to the power of non-conscious evaluation processes, demands and/or resources evaluations can be induced without conscious awareness on the part of individuals; we (Weisbuch-Remington, Mendes, Seery, & Blascovich, 2005) have demonstrated that below-awareness presentation of benign versus threatening visual stimuli (positive and negative Christian icons) induces the appropriate cardiovascular patterns of challenge and threat, respectively, during a perceptual judgment task on the masks overlaying the subliminally presented stimuli.

Implications for Intensive Medical Treatment

When an individual takes on the role of "patient" with a serious malady, the self- or goal relevance for being compliant during prescribed treatments is high. If patients are members of a stigmatized group, their interactions with non-stigmatized individuals in the treatment setting, including other patients or providers, are likely to increase situational threat in terms of the cardiovascular markers described above. Additionally, if patients are not members of a stigmatized group, but interact with healthcare providers who are, the situational setting is more likely to be threatening to those patients.

What makes threat pernicious in terms of medical treatment in which patients take an active coping role psychologically is the cardiovascular activity associated with it. What stands out for our purposes here is the fact that threat is associated with decreased blood flow. Blood flow itself is critical in terms of the delivery of medication such as chemotherapy agents to diseased tissues and the removal of wastes and toxins from the body such as necrotic tissue and unhealthy metabolites produced by pharmaceutical and radiological treatments themselves.

Because threat during treatment is related to reduced rather than increased blood flow, it can affect the delivery and thereby the potency of therapeutic pharmacological agents. This situation can lead to prescriptions for increased dosage levels accompanied by increased side effects and toxicity, or abandonment of an otherwise therapeutic agent. In these situations, if a patient's threat motivation can be reduced or eliminated, or more ideally, replaced by challenge motivation, blood flow should rise, thereby increasing the potency of pharmacological agents. The picture is a bit different regarding radiation therapies. Blood flow is not critical to the delivery of the therapeutic agent, but increased blood flow can decrease the toxicity of its effects by removing destroyed or necrotic cells from the body.

In our opinion, the critical question is not necessarily a question of whether blood flow should be increased during treatments such as chemotherapy and radiation, but rather how. We suggest this can be accomplished non-pharmacologically via virtual reality technology.

In the early 1990s, researchers investigating visual and spatial cognition phenomena discovered that individuals could be perceptually immersed in digitally created, three-dimensional virtual places such as mazes, buildings, the great outdoors, and on precipices. Furthermore, they discovered that such experiences trumped participants' conscious knowledge that they were safe and sound physically in a research facility. They were affected much more by the simulated or virtual world in which they were located than by the physical or grounded world they conterminously inhabited. In other words, they were "immersed" (see Blascovich & Bailenson, 2011).

In the late 1990s, social psychologists began to realize the potential of using immersive virtual environment technology (IVET) to conduct traditional social psychological experiments. Using IVET, they began by replicating well-known social influence processes such as social facilitation, interpersonal distance, conformity, leadership styles, social comparison, etc., establishing the viability and advantages of IVET for conducting social psychological research (Blascovich et al., 2002). Since these beginnings, a theoretical structural model of social influence in virtual environments evolved (Blascovich, in press, a), guiding the development of socially dynamic digital virtual worlds on the basis of critical factors including theory of mind, communicative realism, response-level system, self-relevance, and context.

At this point, it is important to point out that the concept of "virtual reality" should not be confounded with any particular technology. Humans can enter virtual worlds via endogenous processes such as dreams and daydreams. Facilitating technologies include nearly all, if not all, exogenous media technologies that have evolved throughout human history from language and story telling, through the early visual arts, manuscripts and the printing press, photography and cinematography, radio and television, and now digital media. Indeed, most social psychology experiments utilize what are virtually realities void of digital technology, and the most famous of these (e.g., Haney, Banks, & Zimbardo, 1973; Milgram, 1963) are quite immersive.

Hence, though modern virtual media technologies appear to induce immersion in virtual realities more easily than their predecessors, substantial immersion is in large part a function of individuals' minds. In particular, humans' low-level, automatic, or unconscious processes do not typically react differently to virtual versus physically grounded stimuli. These low-level bottom-up processes strongly influence human behavior on both automatic and deliberate levels, with the former being more compelling than the latter.

However, top-down processes are often used by people motivated for one reason or another to seek immersion in alternate realities. Hence, digital virtual reality-based games and worlds have become quite popular. Many allow for a change of personal or social identity and demographic characteristics such as dispositional ones and race via avatar (digital representations of people in real time; Bailenson, Beall, Blascovich, Weisbuch, & Raimmundo, 2001) appearance, communications, and behaviors.

Theorists (Blascovich & Bailenson, 2011) maintain that virtual reality media provide an escape from the vicissitudes of human life ranging from existential anxiety or death fears to the normal ups and downs of daily activity. However, there is a yin and yang to digital virtual reality technologies. Such escapes can be healthy and provide welcome respite from one's troubles. They also can become pathological. For example, data suggest that modern digital virtual realities have spawned a number of addictions (Aboujaoude, 2011) including gaming, sex, gambling, and social networking.

It is our thesis that one can capitalize on the notion of psychological relocation via digital virtual reality technologies to reduce the anxieties and threats inherent in everyday life in the physical world, and the pernicious cardiovascular effects associated with them. Specifically, we propose that immersion in digital virtual worlds during intensive facility-bound medical treatments such as chemotherapy and radiation therapy may ameliorate the threat and remove or even reverse the pernicious cardiovascular effects.

In support of this position, research (Hoffman, 2004) has shown, for example, that severe burn patients immersed in passive virtual worlds experienced decreases of as much as 70–80% in self-reported pain during skin ablation procedures. Typically such pain is regarded as among the highest in terms of intensity. Research (e.g., Schneider, Ellis, Coombs, Shonkwiler, & Folsom, 2003) has also shown that immersing individuals in virtual reality games decreases the perceived stress of chemotherapy sessions.

Importantly for our thesis, existing research demonstrates that even the threat associated with interactions between non-stigmatized individuals and members of stigmatized groups can be overcome. In an exploratory study described in Blascovich and Bailenson (2011) that included cardiovascular measurement, participants met with a person (a confederate) who was either non-stigmatized or stigmatized physically via a port wine facial birthmark (a stigma that had experimentally been shown to induce threat for both individuals in a dyadic interaction during a cooperative word-finding game). In this exploratory study, both the participant and the confederate were then immersed via avatars in a digital virtual environment in which they subsequently played a cooperative word-finding game. The avatar of the confederate either bore or did not bear the facial stigma independently of whether she bore it physically before the dyad was immersed in the virtual world, producing four experimental conditions.

The results of the study indicated that during the first minute of the virtual word-finding game, participants exhibited the cardiovascular threat pattern only if the physical person had borne the facial birthmark prior to the immersion in the virtual world; otherwise they were challenged. However, by the fourth minute of the virtual word-finding game, participants exhibited the cardiovascular threat pattern only if the virtual avatar had borne the facial birthmark.

These studies point to the conclusion that threat associated with stigmas can be ameliorated using immersive virtual reality technology. Such threat reduction should result in increases in blood flow at least to baseline homeostatic levels, thereby reducing the likelihood of iatrogenic effects of stigma on individuals being treated intensively via chemotherapy or radiation.

Additionally and importantly, if long-lasting or frequent episodes of challenge can be induced via immersion in virtual reality-based activities, blood flow should actually increase above homeostatic levels. Assuming that increased blood flow will potentiate pharmacological agents and/or remove unwanted metabolites and necrotic tissues during chemotherapy or radiation therapy, these intensive treatments should be potentiated.

Summary and Conclusions

Upon reflection, my journey into social psychophysiology has been a long one—even the label has changed, being subsumed by the more general "social neuroscience." Yet it was my training in social psychological methods coupled with a fortunate acquaintance with philosophy of science, particularly as expounded by Kaplan (1964), that allowed the development of the BPM-CT and the validation of the cardiovascular signatures of challenge and threat themselves.

Although my students, postdoctoral fellows, and I have examined the underlying motivational states of many social psychological phenomena, probably the most socially important of the latter involve social stigma because of the malignant prejudices and discrimination that stem from stigma. That direction for my work was stimulated not only by my close relationship with Brenda Major, but also by members of my first graduate student group, particularly Wendy Berry Mendes, after I arrived permanently at the University of California, Santa Barbara in 1995. To read and hear stigma work informed by the BPM-CT being conducted by colleagues around the world is quite gratifying—see also the chapters by Major and Kunstman (17), Mendes (16), Mendoza-Denton and Page-Gould (14), and Scheepers (13) in this volume.

At this stage of my career, I have decided to combine my BPM-CT work on stigma with my work on social influence within digital immersive virtual environments. The initial ideas that are being pursued at the present time are the ones about which I speculated in the preceding section; that is, reducing threat during medical treatment by digitally immersing patients in a benign and challenging task environment at which they excel. I believe that such a process can lessen the burden that members of stigmatized groups may suffer during pharmacological treatment in typical treatment environments; a motivational state of challenge may provide remediation of this burden via the increased blood flow that accompanies it. In Dienstbier's (1989) terminology, such "physiological toughness" can help humans as well as rodents survive in potentially threatening situations.

Suggestions for Further Reading

Theoretical

Blascovich, J. (2008). Challenge and threat. In A. J. Elliot (Ed.), *Handbook of approach and avoidance motivation* (pp. 431–446). New York: Erlbaum.

This article provides the most up-to-date description of the biopsychosocial model of challenge and threat.

Dienstbier, R. A. (1989). Arousal and physiological toughness: Implications for mental and physical health. *Psychological Review, 96*, 84–100.

Dienstbier's classic article is the neuroendocrine basis for the development of cardiovascular indexes associated with the biopsychosocial model.

Kaplan, A. (1964). *The conduct of inquiry: Methodology for behavioral science.* San Francisco: Chandler.

Kaplan provides a lucid treatment of the nature of scientific inquiry in plain language.

Methodological

Blascovich, J. (in press). Psychophysiological methods. In H. T. Reis & C. M. Judd (Eds.), *Handbook of research methods in social and personality psychology* (2nd ed.). Cambridge, UK: Cambridge University Press.

An updated general treatment of psychophysiological methods is provided.

Blascovich, J., & Mendes, W. B. (2010). Social psychophysiology and embodiment. In D. Gilbert, S. Fiske & G. Lindzey (Eds.), *The handbook of social psychology* (5th ed., pp. 194–227). New York: Wiley.

This reference work, only the first such chapter since the second edition of the handbook in 1969, puts the challenge/threat work in the context of social psychophysiological work in general.

Blascovich, J., Mendes, W. B., Vanman, E., & Dickerson, S., (2011). *Social psychophysiology.* Newbury Park, CA: Sage Publications.

This is a "how to" book on physiological measurement and methods in social psychology including cardiovascular, electromyographic, and endocrine approaches.

References

Aboujaoude, E. (2011). *Virtually you: The dangerous powers of the e-personality.* New York: W. W. Norton & Company.

Bailenson, J. N., Beall, A. C., Blascovich, J., Weisbuch, M., & Raimmundo, R. (2001). Intelligent agents who wear your face: Users' reactions to the virtual self. *Lecture Notes in Artificial Intelligence, 2190*, 86–99.

Blascovich, J. (2008). Challenge and threat. In A. J. Elliot (Ed.), *Handbook of approach and avoidance motivation* (pp. 431–446). New York: Erlbaum.

Blascovich, J. (in press, a). Challenge, threat and social influence in digital immersive virtual environments. In J. Gratch & S. Marsella (Eds.), *Social emotions in nature and artifact: Emotions in human and human-computer interaction.* Oxford, UK: Oxford University Press.

Blascovich, J. (in press, b). Psychophysiological methods. In H. T. Reis & C. M. Judd (Eds.), *Handbook of research methods in social and personality psychology* (2nd ed.). Cambridge, UK: Cambridge University Press.

Blascovich, J., & Bailenson, J. (2011). *Infinite reality: Avatars, eternal life, new worlds and the dawn of the virtual revolution*. New York: Wiley.

Blascovich, J., Loomis, J., Beall, A., Swinth, K., Hoyt, C., & Bailenson, J. (2002). Immersive virtual environment technology as a research tool for social psychology. *Psychological Inquiry, 13*, 103–125.

Blascovich, J., & Mendes, W. B. (2000). Challenge and threat appraisals: The role of affective cues. In J. Forgas (Ed.), *Feeling and thinking: The role of affect in social cognition* (pp. 59–82). Cambridge, UK: Cambridge University Press.

Blascovich, J., & Mendes, W. B. (2010). Social psychophysiology and embodiment. In D. Gilbert, S. Fiske & G. Lindzey (Eds.), *The handbook of social psychology* (5th ed., pp. 194–227). New York: Wiley.

Blascovich, J., Mendes, W. B., Hunter, S. B., Lickel, B., & Kowai-Bell, N. (2001). Perceiver threat in social interactions with stigmatized others. *Journal of Personality and Social Psychology, 80*, 253–267.

Blascovich, J., Mendes, W. B., & Seery, M. (2002). Intergroup encounters and threat: A multi-method approach. In D. Mackie & E. Smith (Eds.), *From prejudice to intergroup emotions: Differentiated reactions to social groups* (pp. 89–110). New York: Psychology Press.

Blascovich, J., Mendes, W. B., Vanman, E., & Dickerson, S. (2011). *Social psychophysiology*. Newbury Park, CA: Sage Publications.

Blascovich, J., Nash, R. F., & Ginsburg, G. P. (1978). Heart rate and competitive decision making. *Personality and Social Psychology Bulletin, 4*, 115–118.

Blascovich, J., Spencer, S., Quinn, D., & Steele, C. (2001). African-Americans and high blood pressure: The role of stereotype threat. *Psychological Science, 12*, 225–229.

Blascovich, J., & Tomaka, J. (1996). The biopsychosocial model of arousal regulation. In M. P. Zanna (Ed.), *Advances in experimental social psychology* (Vol. 28, pp. 1–51). New York: Academic Press.

Crocker, J., Major, B., & Steele, C. (1998). Social stigma. In D. Gilbert, S. T. Fiske & G. Lindzey (Eds.), *The handbook of social psychology* (4th ed., Vol. 2, pp. 504–553). New York: McGraw Hill.

Dienstbier, R. A. (1989). Arousal and physiological toughness: Implications for mental and physical health. *Psychological Review, 96*, 84–100.

Eagleman, D. (2011). *Incognito: the secret lives of the brain*. New York: Vintage Books.

Goffman, E. (1963). *Stigma: Notes on a spoiled identity*. New York: Simon & Schuster.

Haney, C., Banks, C., & Zimbardo, P. (1973). Interpersonal dynamics in a simulated prison. *International Journal of Criminology and Penology, 1*, 69–97.

Hoffman, H. G. (2004). Virtual reality therapy. *Scientific American, 291*, 58–65.

Kaplan, A. (1964). *The conduct of inquiry: Methodology for behavioral science*. San Francisco: Chandler.

Kleck, R. E., & Strenta, A. (1980). Perceptions of the impact of negatively valued physical characteristics on social interaction. *Journal of Personality and Social Psychology, 39*, 861–873.

Lazarus, R. S., & Folkman, S. (1984). *Stress, appraisal and coping*. New York: Springer.

Mendes, W. B., Blascovich, J., Lickel, B., & Hunter, S. (2002). Challenge and threat during interactions with White and Black men. *Personality and Social Psychology Bulletin, 28*, 939–952.

Milgram, S. (1963). Behavioral study of obedience. *The Journal of Abnormal and Social Psychology, 67*, 371–378.

Obrist, P. (1981). *Cardiovascular psychophysiology: A perspective*. New York: Plenum.

Schneider, S. M., Ellis, M., Coombs, W. T., Shonkwiler, E. L., & Folsom, L. C. (2003). Virtual reality intervention for older women with breast cancer. *Cyberpsychology and Behavior, 6*, 301–307.

Shapiro, D., & Crider, A. (1969). Psychophysiological approaches to social psychology. In G. Lindzey & E. Aronson (Eds.), *The handbook of social psychology* (2nd ed., Vol. 3, pp. 1–49). Reading, MA: Addison-Wesley.

Steele, C. M. (1997). A threat in the air: How stereotypes shape intellectual identity and performance. *American Psychologist, 52*, 613–629.

Steele, C. M., & Aronson, J. (1995). Stereotype threat and the intellectual test performance of African Americans. *Journal of Personality and Social Psychology, 69*, 797–811.

Tomaka, J., Blascovich, J., Kelsey, R. M., & Leitten, C. L. (1993). Subjective, physiological, and behavioral effects of threat and challenge appraisal. *Journal of Personality and Social Psychology, 65*, 248–260.

Vick, S. B., Seery, M. D., Blascovich, J., & Weisbuch, M. (2008). The effect of gender stereotype activation on challenge and threat motivational states. *Journal of Experimental Social Psychology, 44*, 624–630.

Weisbuch-Remington, M., Mendes, W. B., Seery, M. D., & Blascovich, J. (2005). The non-conscious influence of religious symbols in motivated performance situations. *Personality and Social Psychology Bulletin, 31*, 1203–1216.

13

STUDYING SOCIAL IDENTITY-BASED THREATS AND CHALLENGES USING CARDIOVASCULAR MEASURES

Daan Scheepers

Where Is the Threat in Social Identity Threat?

Social identity is the part of one's identity derived from group membership (e.g., identity based on ethnicity, gender, occupation, sports team fandom). Social identity theory specifies how people search for a positive social identity, a need that is met through inclusion in groups that are in a positive way distinct from relevant out-groups (i.e., indicating a relatively high rank in the intergroup status hierarchy; Ellemers & Haslam, 2012; Tajfel & Turner, 1979). Representing a "cognitive-motivational" perspective on intergroup relations, work in the social identity tradition is based on the assumption that this positive group distinctiveness, and the positive social identity emerging from it, in turn serve core human motives for self-esteem, certainty, and meaning (Abrams & Hogg, 1988).

In the mid-1990s, the concept of threat was seen as a key motivational construct within the psychology of social identity (Branscombe, Ellemers, Spears, & Doosje, 1999; Ellemers, Spears, & Doosje, 2002). Interest in threat had developed in attempts to explain a variety of important consequences of situations where the value of one's in-group, and thereby the value of one's social identity, was undermined—as when one's favorite soccer teams loses a game, or when one is a member of a stigmatized group (Ellemers, 1993; Branscombe & Wann, 1994). Examples of the consequences of these situations include diminished self-esteem, strengthened in-group identification, and increased discrimination against out-groups. Could social identity threat explain these important phenomena?

It was around this time that I wrote my Master's degree thesis under the supervision of Russell Spears at the University of Amsterdam. The topic of my thesis was social identity threat, and the aim was to provide a taxonomy of its different forms. Despite the popularity of the concept of threat among social identity theorists, the question remained whether these assumed threats were one and the same thing. In my thesis (Scheepers, 1997), I made distinctions between different forms of threat (e.g., threat to the value of the group, threats to the distinctiveness of the group, the threat of being categorized as a member of the group, the threat of being excluded from the group), that at the time also emerged in other places in the literature (Branscombe et al., 1999; Ellemers et al., 2002).

However, when reviewing the literature on this topic, there was a thing that struck me: Social identity threat was often assumed to be a motivational process, but was seldom directly measured. It was noted that self-report measures of threat might not be ideal because they could lead to defensive responding ("Me, threatened? Not at all!"), making those who are the most threatened indicate this to the least extent (Bettencourt, Miller, & Hume, 1999). To deal with the absence of more direct measures, the state of social identity threat was often inferred from other measures like (lowered) self-esteem, or even from the outcome variable it was hypothesized to explain (e.g., "They show discrimination, so they must have been threatened."). However, this strategy seemed problematic given the diversity of responses to which social identity threat was linked, like working harder at versus turning away from group status-relevant tasks (e.g., Ouwerkerk, De Gilder, & De Vries, 2000; Steele, Spencer, & Aronson, 2002). As a consequence, in the discussion section of my thesis, I concluded that although more insight might have been obtained on the *content* of threat, the *process* of threat remained in a black box. Put differently: Where was the threat in social identity threat?

This question became the start of a journey in which I have sought to measure the motivational state of threat— and its motivational counterpart, challenge—in relation to social identity in a more direct, continuous, and online way by means of the cardiovascular (CV) response patterns described in the biopsychosocial model of challenge and threat (BPM-CT; Blascovich & Mendes, 2010). This chapter provides a "ship's journal" of this journey.

A Biopsychosocial Approach for Studying Social Identity Threat

It has often been noted that coincidences are a main force driving scientific progress. Although this notion has mainly been made at the macro level, in terms of great scientific discoveries, coincidence also played a major role at the micro level, in my case, with the events leading up to my decision to use physiological measures in my work on intergroup relations.

During the research I conducted for my PhD, we repeatedly found that after intergroup comparisons, members of low-status—or otherwise disadvantaged—groups expressed in-group bias mainly for instrumental reasons (i.e., to change the relations between groups), while members of high-status groups expressed in-group bias as a form of identity expression. We found this in different contexts; e.g., under controlled laboratory settings (minimal groups), but also in the out-group derogatory chants sung by die-hard soccer fans (see Scheepers, Spears,

Doosje, & Manstead, 2006, for an overview). Because we hypothesized that group threat (versus affirmation) would be the operating principle underlying these different functions of in-group bias, in one of our studies (Scheepers, Spears, Doosje, & Manstead, 2003, Study 1) we measured threat using a self-report measure. More specifically, by means of a single item, participants indicated on a 7-point Likert scale the extent to which they had "felt threatened" when they had expressed in-group bias. Somewhat to our surprise, this single item predicted the instrumental function of in-group bias quite well. Would it be so easy then, after all? Would defensiveness not play such a big role as we had thought?

However, at the same time, other issues regarding the assessment of threat remained, and new ones came up. For example, as we measured threat retrospectively, it was unclear whether participants had actually felt threatened at the moment they had expressed in-group bias. In addition, we wondered whether our participants actually had full conscious access to a complex motivational state such as threat. We also did not know whether all participants interpreted our question regarding threat in the same way; did they even understand what we meant by "threat"? (and did we actually understand this ourselves?). And finally, because low group status triggered rather different forms of in-group bias under different circumstances (more out-group derogatory forms when status was stable, more in-group favoring forms when status was unstable), we wondered whether threat would explain all of these different forms of bias. For example, might the more positive in-group-oriented response when status was unstable (directed at status improvement) be explained by a more positive, constructive motivational state, like challenge? How nice it would be if we could measure motivational states of threat (and possibly also challenge), in a more direct, online, continuous, and unambiguous way. . .

It was around the time when we conducted our studies on group status and in-group bias that I participated in the summer school of the European Association of Social Psychology in Clermont-Ferrand in France (2000). Although Jim Blascovich was not a teacher at the summer school, he was visiting and delivered a plenary address on how to measure the motivational states of challenge and threat in an online, continuous, direct, and unambiguous way by means of specific patterns of cardiovascular reactivity. I remember this presentation as the moment when I thought, "This is what work on social identity threat really needs."

During the presentation I also sketched a 2 × 2 design of a study on group status, status stability, and CV markers of challenge and threat; this became one of the first studies I conducted on social identity-based challenge and threat (Scheepers, 2009). For this study, which will be described in more detail below, I hypothesized—in line with our work on in-group bias—that although low group status will be rather threatening when intergroup status differences are stable and unlikely to change, when intergroup status differences become unstable, this threat will turn into a challenge because status improvement seems possible. At the same time, members of high-status groups might respond in a threatened manner to unstable status differences because for them their group's position could only become worse.

The Biopsychosocial Model of Challenge and Threat

The model Blascovich described during his presentation is the biopsychosocial model of challenge and threat (BPM-CT) (Blascovich, 2008; Blascovich & Mendes, 2010; Blascovich & Tomaka, 1996; see also Blascovich, Chapter 12 in this volume). The BPM-CT describes specific CV markers of the motivational states of threat and challenge during so-called motivated performance situations (e.g., a job interview, bargaining, giving a speech, performing a group task, intergroup competition, inter-ethnic interactions). The appraisal part of the BPM-CT describes challenge and threat as the outcome of an evaluation of a motivated performance situation in terms of the *demands* of the situation (e.g., required effort, danger, uncertainty) and the *resources* the person brings into the situation in order to deal with these demands (e.g., skills, support, dispositions). When demands outweigh resources, a threat motivational state emerges; whereas when resources approach or exceed demands, a challenge motivational state emerges.

Importantly, the BPM-CT also describes how these motivational states can be measured in a continuous, online, and non-reactive way using a combination of four CV measures: heart rate (HR); pre-ejection period (PEP: an index of left ventricular contractile force); cardiac output (CO: the amount of blood pumped by the heart, in liters per minute); and total peripheral resistance (TPR: an index of net constriction versus dilation in the arterial system). Motivated performance requires a certain level of task engagement, which is indicated by decreased PEP and increased HR. Challenge is in turn marked by relatively high levels of CO and low levels of TPR, enabling the efficient mobilization and transportation of bodily resources during motivated performance. By contrast, threat is marked by relatively low CO and high TPR. In addition, and important for our initial studies, the CV pattern indicative of threat is typically accompanied by high blood pressure (Blascovich & Mendes, 2000).

Intergroup Status Differences as Challenge and Threat: A Research Program

After obtaining my PhD degree in 2002, I moved to Leiden University, to join the lab of Naomi Ellemers. Naomi had some experience with measuring threats in relation to social identity using physiological measures (Van Rijswijk & Ellemers, 1998), and was very keen to extend this work.

In our initial research, we focused on the influence of intergroup status differences on the state of threat (and more recently, also challenge). Differences between groups in terms of status and prestige are key within the psychology of social identity, as they form a proxy of the value of one's social identity, which is in turn an important factor in actions to improve or preserve the group's position. Based on this reasoning, the assumption has often been made that inclusion in a group with a relatively low status is threatening, whereas inclusion in a group with a relatively high status is not. A first study in this line of research (Scheepers & Ellemers, 2005) challenged both these assumptions.

In our study, participants were categorized in minimal groups (ostensibly on the basis of whether they had a more "detail-driven" or "holistic" association style), after which they performed on two rounds of a task. After the first round of the task, we provided group-level status feedback, as a means to induce group status differences. After this we (unexpectedly) announced a second round of the status-defining task. During the experiment we measured each participant's blood pressure three times: at the start of the experiment (Time 1; baseline); after we provided group-level performance feedback on the first round of the task (Time 2); and just after we announced, but before they performed on, the second round of the task (Time 3). We created reactivity scores by subtracting baseline blood-pressure levels from blood pressure at Time 1 and Time 2. In addition to blood pressure, we also measured the participant's subjective identification with the minimal group using a standard group identification scale.

The results indicated that just after receiving the group status feedback (Time 2), participants in the low-status condition had higher blood pressure than participants in the high-status condition. However, at Time 3—i.e., just before the second round of the task— this pattern turned around so that participants in the high-status condition displayed higher blood pressure than participants in the low-status condition. We explained this latter finding by arguing that the inter-group status relations, and consequently the social identity of the group members, could possibly change during the second round of the task. This should be rather threatening for members of the high-status group, as from their superior position things could only go downhill. Thus, reflecting on *existing* status differences is more threatening for members of low-status groups, while anticipating possibly *changing* status differences is more threatening for members of high-status groups.

Follow-up analyses revealed that the effects described above were moderated by group identification such that the pattern described above only emerged for those who were relatively highly identified with their group; for low identifiers, blood pressure did not change as a function of group status feedback. These latter effects are important because they directly show that social identity is implied in the effects of group status on identity threat.

In a follow-up study (Scheepers, Ellemers, & Sintemaartensdijk, 2009, Experiment 2), we conceptually replicated this first study in the context of existing social categories, namely gender categories. In addition, while in our first study a sense of status instability was inferred from the announcement of a second round of the group-status defining task, in this study we sought more directly to manipulate the *reflection* on existing intergroup status differences versus the *anticipation* on possible changing intergroup status differences. To this end, we had males (traditionally the dominant group) and females (traditionally the superordinate group) debating in either an intragroup context (male against male and female against female) or an intergroup context (male against female). We varied the topics of the debate to make the debaters focus on either the existing disparities between men and women (an example of such a topic was "After a child is born the mother [rather than the father] should stay at home to care for it") or on changing intergroup differences (e.g., "More subsidies for day care centers should become available so that women can keep working when they have a child"). We again took blood-pressure assessments at the start of the experiment (baseline) and at each of the two debates, and created reactivity scores by subtracting baseline blood-pressure levels from blood-pressure levels during each of the debates.

Analyses of the reactivity scores revealed that the strongest effects took place in the intergroup contexts. In addition, and as predicted, in this context, women had higher blood pressure than men when debating *existing* gender disparities, while men had higher blood pressure than women when debating *changing* gender relations. Interestingly, on an explicit measure of sexism, males in the intragroup condition (i.e., being only with other men around) scored higher than females; but in the intergroup condition, this difference disappeared. That is, although at the implicit level, members of the dominant group were showing displays of threat when changes in the status quo were made salient, at the explicit level, they seemed strategically to attenuate expressions of sexism, a variable that has often been linked to threats to masculine identity (Maass, Cadinu, Guarnieri, & Grasselli, 2003).

In summary, two important conclusions can be drawn on the basis of these initial studies. First, they illustrate the usefulness of physiological indicators for measuring psychological constructs (threat in this case) that are difficult to assess with explicit measures. For example, in the gender study, the divergence between physiological signs of threat on the one hand and out-group prejudice (sexism in this case) on the other hand, illustrates the danger of solely relying on explicit measures of prejudice, for example, to indicate a state of threat. More specifically, the self-report measure of prejudice would (erroneously) have led to the conclusion that males in the intragroup condition were more threatened because they were most inclined to display sexism. Second, these studies also help to make a more theoretical point; namely, that being a member of a low-status group is not always threatening, while being a member of a high-status group can be rather threatening, depending on the security of status differences (see also Sapolsky, 2005).

There were, however, a couple of limitations to these first studies. For example, although we took repeated measurements, we still did not assess physiological states in an online and continuous way. More importantly, however, although blood pressure has been related to threat (Blascovich & Mendes, 2000; Mendes, Blascovich, Lickel, & Hunter, 2002), when measured in isolation, blood-pressure readings suffer from some interpretational difficulties. Indeed, there is also a rich literature linking blood pressure (in particular systolic blood pressure) to effort mobilization (Wright & Kirby, 2003). In relation to this, by relying solely on blood-pressure assessments, it was not possible to discern cardiovascular correlates of threat from cardiovascular correlates of challenge. Extending the work on social identity threat to challenge motivational states would not only be theoretically innovative, and lead to more motivational specificity regarding the consequences of intergroup status differences; there would also be a practical gain in finding out, for example, how threats to social identity can be turned into challenges. Therefore, in follow-up studies we sought to measure motivational states of challenge (as well as threat) in relation to social identity in a more specific, direct, online, and continuous way by using more sophisticated CV indices.

Turning Social Identity Threat into Challenge

Around the same time that we started experimenting with measuring the blood-pressure correlates of threats to social identity, a second "coincidence" took place that heavily influenced the development of the current line of research; namely, the funding that became available at my university (the so-called "beta-funding") for behavioral scientists who wanted to extend their research by incorporating physiological and neuroscientific measures. By means of this money we were able to buy equipment to take the more sophisticated CV measurements (HR, PEP, CO, TPR) that are necessary to distinguish challenge from threat. These measurements (electrocardiographic measurements [ECG], continuous blood pressure, and impedance cardiographic measurements [ICG]) are described in more detail by Mendes, Chapter 16 in this volume. In addition to this equipment, lab space was made available, and a research technician was hired. Around the same time I also acquired a large grant from the Netherlands Organisation for Scientific Research which made it possible to make social identity-based challenge and threat the main focus of my research for the next 3 years.

When building our more advanced cardiovascular lab, it also became apparent how difficult and time-consuming it is to start up such a lab from scratch. Finding the right equipment, renovating a lab to incorporate this equipment, pilot testing the set-up are all activities that are rather time-consuming (and sometimes stressful) and do not directly produce results, while in the background, teaching obligations and other research lines also continuously call for attention.

During this process there were three things that I learned in particular. First, there is great value in having a research technician/programmer who can create custom-based solutions for equipment and software. The second thing I learned is how important and useful it is to have good relations with other researchers in your department (clinical psychologists, psychophysiologists) who have experience with the type of physiological measures you want to start using. Not only can they teach you tips and tricks to take these measurements, they can also help you out when you run out of electrodes or when that crucial cable breaks down halfway through your experiment. . . Related to this, the third thing I learned is that when building your lab, there is immeasurable value in seeing things really happen in motion in another lab. That is, regardless of how many journal articles and handbook chapters you read, seeing the practices of working with participants (and being one yourself) is of great importance when learning to take psychophysiological measures. I was very fortunate to be able to visit Jim Blascovich's lab and to look over the shoulders of his students (for example, when scoring impedance cardiograms) and to function as a guinea pig when a research assistant was taught to apply band-electrodes for impedance cardiographic measurements.

After visiting Jim's lab, when our own lab was up and running, I was finally able to run the 2 × 2 design (status × stability) that I had sketched during Jim's presentation in Clermont-Ferrand, 5 years earlier. In this experiment, I further examined social identity threat (and this time, also challenge) as a function of intergroup status differences, and the stability of these differences.

In this study (Scheepers, 2009), baseline measurements of cardiovascular indices (HR, PEP, CO, TPR) were taken, after which participants were categorized in minimal groups. Then they performed on a task (a number-counting task) on which they received group-level status feedback (group status manipulation). After this, a second task was announced (a letter-counting task). Participants were led to believe that their performance on this task was typically well-predicted from performance on the number-counting task. By means of this information, a stable intergroup situation was created. I predicted that motivated performance in the context of such stable status differences would be more threatening for members of low-status groups than for members of high-status groups. After this second task, a third task was introduced (a word-searching task). This time, participants were informed that performance on this task was only poorly predicted by performance on the first two tasks. By means of this manipulation, an unstable intergroup situation was created. Motivated performance in the context of such unstable intergroup status differences was expected to be challenging for members of low-status groups, as it made possibilities for status improvement salient, while being threatening for members of high-status groups, for whom things could only go downhill.

The results provided support for these predictions. After calculating CV reactivity scores by subtracting mean baseline levels from task levels, the expected pattern emerged: In the stable situation, participants in the low-status condition were relatively more threatened (lower CO, higher TPR) than participants in the high-status condition. However, when the situation was unstable, this pattern turned around, in the sense that members of the low-status group were relatively more challenged (higher CO, lower TPR) than participants in the high-status condition. It is noteworthy that the strongest within-subjects effect was found for participants in the low-status condition: They were threatened during the stable situation, but this threat turned into challenge when status differences became unstable.

In a follow-up study (Scheepers, 2013b), I addressed another classic social-structural variable that shapes the meaning of status differences, namely the *legitimacy* of these differences. There is great variability in the extent to which status differences between groups (e.g., gender, ethnicity) are seen as fair, just, and legitimate (Jost & Major, 2001). Similar to differences in perceived stability, differences in perceived legitimacy signal the *security* of these differences as they make cognitive alternatives for the current state of affairs salient (Tajfel & Turner, 1979). Legitimate intergroup status differences are a proxy for secure status differences. Therefore, intergroup competition in the context of legitimate status differences should be more threatening for members of low-status groups than for members of high-status groups. By contrast, motivated performance in the context of illegitimate status differences should be particularly threatening for members of high-status groups because this illegitimacy will increase uncertainty and, beyond that, constitutes a direct threat to the value of the group (Doosje, Branscombe, Spears, & Manstead, 1998).

These hypotheses were tested in a study using a similar set-up to the stability study described above. After obtaining baseline levels of CV indices, participants were categorized in minimal groups, after which they received group-level performance feedback on a first round of a task (group status manipulation). The legitimacy of the intergroup status relations was then manipulated by means of a message, ostensibly sent by a fellow in-group member, which claimed that group performance was assessed in either a correct (legitimate status differences) or incorrect (illegitimate status differences) way (see Scheepers, Branscombe, Spears, & Doosje, 2002, for a similar procedure). After this, a second round of the group task followed, which was the motivated performance situation of particular interest to assess the CV indices of challenge and threat. In addition to CV responses, the participant's subjective identification with the group was assessed via a self-report questionnaire.

In line with expectations, we found that motivated performance in the context of legitimate status differences led to more threat in members of low-status groups than in members of high-status groups, while these differences disappeared when status differences were illegitimate. Indeed, members of high-status groups displayed a greater tendency toward threat when the status differences were illegitimate than when they were legitimate. Importantly, however, and in line with the social identity rationale, the interaction between group status and status legitimacy was only significant for those who were relatively highly identified with their group; for those who were only weakly identified, the group status manipulations did not significantly affect their CV responses. In addition, we found a correlation between the CV challenge pattern and performance during the second round of the task, indicating that those who were more challenged during the task performed better on it.

It is particularly noteworthy that among those highly identified with the group, we found a stronger challenge response when low status was illegitimate than when it was legitimate. This is interesting because—although illegitimate status differences signal status insecurity—it is not at all evident that actual change will occur. That is, unlike unstable intergroup status differences which directly imply the opportunity for change, status differences that are illegitimate might nevertheless be rather stable (as is the case for oppressed groups in certain societies). However, the observation that high identifiers respond to illegitimate status differences with challenge is perfectly in line with work showing that high identifiers are less strategic than those who feel more weakly identified with their group; high identifiers tend to work toward group status improvement even when the chances of success are unclear (Ellemers, Spears, & Doosje, 1997). In a final set of studies we tested this effect more directly by examining whether highly identified members of low-status groups respond with challenge to negative feedback about their group, even in the absence of further information about the social structure.

In one study (Scheepers, 2013a, Study 1), after obtaining baseline levels of CV responses and categorization in minimal groups, the participant's identification with the group was manipulated by providing bogus feedback on an implicit association task. Participants in the low identification condition learned that their responses during the test had revealed that they felt only weakly connected to their group, while participants in the high identification condition learned that their responses during the test had revealed that they felt strongly connected to their group. A manipulation check in the form of a standardized identification questionnaire confirmed that the manipulation had been successful. Then, after performing on a group task, all participants received negative group-level performance feedback. In line with expectations, during a second round of the task, participants in the high identification condition showed stronger CV reactivity indicative of challenge than participants in the low identification condition.

This effect was replicated in a second study in a real group context (Scheepers, 2013a, Study 2). After obtaining baseline measures of CV responses and measuring identification with their university, undergraduate students read a series of newspaper articles about a recent examination of the quality of Dutch universities from which it appeared that their university was among the lowest ranked. Immediately after reading these articles, the students worked on a series of assessment-center tests. Beforehand the students were told that from these tests we could predict their future career success. This was the motivated performance situation we focused on with regard to the CV indices of challenge and threat motivational states. We calculated CV reactivity scores by subtracting baseline levels of CV responses from CV responses when working on the assessment-center tests. On the basis of a median split on their group identification scores, the sample was divided into two groups—a group of students who were relatively weakly identified with their university, and a group of students who were relatively strongly identified with their university. Analyses of the reactivity scores showed that, in line with expectations, the group of students that was only weakly identified with their university showed a CV pattern indicative of threat (low CO, high TPR), whereas the group of students that was relatively strongly identified showed a CV pattern indicative of challenge (high CO, low TPR).

Summary, Implications, and Future Challenges

The research described above demonstrates how social identity concerns, stemming from socio-structural factors, shape basic psychophysiological responses that are indicative of challenge and threat motivational states. The results clearly indicate that being a member of a low-status group is not always threatening, but can also be challenging—for instance, when there are signs that the group's position might improve. On the other hand, being a member of a high-status group can be rather threatening—for example, when there are signs that the group's relative favorable position is not legitimate.

The current results are in line with social identity theory, but extend this perspective in important ways. Most importantly, these cardiovascular data provide a new motivational perspective on social identity processes. Previous work on the motivational basis of social identity has mainly focused on motivational end-states, such as enhanced self-esteem or reduced uncertainty (Abrams & Hogg, 1988). More recently, however, researchers have started to apply self-regulation models to the study of social identity and intergroup relations (see Jonas, Sassenberg, & Scheepers, 2010, for an overview). These self-regulation perspectives focus on motivational processes underlying goal pursuit, rather than on ultimate goals or motives themselves. The use of cardiovascular indicators of challenge versus threat fits with this development, as it elucidates the emergence of different motivational status.

The results of this research program also have implications for the relationship between social identity and health. Previous work in this area has demonstrated that a salient social identity can form a buffer against stress (Jetten, Haslam, C., & Haslam, S., 2011). This stress-buffering function of social identification has even been observed at the physiological level in terms of diminished cortisol reactivity during group-demanding situations (Haslam & Reicher, 2006). The current research on group identification and challenge is in line with prior work, but also extends existing insights by showing that social identity not only functions as a buffer against maladaptive physiological stress responses (cortisol), but can even elicit more benign and constructive physiological responses during demanding situations (i.e., challenge; see also Blascovich, Chapter 12 in this volume).

Finally, there are parallels between the current work on psychophysiological effects of group status and the work by Sapolsky on the relationship between social rank in primate hierarchies and physiological stress responses (see Sapolsky, 2005, for an overview). One of the key messages of this prior work is that rank per se is only a weak predictor of maladaptive physiological stress responses (cortisol reactivity). Instead, the relation between rank and stress depends on the *meaning* of rank; for example, as a function of the stability of the hierarchy. Specifically, Sapolsky reports evidence that while lower-ranked primates show the strongest stress response in stable hierarchies, in unstable hierarchies, the higher-ranked primates show the strongest cortisol reactivity. The current research extends this previous work to a human population, and examines intergroup rather than intra-group status hierarchies, addressing social identity-based, rather than realistic conflict-based, sources of group threat. Despite these differences, the main message of the current work is similar to that of Sapolsky: It is much too simple to assume a straightforward relationship between social status and maladaptive physiological stress responses. Rather, what matters is what status *means* as a function of the broader socio-structural context. For example, when (low) status means opportunity for improvement, this leads to a more benign CV response-profile than when status differences are stable. Similarly, when (high) status means insecurity, this leads to a more maladaptive CV response-profile than when status differences are secure.

In terms of future challenges, two issues should be noted. First, future research should more systematically study the consequences of social identity-based challenge and threat for performance and other outcome measures (e.g., well-being). Although in some of our experiments, we found evidence for a relation between the CV challenge pattern and performance (Scheepers, 2013b), in other experiments, we were not able to demonstrate this relationship (e.g., Scheepers, 2009). There are methodological as well as conceptual reasons why the relation between challenge and performance does not always appear to be straightforward. For example, because physiological, self-report, and behavioral measures are often based on different measurement systems, the increased error variance that is associated with the combination of these measures might make it more difficult to establish statistically reliable relationships between different types of measures. At the more conceptual level, it is also important to note that behavior is the outcome of a complex interplay between affective, cognitive, and motivational factors, while the CV indices described by the BPM-CT only mark this latter factor (Blascovich, Mendes, Hunter, & Salomon, 1999). Still, because the ultimate aim of psychology is to explain behavior, future research should address how the predictive value of CV challenge and threat responses for task behavior can be further optimized.

A second challenge for future research concerns precisely the other side of the causal chain, namely the appraisal processes, in terms of the demands and resources leading to social identity-based challenge and threat. Uncertainty, which is one of the main demand components in the BPM-CT, is likely to play a major role here, as the reduction of uncertainty has also been identified as a main factor in defining and defending a positive social identity (Hogg, 2009). For example, the uncertainty of a possibly changing status quo might have contributed to the threat in members of high-status groups when status differences were unstable or illegitimate. In addition, the certainty that a strong sense of social identity yields may have contributed to the positive relation between group identification and challenge (Scheepers, 2013a).

Added Value, Strengths, and Limitations

Cardiovascular indices of challenge and threat seem particularly well-suited to address a dynamic and "elusive" concept such as social identity because these indices can be measured in an online and continuous way. These particular features are useful for studying the motivational implications of social identity because social categorization—and the social identity-related concerns stemming from this—change from situation to situation. In addition, and as indicated above, the state of threat can also elicit defensive responses, making it necessary to measure the occurrence of threat unobtrusively. Thus, the integration of principles derived from the BPM-CT with insights from social identity theory greatly enriches the study of social identity threat, a phenomenon that is of great theoretical as well as practical importance.

Despite these advantages, physiological measures are by no means "better" than cognitive, self-report, or behavioral measures; neither do they provide the "royal road to truth." Physiological measures are just another source of information. For some questions, like those concerning the motivational basis of social identity phenomena, physiological measures have revealed important new insights; for other questions, they might not necessarily be needed (see also Mendes, Chapter 16 in this volume).

In some cases, a combination of different measures may be a particularly fruitful way of addressing social psychological themes. For example, in the gender study discussed above, we combined a physiological index of threat (blood pressure) with a self-report measure of sexism (Scheepers et al., 2009). Had we only measured sexism, it would have remained unclear if and when threat had occurred. However, if we had only included physiological measures, we would have missed how our male participants strategically adapted their expressed sexism toward a female audience, despite blood-pressure evidence indicating the occurrence of threat.

Despite the advantages that physiological measures offer, there are also certain costs associated with the use of these measures. For example, the use of these measures may be more expensive and time-consuming, and may require additional training in comparison to more traditional methodology. In addition, applying physiological measures to social psychological research requires the development of expertise in two areas: On the one hand, it is necessary to learn about the principles of psychophysiology (e.g., about bio-potentials, signal sampling, amplifying, filtering); on the other hand, it is important to be able to formulate new and important questions that help to move forward the domain of social psychology.

The use of psychophysiological measures places demands on specific requirements of the paradigm you use and is typically also more involving for research participants than traditional methods. In our studies, participants are hooked up to six electrodes on their chest and back; a blood-pressure sensor (which makes mild, but continuous contractions) is attached to a finger while participants are required to sit as still as possible. Although this set-up is still less involving for the participant than, for example, undergoing a functional magnetic resonance imaging (fMRI) study, it is still more involving than the average behavioral study. Furthermore, even though physiological measures can be unobtrusive in the sense that people cannot exactly figure out what is being measured, the use of these measures can result in an increased focus on being the "research object," which in itself might cause increased self-awareness and even induce stress.

Neuroendocrine measures, such as cortisol, form a class of alternative physiological measures to examine threats to social identity (Matheson & Cole, 2004). Although it is not possible to measure the "challenge" response with this type of measure, an advantage of cortisol, compared to the CV measures, is that it offers a direct measure of activation of the hypothalamic–pituitary–adrenal (HPA) axis. In addition, cortisol might be relatively simply derived from saliva samples, which makes it also more easily applicable in more ambulatory settings. A disadvantage of using cortisol as a measure of threat is that there are many restrictions (regarding e.g., substance use, ovulatory cycle, time of day) for obtaining reliable measurements. In addition, it is not possible to measure cortisol in a continuous way, making it less suitable as a measure to examine how certain changes in physiological and motivational processes occur.

Conclusion

The BPM-CT provides a new, important, and exciting motivational framework for social identity theory and the phenomena that this theory aims to explain. The CV profiles described by the BPM-CT can offer more direct insight into specific motivational processes, and can be assessed in an unobtrusive, online, and continuous way, making it possible to measure changes in motivational states, which are characteristic for the dynamic quest for a positive social identity. Cardiovascular indices of challenge and threat can also help to provide more direct evidence for the consequences of social identity threat for health and well-being. The research presented in this chapter has shown how threat and challenge emerge as a function of social identity concerns. Future research should examine the precise determinants (i.e., appraisals) of these states, as well as the consequences for performance and well-being.

Suggestions for Further Reading

Blascovich, J., & Mendes, W. B. (2010). Social psychophysiology and embodiment. In S. T. Fiske, D. T. Gilbert & G. Lindzey (Eds.), *The handbook of social psychology* (5th ed., pp. 194–227). New York: Wiley.

Most recent statement and overview of the biopsychosocial model of challenge and threat.

Blascovich, J., Mendes, W. B., Vanman, E., & Dickerson, S. (2011). *Social psychophysiology for social and personality psychology*. London, UK: Sage.

Excellent primer on the application of psychophysiological measurements to social psychological research.

Branscombe, N. R., Ellemers, N., Spears, R., & Doosje, B. (1999). The context and content of social identity threats. In N. Ellemers, R. Spears & B. Doosje (Eds.), *Social identity: Context, commitment, content* (pp. 35–58). Oxford, UK: Blackwell.

Classic overview of different forms and contextual determinants of social identity threat.

Ellemers, N., & Haslam, S. A. (2012). Social identity theory. In P. Van Lange, A. Kruglanski & T. Higgins (Eds.), *Handbook of theories of social psychology* (pp. 379–398). London, UK: Sage.

Excellent recent overview of the social identity approach.

References

Abrams, D., & Hogg, M. A. (1988). Comments on the motivational status of self-esteem in social identity and intergroup discrimination. *European Journal of Social Psychology, 18*, 317–334.

Bettencourt, B. A., Miller N., & Hume, D. L. (1999). Effects of numerical representation within cooperative settings: Examining the role of salience in in-group favouritism. *British Journal of Social Psychology, 38*, 265–287.

Blascovich, J. (2008). Challenge, threat, and health. In J. Y. Shah & W. L. Gardner (Eds.), *Handbook of motivation science* (pp. 481–493). New York: Guilford Press.

Blascovich, J., & Mendes, W. B. (2000). Challenge and threat appraisals: The role of affective cues. In J. Forgas (Ed.), *Feeling and thinking: The role of affect in social cognition* (pp. 59–82). Cambridge, UK: Cambridge University Press.

Blascovich, J., & Mendes, W. B. (2010). Social psychophysiology and embodiment. In D. Gilbert, S. T. Fiske & G. Lindzey (Eds.), *The handbook of social psychology* (5th ed., pp. 194–227). New York: Wiley.

Blascovich, J., Mendes, W. B., Hunter, S. B., & Salomon, K. (1999). Social "facilitation" as challenge and threat. *Journal of Personality and Social Psychology, 31*, 422–429.

Blascovich, J., & Tomaka, J. (1996). The biopsychosocial model of arousal regulation. In M. P. Zanna (Ed.), *Advances in experimental social psychology* (Vol. 28, pp. 1–51). New York: Academic Press.

Branscombe, N. R., Ellemers, N., Spears, R., & Doosje, B. (1999). The context and content of social identity threats. In N. Ellemers, R. Spears & B. Doosje (Eds.), *Social identity: Context, commitment, content* (pp. 35–58). Oxford, UK: Blackwell.

Branscombe, N. R., & Wann, D. L. (1994). Collective self-esteem consequences of out-group derogation when a valued social identity is on trial. *European Journal of Social Psychology, 24*, 641–657.

Doosje, B., Branscombe, N. R., Spears, R., & Manstead, A. S. R. (1998). Guilty by association: When one's group has a negative history. *Journal of Personality and Social Psychology, 75*, 872–886.

Ellemers, N. (1993). The influence of socio-structural variables on identity management strategies. In W. Stroebe & M. Hewstone (Eds.), *European review of social psychology* (Vol. 4, pp. 27–57). Hove, UK: Psychology Press.

Ellemers, N., & Haslam, S. A. (2012). Social identity theory. In P. Van Lange, A. Kruglanski & T. Higgins (Eds.), *Handbook of theories of social psychology* (pp. 379–398). London, UK: Sage.

Ellemers, N., Spears, R., & Doosje, B. (1997). Sticking together or falling apart: Group identification as a psychological determinant of group commitment versus individual mobility. *Journal of Personality and Social Psychology*, *72*, 123–140.

Ellemers, M., Spears, R., & Doosje, B. (2002). Self and social identity. *Annual Review of Psychology*, *53*, 161–186.

Haslam, S. A., & Reicher, S. D. (2006). Stressing the group: Social identity and the unfolding dynamics of stress. *Journal of Applied Psychology*, *91*, 1037–1052.

Hogg, M. A. (2009). Managing self-uncertainty through group identification. *Psychological Inquiry*, *20*, 221–224.

Jetten, J., Haslam, C., & Haslam, S. A. (Eds.). (2011). *The social cure: Identity, health and well-being*. New York & Hove, UK: Psychology Press.

Jonas, K. J., Sassenberg, K., & Scheepers, D. (2010). Self-regulation within and between groups. *Group Processes and Inter-group Relations*, *13*, 131–136.

Jost, J. T., & Major, B. (Eds.). (2001). *The psychology of legitimacy: Emerging perspectives on ideology, justice, and intergroup relations*. New York: Cambridge University Press.

Maass, A., Cadinu, M., Guarnieri, G., & Grasselli, A. (2003). Sexual harassment under social identity threat: The computer harassment paradigm. *Journal of Personality and Social Psychology*, *85*, 853–870.

Matheson, K., & Cole, B. (2004). Coping with a threatened group identity: Psychological and neuroendocrine responses. *Journal of Experimental Social Psychology*, *40*, 777–786.

Mendes, W. B., Blascovich, J., Lickel, B., & Hunter, S. (2002). Challenge and threat during interactions with White and Black men. *Personality and Social Psychology Bulletin*, *28*, 939–952.

Ouwerkerk, J. W., De Gilder, D., & De Vries, N. K. (2000). When the going gets tough the tough gets going: Social identification and individual effort in intergroup competition. *Personality and Social Psychology Bulletin*, *26*, 1550–1559.

Sapolsky, R. M. (2005). The influence of social hierarchy on primate health. *Science*, *308*, 648–653.

Scheepers, D. (1997). *Wat is dreigend? Een indeling van de soorten identiteitsdreiging in intergroepssituaties*. [What is threatening? A taxonomy of identity threats in intergroup contexts]. Unpublished Master's thesis, University of Amsterdam, the Netherlands.

Scheepers, D. (2009). Turning social identity threat into challenge: Status stability and cardiovascular reactivity during intergroup competition. *Journal of Experimental Social Psychology*, *45*, 228–233.

Scheepers, D. (2013a). Active coping with social identity threat: The importance of a strong social identity. Manuscript in preparation.

Scheepers, D. (2013b). Inter-group status differences as challenge and threat: The moderating role of legitimacy. Manuscript submitted for publication.

Scheepers, D., Branscombe, N. R., Spears, R., & Doosje, B. (2002). The emergence and effects of deviants in low and high status groups. *Journal of Experimental Social Psychology*, *38*, 611–617.

Scheepers, D., & Ellemers, N. (2005). When the pressure is up: The assessment of threats to social identity in low and high status groups. *Journal of Experimental Social Psychology*, *41*, 192–200.

Scheepers, D., Ellemers, N., & Sintemaartensdijk, N. (2009). Suffering from the possibility of status loss: Physiological indicators of social identity threat in high status groups. *European Journal of Social Psychology*, *39*, 1075–1092.

Scheepers, D., Spears, R., Doosje, B., & Manstead, A. S. R. (2003). Two functions of verbal inter-group discrimination: Instrumental and identity motives as a result of group identification and threat. *Personality and Social Psychology Bulletin, 29*, 568–577.

Scheepers, D., Spears, R., Doosje, B., & Manstead, A. S. R. (2006). The social functions of in-group bias: Creating, confirming, or changing social reality. *European Review of Social Psychology, 18*, 359–396.

Steele, C. M., Spencer, S. J., & Aronson, J. (2002). Contending with group image: The psychology of stereotype and social identity threat. In M. P. Zanna (Ed.), *Advances in experimental social psychology* (Vol. 34, pp. 379–440). New York: Academic Press.

Tajfel, H., & Turner, J. C. (1979). An integrative theory of intergroup conflict. In W. G. Austin & S. Worchel (Eds.), *The social psychology of intergroup relations* (pp. 33–47). Monterey, CA: Brooks/Cole.

Van Rijswijk, W., & Ellemers, N. (1998). Emotionele reacties op feedback over de eigen groep. [Emotional response to feedback about one's group]. *Fundamentele Sociale Psychologie, 12*, 86–96.

Wright, R. A., & Kirby, L. D. (2003). Cardiovascular correlates of challenge and threat appraisals: A critical examination of the biopsychosocial analysis. *Personality and Social Psychology Review, 7*, 216–233.

14

PHYSIOLOGICAL AND SELF-REPORT MEASURES OF STRESS AND COPING IN THE STUDY OF STIGMA

Rodolfo Mendoza-Denton and Elizabeth Page-Gould

In *The Nature of Prejudice*, Gordon Allport's (1954) book that has inspired several generations of researchers on the psychological roots and consequences of prejudice and stigma (e.g., Dovidio, Glick, & Rudman, 2005), Allport (p. 42) posed this poignant question: "What would happen to your personality if you heard it said over and over again that you are lazy and had inferior blood?"

We have been drawn to this question (e.g., Mendoza-Denton, Page-Gould, & Pietrzak, 2006) because it suggests that individual differences play an important role in the psychology of stigmatization. This idea, as we describe below, has received strong support in our program of research. Further, when it comes to the question of how targets of stigma are affected by prejudice (e.g., being told one is lazy and has inferior blood), the recent introduction of physiological approaches has breathed new life into a question that historically has been addressed through self-report measures.

One of the primary reasons that psychophysiological measures have been embraced by intergroup researchers is that they help us address some of the self-presentational biases that affect self-report measures. Although self-presentational concerns are traditionally associated with prejudice from the perpetrator's perspective (e.g., American Psychological Association [APA], 2012; Butz & Plant, 2009; Plant & Devine, 2009), there is reason to believe that self-presentational concerns are also relevant to the stigmatized, such that people who are stigmatized may be motivated not to show how much prejudice affects or hurts them (Barger, Marsland, Bachen, & Manuck, 2000; Mendes, Major, McCoy, & Blascovich, 2008; N. Mendoza-Denton, 2009). To the degree that physiological stress is difficult to control, measures that capture physiological responses may facilitate the interpretation of stigmatized individuals' self-report responses.

While the effects of prejudice on targets are well documented (Crocker, Major, & Steele, 1998) and the physiological consequences of stigmatization are increasingly well understood (e.g., Ratner, Halim, & Amodio, 2013), less is known about the specific *interrelationships* between self-report and physiological measures from the perspective of targets. In other words, little research has specifically addressed the correspondence (or lack thereof) between self-report and physiological measures with the same specificity that the topic has garnered from the perpetrator's perspective. Do self-report and uncontrollable measures converge or diverge in general among targets of prejudice? Are there certain contexts or groups for whom they diverge and for whom they converge? What does it mean for the subjective and physiological levels of experience to coalesce or diverge? These questions are only beginning to be addressed in the literature (e.g., Mendes et al., 2008), but much remains to be learned.

There is reason to suspect that comparing subjective and physiological measures would not only be interesting, but that altogether, measuring *multiple* levels of experience will provide a broader picture of the coping mechanisms underlying stigmatization. In this chapter, we explore one domain of research on the experience of stigma that seems ripe for such comparisons. This domain centers around self-protective processes among African Americans specifically, and members of minority groups more generally.

A Brief History of Research on Self-protection Among African Americans

"There is ample evidence of inferiority feelings and morbid self-hate in all minority groups," wrote Erik Erikson in 1956 (p. 115), and Cartwright (1950) noted that "self-hatred and feelings of helplessness tend to arise from membership in under-privileged or outcast groups" (p. 440). Implicit in these statements (as well as Allport's [1954] own discussions) was the dominant belief at the time that targets of prejudice *internalize* the negative attitudes and stereotypes about them, thus leading to feelings of low self-worth. Evidence for this proposition came from clinical observations (e.g., Erikson, 1956), as well as from empirical demonstrations. Kenneth and Mamie Clark (1947), in what is now famously known simply as their "doll study," asked young Black and White children in the United States to choose between two dolls—one that looked White, and another that looked Black. When asked to point to the doll that "looks like you," the children overwhelmingly chose the doll that indeed corresponded to their own racial group, suggesting that the children understood the concept of race, and categorized both themselves and the dolls accordingly. However, when asked to choose the doll that "looks bad," the children overwhelmingly chose the Black doll, and when asked to choose the doll that "you like best," most children chose the White doll.

Most poignant were the reasons that the Black children gave for rejecting the Black doll. "Cause he's ugly," reasoned one child, while another was more direct: "cause he's got black on him." As Clark and Clark (1947) wrote, one light-skinned, Black 7-year-old child was at pains to convince the researchers he was

White, but looked "brown because [he] got a suntan in the summer." (p. 178). The conclusion that negative stereotypes are internalized by members of minority groups, with adverse consequences for the self-concept, seemed self-evident.

For one of the authors (Mendoza-Denton), the notion of internalization resonated with personal experience. When he was 10, his father got a job with UNICEF, the United Nations International Children's Emergency Fund. Although his father's assignment was eventually to be in Abidjan, Ivory Coast (an experience that helped shape Mendoza-Denton's interest in intergroup relations; see Mendoza-Denton & Hansen, 2007), a year of training at the UN headquarters in New York City was required. The family moved for a year to Westport, Connecticut, chiefly because it was a town directly located along the Red Line of the Amtrak train on which his father could commute. Little did the family anticipate the chilly welcome that awaited them: While Mendoza-Denton and his sister were playing in the yard of the house they rented, a neighbor walked up and scolded his sister for being where she did not belong, while little brother hid in a nearby bush. Over the course of that year, he recalls desperately wanting to fit in, wishing he were White, while his sister went as far as to scrub her skin red with a hard sponge so as to appear to have "rosy cheeks." Thankfully, responses to stigma change and grow along with the individual who experiences them.

Years later, in 1989, Jennifer Crocker and Brenda Major provided a seminal challenge to the notion of internalization—and it is here that the potential to compare self-report versus psychophysiological measures takes seed. According to Crocker and Major's (1989) hypothesis of the self-protective properties of stigma, when targets of stigma or prejudice encounter negative messages about themselves or their groups, targets may be motivated to *discount* this negative information, thereby rendering such information non-informative about one's qualities. The protection of the self occurs specifically in reaction to negative outcomes for which there is *attributional ambiguity*—that is, ambiguity as to whether one should attribute negative outcomes to one's abilities or to societal prejudice (see also Major & Kunstman, Chapter 17 in this volume).

Crocker, Voelkl, Testa, and Major (1991) provided an early empirical test of how attributional ambiguity is resolved by some minority group members in favor of self-protection. African American participants were invited to partake in a study on impression formation, and were led to believe that another person (whom participants thought was White) was already sitting in another room on the other side of a one-way mirror that participants could not see through. In one condition, the "blinds up" condition, participants were told that the blinds would remain up because visual cues are a critical part of the impression-formation process (thereby reinforcing the notion that the purported "other participant" could see them). In the "blinds down" condition, participants were given the equally plausible explanation that the researchers were interested in the impression-formation process independently of physical attributes. The (real) participants filled out some brief attitudinal surveys that were then "taken" to the other person for examination, and subsequently the experimenter returned with a similar attitudinal survey from the purported other participant. This survey revealed that the other person was either quite willing, or quite unwilling, to meet the participant face-to-face. The participants' self-esteem was measured both before and after the manipulation.

The results confirmed that in the "blinds down" condition—that is, when participants could not attribute the interpersonal feedback that they received to their race—self-esteem rose with positive feedback and fell with negative feedback. These findings are as one might expect, particularly given advances in theory and research suggesting that self-esteem acts as a "sociometer" of whether one is accepted or rejected by others (Leary, 2000). In the "blinds up" condition, however, the data suggested that participants experienced attributional ambiguity, not knowing whether their partner disliked them because of prejudice, or because of something the participant him/herself had indicated on the attitude survey. As predicted, when given negative feedback, participants' self-esteem did not change—that is, when faced with attributional ambiguity, they discounted the negative feedback, leaving self-esteem intact.[1] The hypothesis of the self-protective properties of stigma has generated an impressive amount of research (e.g., Aronson & Inzlicht, 2004; Hoyt, Aguilar, Kaiser, Blascovich, & Lee, 2007; Major, Quinton, & McCoy, 2002; Mendoza-Denton, Goldman-Flythe, Pietrzak, Downey, & Aceves, 2010; see also Major & Kunstman, Chapter 17 in this volume).

At the level of self-report measures, Major, Kaiser, and McCoy (2003) report evidence that attributions to discrimination, while protecting self-esteem, continue to leave targets vulnerable to anger. Crocker, Cornwell, and Major (1993) have identified the critical moderating role of perceived responsibility in the link between discrimination and self-esteem protection, such that those stigmatized targets who may blame themselves for their condition (e.g., overweight individuals; see Crocker et al., 1993) do not experience self-esteem protection. Nevertheless, in a comprehensive meta-analysis, Twenge and Crocker (2002) confirmed that, consistent with the hypothesis of the self-protective properties of stigma, African Americans today reliably report higher self-esteem relative to Whites in the United States.

Nevertheless, these higher ratings of *subjective* well-being among African Americans seem to fly in the face of well-documented disparities in physical health (see APA, 2012). Moreover, meta-analytic results show that the burden of ethnic health disparities is not ubiquitous across the African American population, but rather is the greatest among the African Americans who report experiencing the most discrimination (Williams, Neighbors, & Jackson, 2003). Thus the question arises: Are perceptions of discrimination a *risk* factor—as would be suggested by research on discrimination and health; or a *buffering* factor—as would be suggested by attributional ambiguity?

The Self-protective Qualities of Stigma: New Questions and Opportunities

Recent developments examining physiological responses among the stigmatized underscore the complexity of people's reactions to discrimination, while also inviting new questions about the interpretation of the extant self-report data. Sometimes self-report measures converge with physiological measures and sometimes they do not, suggesting that the physiological and subjective levels of experience are at least partially distinct. In our own work, we first turned to psychophysiological measures out of a desire to capture a richer, more nuanced image of intergroup experiences. In the service of this goal, we strive to compare between and integrate across different modes of experience (e.g., subjective, visceral). As psychophysiological measures become more mainstream, we hope that social psychologists will not abandon classic self-report and reaction-time (RT) methods, as these metrics, together, give us a more comprehensive view for the interpretation of physiological responses.

We have drawn our work primarily from the biopsychosocial model of challenge and threat (see Blascovich, Chapter 12 in this volume; Blascovich, Mendes, Tomaka, Salomon, & Seery, 2003), drawing heavily on stress and coping theory (see Folkman, Lazarus, Dunkel-Schetter, DeLongis, & Gruen, 1986; Trawalter, Richeson, & Shelton, 2009). From this perspective, people evaluate the demands of a given situation and then determine whether they have what is needed (i.e., the "resources") to overcome the situational demands. If you appraise your resources as exceeding the situational demands, then you respond with a pattern of responses across the cardiovascular and hormonal stress systems that is referred to as "challenge." Conversely, you respond with a pattern of responses termed "threat" when the demands appear to outweigh your resources. At a physiological level, only one of the stress systems, the sympathetic–adrenal–medullary (SAM) axis, is activated during challenge responses, but both the SAM and the hypothalamic–pituitary–adrenal (HPA) axes are activated during threat. As such, physiological changes associated with the activation of the SAM and the HPA axes allow for inferences related to challenge and threat.

The SAM and HPA axes trigger both hormonal and autonomic stress responses, which allows for nice empirical flexibility. Although the ideal psychophysiological studies of challenge and threat would look for patterns of physiological data to evaluate challenge and threat, neuroendocrine approaches to challenge and threat must rely on one metric for each stress axis that can be measured non-invasively in human saliva. That being said, SAM activation is reflected by the presence of an enzyme known as α-amylase approximately 5 minutes after the onset of a stressor. Hypothalamic–pituitary–adrenal activation is reflected by the presence of the glucocorticoid cortisol approximately 18–20 minutes after the onset of a stressor. The important thing to note is that you must collect one saliva sample 5 minutes after the onset of the stressor to measure the α-amylase response, and one saliva sample 20 minutes after the onset of the stressor to measure the comparable cortisol response. Although they are measured close to 15 minutes apart, the values of each will reflect participants' stress responses at approximately the same moment in time. The method is simple because participants simply expectorate through a straw into a small, 2-ml vial at the key times after onset.

If you have access to physiological hardware that measures the autonomic nervous system—the peripheral nervous system regulating your visceral organs like the heart, lungs, and skin—then the key measures of interest for challenge and threat patterns are heart rate (HR), the strength of the heart contraction (measured by and inversely related to the pre-ejection period [PEP]), the volume of blood in the torso (cardiac output [CO]), and either blood pressure (BP) or pulse plethysmograph (PPG). Heart rate can be measured with electrocardiography (ECG), and both CO and PEP can be captured with vascular impedance. You will also want to collect either BP or PPG to estimate vascular constriction.

As described in detail by Blascovich and colleagues (Blascovich, Spencer, Quinn, & Steele, 2001; Blascovich et al., 2003, see also Blascovich, Chapter 12 in this volume), two key cardiovascular measures differentiate challenge and threat states: the volume of blood circulated by the heart and the relative dilation or constriction of the vascular system (e.g., arteries, capillaries). Cardiac output increases significantly above baseline with challenge and either stays the same or decreases with threat. However, with threat, the vasculature constricts, restricting the flow of blood throughout the body (thus the feeling of cold, clammy extremities); but with challenge, the vasculature dilates, reducing resistance to the flow of blood throughout the body (akin to when one feels, literally, "pumped"). Vascular constriction can be estimated in two ways, either by deriving total peripheral resistance (TPR) from CO and BP with the formula $TPR = 80 \times \text{mean arterial pressure} \div CO$, or by calculating pulse amplitude from the pulse plethysmograph. Altogether, when a challenge pattern is observed, the heart is performing efficiently by beating quickly, strongly, and circulating large volumes of blood. When a pattern of threat is observed, the heart is also beating faster, but is less efficient, emitting less blood on each beat and restricting the flow of blood through the vasculature.

In a recent, important addition to this literature, Mendes et al. (2008) have further clarified that activational challenge states can be further differentiated into those that are associated with positive emotions, and those that are associated with what may not be a "positive" response—anger. While both of these challenge states facilitate performance, both ventricular contractility and heart rate seem to be reliably higher when an individual is experiencing anger as opposed to a more positive challenge state. This distinction is consistent with other conceptualizations of anger as an approach-related response (Chan & Mendoza-Denton, 2008), and may help illuminate, in future research, the role of anger in discrimination alongside the role of threat.

The role of cognitive appraisals in predicting responses to stress introduces a key pathway through which social cognitive factors may affect bodily responses. Our own work on race-based rejection sensitivity (Mendoza-Denton, Downey,

Purdie, Davis, & Pietrzak, 2002; see also below) has shown how people who are high in race-based rejection sensitivity are vigilant for discrimination in intergroup interactions. Thus their stress responses can be understood in part by an amplification of the perceived demands of intergroup interaction.

Among those high in race-based rejection sensitivity, being at risk for experiencing discrimination (i.e., in an intergroup interaction) is associated with greater cortisol reactivity (Mendes, Blascovich, Lickel, & Hunter, 2002; Page-Gould, Mendoza-Denton, & Tropp, 2008). Barger et al. (2000) suggest that benign self-reported reactions to negative events (such as discrimination) may signal a repressive coping style that exacts an endocrine and immunological toll. This idea is supported by data (mentioned above) showing marked health disparities in cardiovascular health and other ailments between stigmatized and non-stigmatized groups (e.g., APA, 2012).

Mendes et al. (2008) succinctly phrase the potential of using both physiological and self-report measures to further our understanding of targets' reactions to stigma:

> Are participants simply disengaging from the social interaction with a prejudiced person? Are the self-esteem effects observed due simply to individuals masking the pain of being a target of discrimination and self-reporting that they are not affected by the discrimination when, in reality, they are feeling anxiety and threat? Is buffered self-esteem a defensive reaction?
>
> *(Mendes et al., 2008, p. 280)*

These kinds of questions seem ripe for a multi-method approach where self-report and physiological data inform and constrain the interpretation of one another. Mendes et al. (2008), for example, were able to use physiological markers of challenge and threat states as described above to disentangle two interpretations of self-esteem protection following attributions to discrimination. One interpretation suggests that protected self-esteem results from a shifting of blame from oneself to another (see e.g., Major et al., 2003), while another suggests that protected self-esteem is a masking response that hides underlying distress (e.g., Barger et al., 2000). Mendes et al. (2008) found that participants who experienced discrimination showed an activational anger challenge response (see discussion above), consistent with the shifting-of-blame explanation. For positive feedback, Mendes et al. (2008) observed a cardiovascular pattern of threat, suggesting that positive feedback in the face of minority status may be met with suspicion by targets of stigma.

A Role for Individual Differences: Race-based Rejection Sensitivity

Mendoza-Denton et al. (2010) have similarly found evidence of self-esteem protection while also considering the effect of individual differences in *race-based rejection sensitivity* (RS-race, Mendoza-Denton et al., 2002). Race-based rejection sensitivity refers to a cognitive–affective dynamic whereby direct or vicarious experiences of race-based discrimination, mistreatment, or prejudice lead people to anxiously expect similar treatment in future applicable situations, such as interactions with out-group members or contexts in which one's identity is undervalued. These anxious expectations have been hypothesized to activate what Lang, Bradley, and Cuthbert (1990) term the defensive motivational system, whereby the threshold for perceiving the noxious stimulus is lowered as a coping response against threat. Perceptions of discrimination are then followed by intense cognitive and affective reactions to the perceived discrimination. Physiological correlates of RS-race include cortisol reactivity in the short term (Page-Gould et al., 2008) as well as more chronic physical symptoms associated with stress (Page-Gould, Mendoza-Denton, & Mendes, in press).

Mendoza-Denton et al. (2010), in a similar way to Crocker et al. (1991), examined African American students' reactions to negative versus positive feedback under conditions in which they either disclosed or did not disclose their race to a purported evaluator. Drawing on a procedure described by Cohen, Steele, and Ross (1999), participants were asked to write a position essay that they thought would be evaluated by a White teaching assistant at their (predominantly White) university. The race-disclosure manipulation consisted of having participants fill out a demographic form that either included an item about race or did not include this item. Additionally, as discussed above, the researchers measured participants' RS-race, and examined participants' changes in performance self-esteem (Heatherton & Polivy, 1991) following positive or negative feedback.

Analogous to Crocker et al.'s (1991) "blinds down" condition, when participants had not disclosed their race to the evaluator, only a main effect of feedback was observed, such that participants who received positive feedback reported greater performance self-esteem than those who received negative feedback. The fact that no effect of RS-race was observed is consistent with the conceptualization of the dynamic as being activated only in applicable contexts (Mendoza-Denton et al., 2002): In the context of not having disclosed one's race, anxious expectations of race-based rejection did not predict reactions to feedback. A different story emerged when participants did disclose their race, however. Here, for participants higher in RS-race, the type of feedback that they received— either positive or negative— did not affect their self-reported self-esteem. This suggests that among participants highly concerned about discrimination, discounting of the feedback was at work. As expected, for participants low in RS-race, self-esteem continued to rise or fall depending on whether they received positive or negative feedback, respectively.

The findings from Mendoza-Denton et al. (2010) raise the following question: How might negative versus positive feedback affect participants high and low in RS-race at a physiological level? By all accounts, given research on attributional ambiguity as well as the findings from Mendoza-Denton et al. (2010) described above, one might expect that discounting would afford some type of physiological protection against negative feedback. A similar prediction might be made from accounts of disengagement (see Major & Schmader, 1998), which specifically posit that, as a coping mechanism, targets of stigma may psychologically disengage from stressful evaluative tasks pre-emptively, particularly in domains where their identities are not valued. These accounts would predict that participants high in RS-race would show *reduced* physiological reactivity following negative feedback.

And yet, as we have anticipated throughout, other lines of research, particularly those relying on physiological measures, suggest the *opposite* hypothesis—that is, for participants high in RS-race, there should be a distinct lack of correspondence between physiological and self-report measures. Blascovich et al. (2001), in a directly relevant study, measured mean arterial pressure (MAP) among African American and White students over three brief testing sessions, with a minute of rest in between each session. Participants were tested under conditions of high or low stereotype threat, analogous to traditional stereotype-threat manipulations (see Steele, 1997). For the African American participants in the high stereotype-threat condition, but not in the low stereotype-threat condition, MAP increased significantly from baseline over the course of the three testing sessions—even during the break periods. To the degree that participants in the high threat condition are analogous to individuals high in RS-race, it seems reasonable to predict that under similar conditions, people high in RS-race might exhibit similar MAP increases. Further, as mentioned, in our own research (Page-Gould et al., 2008), we have found that participants high in RS-race exhibited greater cortisol reactivity during an initial meeting designed to promote friendship with an out-group partner. Thus, these lines of evidence would suggest that participants high in RS-race might exhibit *the greatest* physiological threat reactivity in response to negative feedback.

To summarize, then, one line of research would suggest that participants high in RS-race should show reduced physiological threat reactivity relative to their low RS-race counterparts following negative feedback; while another would suggest precisely the opposite. Mendoza-Denton, Page-Gould, and Mendes (2009) conducted a preliminary study to address these alternative hypotheses. We conducted a Trier Social Stress Test (TSST) (Kirschbaum, Pirke, & Hellhammer, 1993) among both African Americans and European Americans (we focus on the data for African Americans here). Briefly, the TSST is a well-established procedure for examining stress responses in motivated performance situations, whereby study participants are asked to perform difficult verbal and arithmetic tasks in front of an audience that usually consists of "judges" who are generally confederates or experimenters. One of the advantages of the TSST, for our purposes, is that one can manipulate both the type of feedback that is given, as well as the group membership of the judges to create an intergroup versus intragroup performance context. In this study, we manipulated the race of the judges to be either White or Black, creating an out-group/in-group performance context for our African American participants. The judges in our study were instructed to provide only neutral feedback, consisting of head nods and other signs of listening/engagement, but excluding both smiling or encouragement as well as frowning or negative commentary.

The procedure was as follows: We brought participants into the laboratory, and hooked them up to psychophysiological equipment that allowed us to measure cardiovascular reactivity (see Mendes et al., 2008). Once participants were hooked up to the equipment, participants sat quietly in a room by themselves for 5 minutes. Participants' physiological signals during this time were used as a baseline against which to compare participants' reactions during the other sections of the study. Following this baseline period, an experimenter came into the lab and informed participants that their task would be to give a 3-minute speech in front of two judges, arguing why they were well-suited for a job. The judges came into the room and introduced themselves briefly to the participant. The participant was then left alone for another 5 minutes to prepare for his or her speech. Following this "prep" period, the judges came back to the room and sat down opposite the participant, armed with clipboards and pens. Again, their feedback was neutral while the participant delivered his or her speech. At the end of the speech segment, the experimenter came back into the room and announced a new surprise task: to assess the participant's qualifications more fully, the participant was to count backward in front of the judges. The task was re-calibrated by the judges (e.g., the participant was asked to count backward by 7s or 5s depending on his/her ease with the task) so that all participants experienced a moderate to high feeling of difficulty. Following this mathematical task, a 5-minute "recovery" phase ensued, during which participants filled out various questionnaires, but continued to be hooked up to the psychophysiological equipment. Following Blascovich, Seery, Mugridge, Norris, and Weisbuch (2004), we computed a physiological "threat" index for each segment of the study.

Our initial expectation was that we would observe physiological reactivity consistent with threat responses among the African American participants high in RS-race when specifically paired with White judges; this hypothesis was based on our previous work showing heightened physiological reactivity among high RS-race participants during our friendship induction and their greater likelihood of perceiving discrimination in applicable situations. Consistent with the idea that RS-race becomes applicable only when the possibility of race-based discrimination is present (Mendoza-Denton et al., 2002), no differences as a function of RS-race emerged among participants in the same-race condition (i.e., when both the participant and the judges were African American). However, the findings in the cross-race condition surprised us: Higher threat scores were significantly—but *negatively*—associated with RS-race, such that participants lower in RS-race exhibited the greatest threat reactivity during this TSST. Particularly striking was the finding that the negative association between threat and RS-race was greatest during the recovery phase, suggesting that participants lower in RS-race were, in fact, the ones who were least able to recover.

These findings were contrary to our original hypotheses—and yet, they make sense within the framework of attributional ambiguity. Seen within this context, a relative lack of threat reactivity among high RS-race participants might be interpreted as successful disengagement of the self-concept from this otherwise aversive task—a successful "online" discounting of the out-group judges' reactions.

An informative aspect of the study was the participants' perceptions of racism at the end of the experimental session: Only a main effect of condition was observed, such that all of our African American participants, regardless of RS-race, perceived higher discrimination in the cross-race condition relative to the same-race condition. So, participants' physiological responses reflect the way the body responds during experiences of discrimination. This suggests that the "neutral" feedback that the judges provided tended to be seen as negative, and raises the possibility that participants who were high in RS-race may have been physiologically *prepared to expect and react protectively to this negativity.* This is consistent with prior research suggesting that unexpected stimuli reliably evoke greater threat reactivity compared to expected stimuli (Mendes, Blascovich, Hunter, Lickel, & Jost, 2007). This notion may also help to explain the threat reactivity for lower RS-race participants: To the degree that they are low in race-based rejection expectations, the perception that the out-group judges were prejudiced may have been *unexpected and thus threatening.*

It is important to note that these preliminary findings are far from definitive; they invite more questions than they provide answers. Yet that is precisely the point: The comparison of the psychophysiological measures to participants' self-reports of discrimination helps us interpret the findings and provides possibilities about the phenomenology of people who are high versus low in RS-race in this situation. The findings both refine our theoretical conceptualization of RS-race and provide valuable (albeit not yet conclusive) data on the reach of the self-protective properties of stigma. We believe that future research comparing self-report and biological measures specifically among targets of stigma will help scientists better understand the psychological sequelae of stigmatization, and open up new avenues for research and theory in this area as it has done for research on prejudice from the actor's perspective.

Conclusion

How do we reconcile research suggesting heightened physiological activity among targets of stigma with the well-documented self-protective properties of stigma? We suggest here that such reconciliation has yet to come about—and is an exciting new frontier in the study of prejudice from the target's perspective. Below are some unanswered questions on prejudice from the target's perspective that research employing physiological and self-report measures together might help to address. Note that these questions are interrelated and not mutually exclusive:

1. *The degree to which self-protection is a self-presentational strategy.* Analogous to the dynamic of "smile now, cry later" that has been documented among Latina girl-gang members (N. Mendoza-Denton, 2009), the question of how deeply the self-protective properties of stigma apply, remains, in our view, largely unresolved in the literature on stigma from the target's perspective.
2. *Short-term versus longer-term risk and benefits.* Our understanding of targets' reactions to stigma at both a self-report and psychophysiological level is still in its infancy. One big question that, in our view, deserves serious attention is the longer-term effects of a given coping strategy. For example, if a person "downplays" the effect of a discriminatory comment, but is in a state of high physiological activation, does this place the person at risk for future physical complications? The data on cardiovascular health of African Americans suggests that this may be a possibility. However, controlled, longitudinal studies that incorporate both biological and self-report measures are in order.
3. *The "reach" of self-esteem protection.* Major et al. (2003) postulate that attributional ambiguity does not protect against anger or other highly arousing emotions. Similarly, Mendes et al. (2008) find that intergroup interactions involving attributional ambiguity trigger an activational challenge response, often associated with anger. Research examining physiological outcomes would help to clarify the boundary conditions (in terms of the emotions or processes that are "protected") of the self-protective properties of stigma.
4. *The degree of effort it takes to regulate self-protective processes.* Related to Question 2 above, how cognitively (or even physiologically) taxing is self-protection? How do different depletion manipulations affect physiological versus self-report measures of self-protection?
5. *Is there a developmental trajectory to the self-protective properties of stigma?* Davis (2005) vividly replicated Kenneth and Mamie Clark's (1947) findings of Black doll rejection among Black children in modern-day New York City. Do children not have as sophisticated self-presentational strategies as adults, or does a self-protective capacity only develop at a later age?
6. *How do different modes of experience interact with and affect each other?* To what extent do subjective experiences drive physiological responses? Even more nascent is the "So what?" question of whether physiological responses to intergroup interactions play any role in our experiences of them. On the one hand, ample research has demonstrated that people are more likely to avoid out-group members following a stressful or anxiety-laden intergroup interaction (Dovidio, Gaertner, Kawakami, & Hodson, 2002; Plant & Devine, 2009). This would suggest that awareness of physiological stress during an intergroup interaction would amplify negative experiences. However, recent research suggests that people behave pro-socially when they believe that they are physiologically stressed, even in an intergroup context (Gu, Zhong, & Page-Gould, in press, Study 1). Altogether, it is unclear whether physiological stress affects intergroup experience at all or rather is just a reaction. If

physiological responses *do* affect subjective experiences in intergroup interactions, then it is not clear if the impact is positive or negative.

Currently, these questions are difficult to address, because self-report data and physiological findings have tended to occur in different studies. As mentioned earlier, in research on the perpetrator's perspective, a lack of correspondence between self-report and physiological measures has helped researchers get a better handle on issues of self-report biases as well as unconscious prejudices. In research on the target's perspective, the field stands to gain many similar theoretical and empirical insights from studies that include physiological as well as self-report data.

Note

1 Interestingly, for positive feedback, the results revealed that the African American participants' self-esteem actually plummeted, suggesting that the participants indeed discounted the positive feedback as insincere or paternalistic. At the same time, the drop in self-esteem suggested that something other than self-esteem protection was happening for positive feedback. As research interest on attributional ambiguity grew, however, the focus remained mostly on negative feedback, in part because of an interest in understanding the effects of negative messages (e.g., stereotypes) on targets. See Mendes et al. (2008) and Hoyt et al. (2007) for an expanded discussion of the effect of positive feedback on targets of stigma.

Suggestions for Further Reading

Crocker, J., & Major, B. (1989). Social stigma and self-esteem: The self-protective properties of stigma. *Psychological Review, 96*, 608–630.

Crocker and Major's classic theoretical piece on attributional ambiguity.

Major, B., Kaiser, C. R., & McCoy, S. K. (2003). It's not my fault: When and why attributions to discrimination protect self-esteem. *Personality and Social Psychology Bulletin, 29*, 772–781.

An article that nicely summarizes and expands on debates regarding the boundary conditions of attributional ambiguity.

Mendes, W. B., Major, B., McCoy, S., & Blascovich, J. (2008). How attributional ambiguity shapes physiological and emotional responses to social rejection and acceptance. *Journal of Personality and Social Psychology, 94*, 278–291.

An excellent article that covers some of the basic methodology related to measuring cardiovascular responses, and includes both self-report and physiological measures. Additionally, this research introduces an important distinction between an activational challenge pattern related to positive emotions and an activational challenge pattern related to anger.

Mendoza-Denton, R., Page-Gould, E., & Pietrzak, J. (2006). Mechanisms for coping with status-based rejection expectations. In S. Levin & C. Van Laar (Eds.), *Stigma and group inequality: Social psychological approaches* (pp. 151–169). Mahwah, NJ: Erlbaum.

An introduction to theory and research on race-based rejection sensitivity.

Page-Gould, E., Mendoza-Denton, R., & Tropp, L. (2008). With a little help from my cross-group friends: Reducing intergroup anxiety through cross-group friendship. *Journal of Personality and Social Psychology, 95*, 1080–1094.

A report that outlines our procedures for salivary assays of cortisol.

References

Allport, G. W. (1954). *The nature of prejudice*. Oxford, UK: Addison-Wesley.

American Psychological Association, APA Presidential Task Force on Reducing and Preventing Discrimination Against and Enhancing Benefits of Inclusion of People Whose Social Identities Are Marginalized in U.S. Society. (2012). *Dual pathways to a better America: Preventing discrimination and promoting diversity*. Washington, DC: Author.

Aronson, J., & Inzlicht, M. (2004). The ups and downs of attributional ambiguity: Stereotype vulnerability and the academic self-knowledge of African American college students. *Psychological Science, 15*, 829–836.

Barger, S. D., Marsland, A. L., Bachen, E. A., & Manuck, S. B. (2000). Repressive coping and blood measures of disease risk: Lipids and endocrine and immunological responses to a laboratory stressor. *Journal of Applied Social Psychology, 30*, 1619–1638.

Blascovich, J., Mendes, W. B., Tomaka, J., Salomon, K., & Seery, M. (2003). The robust nature of the biopsychosocial model of challenge and threat: A reply to Wright and Kirby. *Personality and Social Psychology Review*, 7, 234–243.

Blascovich, J., Seery, M. D., Mugridge, C. A., Norris, R. K., & Weisbuch, M. (2004). Predicting athletic performance from cardiovascular indexes of challenge and threat. *Journal of Experimental Social Psychology, 40*, 683–688.

Blascovich, J., Spencer, S. J., Quinn, D., & Steele, C. (2001). African Americans and high blood pressure: The role of stereotype threat. *Psychological Science, 12*, 225–229.

Butz, D. A., & Plant, E. A. (2009). Prejudice control and interracial relations: The role of motivation to respond without prejudice. *Journal of Personality*, 77, 1311–1342.

Cartwright, D. (1950). Emotional dimensions of group life. In M. L. Raymert (Ed.), *Feelings and emotions* (pp. 439–447). New York: McGraw-Hill.

Chan, W., & Mendoza-Denton, R. (2008). Status-based rejection sensitivity among Asian Americans: Implications for psychological distress. *Journal of Personality, 76*, 1317–1346.

Clark, K. B., & Clark, M. (1947). Racial identification and preference in Negro children. In T. M. Newcomb and E. L. Hartley (Eds.), *Readings in social psychology* (pp. 169–178). New York: Holt.

Cohen, G. L., Steele, C. M., & Ross, L. D. (1999). The mentor's dilemma: Providing critical feedback across the racial divide. *Personality and Social Psychology Bulletin, 25*, 1302–1318.

Crocker, J., Cornwell, B., & Major, B. (1993). The stigma of overweight: Affective consequences of attributional ambiguity. *Journal of Personality and Social Psychology, 64*, 60–70.

Crocker, J., & Major, B. (1989). Social stigma and self-esteem: The self-protective properties of stigma. *Psychological Review, 96*, 608–630.

Crocker, J., Major, B., & Steele, C. (1998). Social stigma. In D. T. Gilbert & S. T. Fiske (Eds.), *The handbook of social psychology* (4th ed., pp. 504–553). New York: McGraw-Hill.

Crocker, J., Voelkl, K., Testa, M., & Major, B. (1991). Social stigma: The affective consequences of attributional ambiguity. *Journal of Personality and Social Psychology, 60*, 218–228.

Davis, K. (Director). (2005). *A girl like me* [Short motion picture]. United States: Reel Works Teen Filmmaking.

Dovidio, J. F., Gaertner, S. L., Kawakami, K., & Hodson, G. (2002). Why can't we just get along? Interpersonal biases and interracial distrust. *Cultural Diversity and Ethnic Minority Psychology, 8*, 88–102.

Dovidio, J. F., Glick, P., & Rudman, L. A. (Eds.). (2005). *On the nature of prejudice: 50 years after Allport*. Malden, MA: Blackwell Publishing.

Erikson, E. (1956). The problem of ego-identity. *Journal of the American Psychoanalytic Association, 4*, 56–121.

Folkman, S., Lazarus, R. S., Dunkel-Schetter, C., DeLongis, A., & Gruen, R. J. (1986). Dynamics of a stressful encounter: Cognitive appraisal, coping, and encounter outcomes. *Journal of Personality and Social Psychology, 50*, 992–1003.

Gu, J., Zhong, C.-B., & Page-Gould, E. (in press). Listen to your heart: When false somatic feedback shapes moral behavior. *Journal of Experimental Psychology: General.*

Heatherton, T. F., & Polivy, J. (1991). Development and validation of a scale for measuring state self-esteem. *Journal of Personality and Social Psychology, 60*, 895–910.

Hoyt, C. L., Aguilar, L., Kaiser, C., Blascovich, J., & Lee, K. (2007). The self-protective and undermining effects of attributional ambiguity. *Journal of Experimental Social Psychology, 43*, 884–893.

Kirschbaum, C., Pirke, K. M., & Hellhammer, D. H. (1993). The "Trier Social Stress Test": A tool for investigating psychobiological stress responses in a laboratory setting. *Neuropsychobiology, 28*, 76–81.

Lang, P. J., Bradley, M. M., & Cuthbert, B. N. (1990). Emotion, attention, and the startle reflex. *Psychological Review, 97*, 377–395.

Leary, M. (2000). *The nature and function of self-esteem: Sociometer theory*. Burlington, MA: Elsevier.

Major, B., Kaiser, C. R., & McCoy, S. K. (2003). It's not my fault: When and why attributions to discrimination protect self-esteem. *Personality and Social Psychology Bulletin, 29*, 772–781.

Major, B., Quinton, W. J., & McCoy, S. K. (2002). Antecedents and consequences of attributions to discrimination: Theoretical and empirical advances. *Advances in Experimental Social Psychology, 34*, 251–330.

Major, B., & Schmader, T. (1998). Coping with stigma through psychological disengagement. In J. K. Swim & C. Stangor (Eds.), *Prejudice: The target's perspective* (pp. 219–241). San Diego, CA: Academic Press.

Mendes, W., Blascovich, J., Hunter, S., Lickel, B., & Jost, J. (2007). Threatened by the unexpected: Physiological responses during social interactions with expectancy-violating partners. *Journal of Personality and Social Psychology, 92*, 698–716.

Mendes, W. B., Blascovich, J., Lickel, B., & Hunter, S. (2002). Challenge and threat during social interactions with White and Black men. *Personality and Social Psychology Bulletin, 28*, 939–952.

Mendes, W. B., Major, B., McCoy, S., & Blascovich, J. (2008). How attributional ambiguity shapes physiological and emotional responses to social rejection and acceptance. *Journal of Personality and Social Psychology, 94*, 278–291.

Mendoza-Denton, N. (2009). *Homegirls: Language and cultural practice among Latina youth gangs*. Malden, MA: Blackwell.

Mendoza-Denton, R., Downey, G., Purdie, V. J., Davis, A., & Pietrzak, J. (2002). Sensitivity to status-based rejection: Implications for African American students' college experience. *Journal of Personality and Social Psychology, 83*, 896–918. doi: 10.1037// 0022-3514.83.4.896

Mendoza-Denton, R., Goldman-Flythe, M., Pietrzak, J., Downey, G., & Aceves, M. (2010). Group value ambiguity: Understanding the effects of academic feedback on minority students' self-esteem. *Social Psychological and Personality Science, 1*, 127–135.

Mendoza-Denton, R., & Hansen, N. (2007). Networks of meaning: Intergroup relations, cultural worldviews, and knowledge activation principles. *Social and Personality Psychology Compass, 1*, 68–83.

Mendoza-Denton, R., Page-Gould, E., & Mendes, W. B. (2009, May). Self-protective properties of stigma: Evidence from physiological and self-report measures. In S. S. M. Townsend & B. Major (Chairs), symposium on "Getting under the skin: How psychophysiology is advancing our understanding of intergroup threat," conducted at the annual Association for Psychological Science Convention, San Francisco, CA.

Mendoza-Denton, R., Page-Gould, E., & Pietrzak, J. (2006). Mechanisms for coping with status-based rejection expectations. In S. Levin & C. Van Laar (Eds.), *Stigma and group inequality: Social psychological approaches* (pp. 151–169). Mahwah, NJ: Erlbaum.

Page-Gould, E., Mendoza-Denton, R., & Mendes, W. (in press). Stress and coping in interracial contexts: The influence of race-based rejection sensitivity and cross-group friendship in daily experiences of health. *Journal of Social Issues.*

Page-Gould, E., Mendoza-Denton, R., & Tropp, L. (2008). With a little help from my cross-group friends: Reducing intergroup anxiety through cross-group friendship. *Journal of Personality and Social Psychology, 95*, 1080–1094.

Plant, E. A., & Devine, P. G. (2009). The active control of prejudice: Unpacking the intentions guiding control efforts. *Journal of Personality and Social Psychology, 96*, 640–652.

Ratner, K. G., Halim, M. L., & Amodio, D. M. (2013). Perceived stigmatization, ingroup pride, and immune and endocrine activity: Evidence from a community sample of Black and Latina women. *Social Psychological and Personality Science, 4*, 82–91.

Steele, C. M. (1997). A threat in the air: How stereotypes shape the intellectual identities and performance of women and African Americans. *American Psychologist, 52*, 613–629.

Trawalter, S., Richeson, J. A., & Shelton, J. N. (2009). Predicting behavior during interracial interactions: A stress and coping approach. *Personality and Social Psychology Review, 13*, 243–268.

Twenge, J. M., & Crocker, J. (2002). Race and self-esteem: Meta-analyses comparing Whites, Blacks, Hispanics, Asians, and American Indians and comment on Gray, Little, and Hafdahl (2000). *Psychological Bulletin, 128*, 371–408.

Williams, D. R., Neighbors, H. W., & Jackson, J. S. (2003). Racial/ethnic discrimination and health: Findings from community studies. *American Journal of Public Health, 93*, 200–208.

PART V

Intergroup Interactions

15

USING EEG MU-SUPPRESSION TO EXPLORE GROUP BIASES IN MOTOR RESONANCE

Jennifer N. Gutsell and Michael Inzlicht

The opposite of love is not hate, it is indifference.

(Elie Wiesel, n.d.)

Humans have an incredible capacity to empathize with others, to understand their actions and intentions, and to use this information to coordinate their own behavior with the behavior of these others. When it comes to our loved ones, our family and friends, or our social in-group, we use the full extent of our empathic abilities. We automatically catch their emotions, and the meaning and intentions of their actions, which provide us with an intuitive understanding of their inner states. Thereby, we help them when they are in need, we bond with them, and we can act in unison and close cooperation. With those we hate, our adversaries or competitors, it is different. We still have a need to understand their emotions, actions, and intentions, but for more selfish reasons. Here we use our empathic abilities to anticipate our opponent's next move and to be prepared. Hence, despite our selfish reasons, we still are receptive to the inner states of our enemies. But what about those toward whom we feel nothing but indifference, the homeless person on the street, maybe the nameless service person, or simply those whom we consider part of a social out-group? Could it be that we do not intuitively catch their intentions, needs, and emotions?

We started wondering about this question not, as usually is the case, because we read an interesting article in a social psychology journal, or because we saw an inspiring talk at a social psychology conference. Instead, we came across an interesting phenomenon primarily discussed in the neurosciences—the mirroring process or motor resonance —and were intrigued by its potential implications for social psychology.

Motor resonance is a very basic and simple process: A subset of motor neurons are active not only during motor activity, but also when the perceiver remains completely still and observes someone else engaging in motor activity (Di Pellegrino, Fadiga, Fogassi, Gallese, & Rizzolatti, 1992; Rizzolatti, Fadiga, Fogassi, & Gallese, 1996). Despite its simplicity, motor resonance seems to contribute to many much more complex and important social functions. For instance, motor resonance has now been associated with better performances in tasks that require emotional empathy (Pineda & Hecht, 2009), and it is said to foster an intuitive understanding of the other's actions and emotions (Hurley, 2008; Iacoboni, 2009), potentially facilitating interpersonal coordination and cooperation (Knoblich & Sebanz, 2006).

Reading about motor resonance was fascinating and the phenomenon seemed to be incredibly relevant for social psychology; here was a mechanism that could potentially explain how and why people can so easily and effortlessly connect with each other. We noticed, however, that it was often implied that motor resonance happens relatively automatically, without much influence from top-down processes (e.g., Chong & Mattingley, 2009; Fadiga, Fogassi, Pavesi, & Rizzolatti, 1995; van Schie, Waterschoot, & Bekkering, 2008) and only a few studies have investigated the nature of top-down influences on motor resonance (e.g., Aglioti, Cesari, Romani, & Urgesi, 2008; Calvo-Merino, Glaser, Grèzes, Passingham, & Haggard, 2005; Iacoboni, 2005; Oberman, Ramachandran, & Pineda, 2008; Sartori, Cavallo, Bucchioni, & Castiello, 2011). Moreover, those that did mostly focused on relatively basic processes such as attention or situational context. At the same time, we know from research on social cognition that even very basic perceptional processes are susceptible to higher-order cognitive processes. For instance, face recognition (Bernstein, Young, & Hugenberg, 2007), the interpretation of facial expression (Elfenbein & Ambady, 2002), and memory (Rothbart, Evans, & Fulero, 1979) are all very basic processes that are still influenced by social categorization and prejudice. The same could be true for motor resonance, and finding a group bias in motor resonance could have important implications for both the literature on intergroup processes and the literature on motor resonance. If people indeed resonated less with out-group members, they would potentially have a harder time to predict their actions and intentions and to coordinate their actions with them. Moreover, to show that motor resonance is indeed influenced by social categorization and prejudice would suggest that motor resonance is susceptible to culturally learned attitudes and biases.

We were intrigued by the mirror mechanism and its potential for answering important social psychological questions, and then we came across an article by Pineda (2005) in which he discussed a method that uses electroencephalography (EEG) to investigate motor resonance. Being experienced with EEG, it was clear to us that this technique would enable us to measure motor resonance cost-efficiently and relatively conveniently. In addition, EEG is easily applicable and a relatively unobtrusive measure of brain activity so that participants would be left relatively unconstrained, allowing them to perform complex actions, and potentially even to interact with others during EEG recording.

Here we describe four studies that use EEG to investigate group biases in motor resonance. Specifically, we used EEG to measure neural activity in sensory motor areas during observation of actions performed by in-group and out-group members. We propose that, as a default, people do not resonate with ethnic out-group members, but once out-group members' behavior becomes motivationally relevant, either because group boundaries are blurred, or because the out-group member poses a threat, people become sensitive to out-group members' inner states.

Motor Resonance

A now classic effect in social psychology is that perceiving someone else's behavior increases a tendency in the perceiver to behave similarly (Chartrand & Bargh, 1999; James, 1890). This happens because actions and the mere perception or imagination of the same action share common representations (Preston & de Waal, 2002; Prinz, 1987). In the brain, this action–perception link is said to be implemented by the so-called mirror neurons, a subset of sensory motor neurons that are active during motor activity, but also when the perceiver remains completely still and sees someone else engaging in motor activity (Carr, Iacoboni, Dubeau, Mazziotta, & Lenzi, 2003; Rizzolatti & Craighero, 2004). Mirror neurons have been discovered using single-cell recordings in monkeys, but being an invasive method requiring open-brain surgery, single-cell recordings in humans are hard to get due to ethical considerations; thus far, only one study has shown mirror neurons in humans using single-cell recording (Mukamel, Ekstrom, Kaplan, Iacoboni, & Fried, 2010). Due to the lack of opportunities for single-cell recording studies in humans, mirror neuron activity is usually investigated using corresponding neural activation of sensory motor areas within an observer in response to the mere perception of motor activity in the target as a proxy. This motor resonance is considered to underlie the understanding of actions and intentions, which is in turn essential for interpersonal and social processes such as learning through observation and interpersonal coordination. For example, research shows that motor resonance goes beyond the mere mechanics of a motion, but seems to map the function and goal of the action: Mirror neurons fire, even when the "observer" can only partially see the action (Umilta et al., 2001) or not at all (Kohler et al., 2002). Moreover, motor resonance is more pronounced for goal-directed action (Cattaneo, Caruana, Jezzini, & Rizzolatti, 2009), and specific groups of mirror neurons respond to specific and unique goals of actions (Gallese, Gernsbacher, Heyes, Hickok, & Iacoboni, 2011), such as grasping to put something into a container versus grasping to bring something to one's mouth (Bonini et al., 2010).

Generally, motor resonance has been suggested to be automatic and to occur without conscious effort (Gallese, Fadiga, Fogassi, & Rizzolatti, 1996; Rizzolatti

& Craighero, 2004; Wilson & Knoblich, 2005), but this does not mean that everyone resonates with just anyone in any situation. What it probably comes down to is whether the action in question is motivationally relevant to the observer or not.

The Functional Approach to Motor Resonance

People tend to allocate their limited cognitive and perceptional resources toward others who are motivationally relevant to them (Ackerman et al., 2006). These relevant others are most often members of the social in-group. One reason for this is that from an evolutionary perspective, in-group members are more often beneficial for one's reproductive fitness. According to the concept of inclusive fitness (Hamilton, 1964, as cited in Cialdini, Brown, Lewis, Luce, & Neuberg, 1997), individuals can promote their own evolutionary success by helping others who share some of their genes (Sime, 1983). The degree of genetic overlap, however, is impossible to detect, and thus people have to use proxies of genetic overlap such as the degree of kinship, similarity, familiarity, and affiliation. In-group members are generally perceived as more similar to oneself than out-group members (Turner, Brown, & Tajfel, 1979). Moreover, in-group members are usually the ones with whom we most often interact, and if we interact with out-group members, this mostly happens on a group level rather than in one-on-one intimate interactions (Fiske, 1992). Consequently, intragroup interactions are most important to us, and since cognitive resources are limited, processing out-group members' inner states most often does not take priority (Ackerman et al., 2006).

Indeed, research in social cognition and social neuroscience has revealed several group biases in the processing of others' inner states. Research participants show less neural activity in areas for social perception and social cognition in response to out-group members (Fiske, Ames, Cikara, & Harris, Chapter 5 in this volume; Harris & Fiske, 2006; Van Bavel, Packer, & Cunningham, 2008), and they have a harder time recognizing out-group members' faces and interpreting their facial expressions (Elfenbein & Ambady, 2002; Sporer, 2001). Moreover, they react less strongly to emotions such as happiness and fear when displayed by out-group members (Weisbuch & Ambady, 2008), and they show less neural activity in the neural pain circuit when they see out-group members experiencing physical pain (Xu, Zuo, Wang, & Han, 2009), or emotionally painful situations (Mathur, Harada, Lipke, & Chiao, 2010) than when they see pain in in-group members. Taken together, these findings suggest that across perceptional domains, people are less sensitive and perceptive to the inner states of social out-group members. However, this does not necessarily mean that social perception is entirely reserved for the social in-group. When the out-group becomes motivationally relevant, it is possible that people would be as responsive to out-group members as to in-group members. For example, Ackerman et al. (2009) showed that, although usually people have problems in telling different out-group faces apart, when the faces are angry, research participants performed even better in recognizing out-group members' faces than in-group faces. Hence, when out-group members become motivationally relevant—in this case, because they potentially could pose a threat—group biases in social perception are alleviated or even reversed.

Given the role of motivational relevance and social group membership in social perception, it is likely that group biases also exist in motor resonance. Such a bias could leave people to be less responsive to the inner states of out-group members, potentially hampering cross-group interactions and cooperation. In what follows, we will describe a series of studies that investigate how group categories, prejudice, and motivational relevance influence the amount of motor resonance in response to others. In all studies reported here, we used suppression of electroencephalographic mu oscillations during the observation of actions as an index of motor resonance.

Mu-suppression as a Measure of Motor Resonance

The mu-rhythm is a well-established measure of neural activity in the sensory motor cortex (Kuhlman, 1978; Pfurtscheller & Aranibar, 1979). The alpha component of the mu-rhythm (8–12 Hz) seems to originate from the primary somatosensory cortex (Hari, 2006) and shows a synchronized activity at rest, which is desynchronized both during movement and during passive observation of movement, leading to a suppression of mu (Pineda, 2005).

During execution of movement, neurons in the premotor, motor, and sensorimotor cortices are activated. Therefore, mu-suppression while a person is moving is probably the result of motor neurons and mirror neurons. In the absence of overt movement, motor neurons should no longer be active, leaving only mirror neurons to contribute to mu-suppression (Pineda, 2005). Mu-suppression during action observation should, therefore, be exclusively attributed to the discharge of mirror neurons. Anatomically, mirror neurons located in the ventral premotor cortex could affect neurons in the primary sensorimotor cortex through existing strong connections between these two regions, resulting in mu-suppression (Pineda, 2005). Mu-suppression, thus, seems to be a good indicator of motor resonance and mirror neuron activity (Puzzo, Cooper, Cantarella, & Russo, 2011).

Studies measuring mu-suppression differ in the experimental and control conditions during which mu is measured, but they mostly follow the same general principle. At the beginning of the experimental session, researchers assess participants' baseline brain activity. For example, EEG is recorded while the participant sits completely still either looking at a blank screen or visual white noise. Following this, participants usually are shown videos of others performing a goal-directed action. These videos often only depict the hands of the other person to keep other target-related factors controlled (e.g., Oberman et al., 2008), but richer stimuli can sometimes be beneficial. For example, in our research, we usually choose to present videos depicting large parts of the body and even the face. We do this to create an experience similar to actual social interactions.

Across studies using mu-suppression, the length of the videos varies, but in general, the methods used can be split into two kinds of protocols: the long-exposure protocol and the multiple-repetitions protocol. Long-exposure protocols use longer videos, usually longer than 10 seconds, and each video is shown once or sometimes twice (e.g., Gutsell & Inzlicht, 2010; Oberman et al., 2005). In contrast, in the multiple-repetitions protocol, the videos are much shorter, usually 3 seconds or less, and the same videos are presented multiple times at each trial preceded by a fixation cross (e.g., Puzzo et al., 2011). Research shows that the multiple-repetitions protocol can elicit stronger mu-suppression for relatively minimal stimuli (e.g., a hand squeezing a ball; Puzzo et al., 2011). Nonetheless, because in our research, the stimulus videos are more complex, and because we are interested in social processes, we chose to use a long-exposure protocol in order to create a more realistic experience for the participants.

The fact that it is possible to use relatively rich stimuli depicting complex actions makes mu-suppression a flexible tool that is particularly suitable for answering social psychological questions. Moreover, this method allows participants to be relatively unrestrained by the equipment so that they themselves can move and perform actions. Both these assets could potentially enable researchers to measure motor resonance during actual interactions with real people, and when using a portable EEG system, this interaction could even take place outside the laboratory. This flexibility is certainly an advantage over more cumbersome methods such as magnetic resonance imaging (MRI) or transcranial magnetic stimulation (TMS).

Mu-suppression is usually strongest over the contralateral sensory motor areas (C3 or C4 depending on laterality), but can still be measured in the adjacent more frontal areas (FC-strip) and more posterior areas (CP-strip) (Puzzo et al., 2011). We measured mu-suppression primarily looking at C3—since right-handed participants were presented with right-hand actions.

Biased Motor Resonance

Our first study using mu-suppression to assess motor resonance in an intergroup context was aimed to test the general effect: Is there a group bias in motor resonance (Gutsell & Inzlicht, 2010). We expected that, by default, people would only show motor resonance in response to seeing members of their in-group performing an action, and not in response to the actions of out-group members. Moreover, we expected that such an in-group bias in motor resonance would be strongest for people high in prejudice. To test this prediction, we first assessed participants' level of prejudice in a mass testing session at the beginning of the academic term. Specifically, we used the Symbolic Racism 2000 Scale (Henry & Sears, 2002)—a measure of modern racism, indicative of a subtle form of racism that obscures racist feelings with abstract values such as justice and order. Several weeks later, we invited participants to the laboratory one at a time and showed them videos of in-group and out-group members performing a simple action, while we measured their EEG. Specifically, we used a one-way within-subject design showing each participant videos of in-group members and out-group members. For the purpose of all the studies described here, we defined group in terms of ethnic group membership. Following the video presentation, we asked participants to perform the same action themselves.

The data obtained during the self condition confirmed that we were indeed picking up neural motor activity. When people actually performed the action, they showed a suppression of mu over electrode C3—the contralateral electrode site over motor areas— compared to mu during baseline. Interestingly, and more central to our hypotheses on motor resonance, we did also find mu-suppression when participants were sitting completely still, simply observing in-group members performing the action. These results are indicative of motor resonance in response to the in-group. Critically, the same test for mu-suppression was not significant for EEG activity during the out-group condition. Participants did not show mu-suppression for the out-group. Moreover, mu-suppression for the out-group was significantly less than mu-suppression in response to the in-group, further supporting a group bias in motor resonance. These findings suggest that people do not resonate with the actions of out-groups. We take this as evidence that people experience less vicarious action and its associated somatic and autonomic states when confronted with out-groups than with in-groups (please see Gutsell & Inzlicht, 2010, for a full description of the findings).

Although our findings suggest that people in general show this bias, not everyone seems to show it to the same degree. When we looked at the influence of prejudice, we found a positive correlation between mu-suppression for the out-group (with more negative scores indicating more suppression) and the level of prejudice reflected in a participant's score on the Symbolic Racism 2000 Scale. Thus, the more prejudiced participants were, the less mu-suppression they showed in response to out-groups. In contrast, mu-suppression for the in-group was unrelated to prejudice. The degree of bias in motor resonance is therefore influenced by how much people dislike the respective ethnic group. This is also reflected in the fact that when we break down the omnibus correlation into the specific ethnic groups, we find results that are consistent with a Canadian context: The correlation was strongest for South Asians, followed by Blacks; the correlation for East Asians, however, was not significant. Thus, motor resonance seems indeed to be limited to the in-group and more so when the perceiver is prejudiced or when the group in question is disliked.

According to social identity theory, the psychological self extends to include other people and our social in-groups so that the in-group becomes part of the self (Turner et al., 1979). Hence, if we take a functional approach to motor resonance, it makes sense that people resonate with in-group members, but not with out-group members; since in-group members are the ones people feel closest to, it is important to understand their actions and intentions. In contrast, people might not feel the same need to resonate with out-group members, especially when these groups are disliked. If a lack of closeness and liking indeed, at least partly, explains the in-group bias in motor resonance, then somehow bringing the out-group closer to the self and reducing prejudice should alleviate such biases. One way to increase closeness to social out-groups and to reduce prejudice is cognitive perspective taking.

Facilitating Cross-group Motor Resonance by Blurring Group Boundaries

Cognitive perspective taking refers to the active contemplation of others' psychological experiences; it facilitates altruistic behavior (Batson et al., 1997) and cooperation (Batson & Moran, 1999), and promotes conflict resolution (Galinsky, Maddux, Gilin, & White, 2008). In addition to these more general benefits, perspective taking can be helpful in an intergroup context (Batson et al., 1997; Dovidio et al., 2004, Galinsky & Moskowitz, 2000). For example, perspective taking improves evaluations of specific members of the out-group (Shih, Wang, Trahan Bucher, & Stotzer, 2009) and of the out-group more generally (Dovidio et al., 2004; Galinsky & Ku, 2004). Given all the benefits of perspective taking, a perspective-taking mindset could potentially be a powerful tool to decrease intergroup biases, including biases in motor resonance.

To investigate the effects of cognitive perspective taking on motor resonance in an intergroup context, we used the same design as in the first study, but this time, before participants saw the videos, they completed a mindset manipulation. Specifically, they saw a picture of a young Black university student and were asked to write a short story about a day in his life, either taking his perspective and writing in the first person, or remaining objective and writing in the third person. Following this, we had participants complete the same action video task as in the previous study while we recorded their EEG. Please see Gutsell and Inzlicht (2013) for a more detailed description of the study and findings.

The results suggest that the kind of mindset really does matter, and that a perspective-taking mindset can indeed reduce group biases in motor resonance. When in an objective mindset, participants did not show mu-suppression, either for the in-group or for the out-group. Hence, taking a step back and distancing oneself from the emotions and intentions of others can reduce motor resonance even for in-group members. Similarly, when in a perspective-taking mindset, participants showed mu-suppression in response to in-group and out-group members, reflecting a facilitating effect of a perspective-taking mindset on motor resonance in general, and motor resonance in response to out-group members in particular. Moreover, while participants in the objective mindset condition show an in-group bias in motor resonance—mu-suppression is stronger for the in-group than for the out-group, those in a perspective-taking mindset do not show such group distinction—mu-suppression for the in-group and out-group did not differ.

Taken together, the two studies described so far suggest that as a default, people tend not to resonate with the actions of out-group members. Consequently they might not intuitively grasp out-group members' actions, intentions, and inner states. Study 2 provides evidence for the utility of perspective taking as a strategy for combating this subtle consequence of racial bias. A potential mechanism for the beneficial effects of perspective taking is an increase in self–other overlap with the out-group. Taking the perspective of another person increases the link between the mental representations of the self and the mental representations of the other (Davis, Conklin, Smith, & Luce, 1996; Galinsky & Moskowitz, 2000). It leads people to ascribe more of their own traits to the other (Galinsky & Ku, 2004). Moreover, perspective taking makes people see more of the other in themselves (Ku, Wang, & Galinsky, 2010), such that the self concept extends to include the other; this is thought to assist goals related to social connection (Tiedens & Jimenez, 2003), and to increase social coordination (Galinsky et al., 2008). Through this mechanism, taking the perspective of the Black student in the story-writing task would increase a feeling of connectedness and affiliation with the student, and this affiliation goal would lead to increases in motor resonance. Thus Study 2 suggests that by blurring the boundary between the self and the out-group, people are more likely to resonate with the actions of out-group members. In a third study (Gutsell, Tullett, Inzlicht, & Plaks, 2013) we aimed to test this mechanism once more, this time looking at a more direct route to people's feelings of similarity and connectedness to out-group members— people's beliefs in genetic overlap between ethnic groups.

Despite the fact that people actually share all but a minuscule portion of their genetic information with everyone else on the planet (Feldman, Lewontin, & King, 2003), for centuries, people have been classified into different racial groups (Frederickson, 2003). How does the belief in genetic overlap between people relate to motor resonance? Do people who think that they share more of their genes with any other person tend to be more likely to resonate with the actions of those around them? Once more we used EEG mu-suppression as an indicator of motor resonance. This time, however, we used a correlational design and correlated mu-suppression scores in response to actions of ethnic in-group and out-group members with participants' scores on a self-report measure of beliefs in genetic overlap. Specifically, we asked participants: "If you were to choose two people at random from the entire world, what percentage of genetic material would they have in common? (0–100%)." This measure of genetic overlap has been shown to be correlated with the tendency to see racial groups as clear-cut categories, and when people are led to believe in a high degree of genetic overlap, they expressed less implicit racial bias (Plaks, Malahy, Sedlins, & Shoda, 2011).

The results further support the functional account of motor resonance. We found a negative correlation between beliefs in genetic overlap and mu-suppression (higher negative score indicates more suppression), such that participants of European-Canadian or East Asian-Canadian descent who believe in high genetic overlap show mu-suppression while they watch videos of ethnic in-group members, but also when they watch videos of ethnic out-group members (African-Canadian and South Asian-Canadian descent). Therefore, when people believe themselves to be more genetically related and thus closer and more connected to others in general, they are more likely to resonate with them, potentially leading to richer, more intuitive understanding of their actions and intentions.

Taken together, the findings on a perspective-taking mindset and on beliefs regarding genetic overlap suggest that creating a sense of interconnectedness or closeness can indeed positively influence motor resonance and might potentially be a good way to alleviate the in-group bias in motor resonance. Motor resonance most likely helps people to get an intuitive understanding of other's actions and intentions (Fogassi et al., 2005) and might even be involved in basic empathic processes such as emotional contagion (Pineda & Hecht, 2009). Hence, group biases in motor resonance might ultimately translate to misunderstandings and impaired social rapport, empathy, and social coordination. Perspective taking and beliefs in strong genetic overlap across groups have been shown successfully to reduce prejudice (Batson et al., 1997; Galinsky & Ku, 2004; Plaks et al., 2011), and to create a sense of connectedness to the out-group. Alleviating group biases in motor resonance might yet be another important benefit (Galinsky & Ku, 2004; Plaks et al., 2011) of these techniques.

Other techniques that reduce the psychological gap between different ethnic groups might be similarly effective in reducing group biases in motor resonance. For example, Hogeveen and Obhi (2011) found that when videos of simple hand motions were paired with words that prime an interdependent self-construal such as "social," "family," and "help" (Aron, A., Aron, E., & Smollan, 1992), participants showed an increase in motor resonance (as measured with transcranial magnetic stimulation) in the interdependent trials, as opposed to control trials. Once someone is perceived as similar, and once they are part of our inner circle of close others whose actions, intentions, and goals we care about, it seems that our neural system for action understanding becomes sensitive to out-group members.

Increasing closeness to the target, however, is not the only possible approach to make a target motivationally relevant. If we truly take a functional approach to motor resonance, it does not seem sensible to think that people only resonate with those to whom they feel close. Altruism and reciprocity are not the only reasons why we need to understand others' goals and intentions, or why others' behavior is motivationally relevant. Another reason why a behavior might be relevant is that it poses a potential threat.

Motor Resonance for Threatening Behavior

It is a well-known psychological effect that people are particularly sensitive to negative information. This negativity bias (see reviews by Cacioppo & Berntson,

1994; Peeters & Czapinski, 1990; Taylor, 1991) has been shown in various domains, such as impression formation (Skowronski & Carlston, 1989), person memory (Ybarra & Stephan, 1996), and stimulus processing. For example, people spend more time looking at negative than positive stimuli, and perceive negative stimuli to be more complex than positive stimuli (DuCette & Soucar, 1974). In an intergroup context, the negativity bias is particularly powerful when anger is concerned. For example, research has shown that angry facial expressions capture and hold our attention (Öhman, Flykt, & Esteves, 2001). The reason probably is that anger is an interpersonal emotion and poses a threat if it is directed at the observer (Ackerman et al., 2006). Threatening out-group behavior should therefore be particularly motivationally relevant for the individual, and consequently should be a likely target for motor resonance. In fact, threatening out-group behavior might even override any existing group biases in motor resonance.

In order to test this idea, we invited participants of European-Canadian and East Asian-Canadian descent to the lab and measured their EEG while they watched videos of European-Canadian or East Asian-Canadian actors respectively (in-group condition), and while they watched South Asian-Canadian and African-Canadian actors (out-group condition) (Gutsell et al., 2013). The design differed from the previous studies in that this time, actors did not only display a neutral action (moving their arm up and down with a neutral face); they also displayed a positive action (repeatedly making the "thumbs up" hand gesture with a positive facial expression) and a threatening action (repeatedly doing the "giving someone the finger" hand gesture with an angry facial expression).

Replicating Study 1, participants did not show significant mu-suppression when observing neutral actions performed by out-group members, but they did show significant mu-suppression in response to neutral in-group actions. Once again, these findings show that as a default, people have an in-group bias when it comes to motor resonance. A similar pattern was found for positive actions—participants only showed significant mu-suppression during the observation of positive actions performed by in-group members, but not during the observation of positive actions performed by out-group members. Interestingly, and as predicted, when the observed behavior was threatening, people did show significant mu-suppression in response to all targets—in-group members as well as out-group members, and the amount of mu-suppression in response to out-group members no longer differed from the amount of mu-suppression in response to the in-group. Thus, in the threat condition, participants no longer show an in-group bias in motor resonance. We resonate with threatening out-group members as much as we would usually only resonate with in-group members.

These findings are consistent with a functional perspective on person perception. No matter whether the behavior of out-group members is neutral or positive, people show an in-group bias in motor resonance. In contrast, when out-group behavior is negative and threatening, people seem to allocate cognitive resources to the threatening individual and start processing them as thoroughly as they would in-group members—the in-group bias in motor resonance disappears. Friendly out-group members might be more likable and pleasant, but as long as we do not include them in our inner circle of relevant others, we will not resonate with them. Thus, it is not so much about the valence of the behavior, but about whether or not the behavior in question has implications for one's own goals and behavior.

Added Value

A failure to resonate with out-group members' actions and intentions, and so to speak, spontaneously to "catch" the meaning of their actions and intentions can potentially have serious consequences for intergroup interactions. We presented research suggesting that, as a default, people do not resonate with ethnic out-group members. However, once an out-group member becomes motivationally relevant—either because the observer is in a perspective-taking mindset, or because the out-group member's behavior poses a potential threat—people start resonating, thereby becoming more sensitive to the inner states of out-group members.

Our findings advance current theoretical understanding of cross-group interactions, showing that one of the most basic processes underlying action perceptions is affected by group biases. Motor resonance and the sharing of somatic, autonomic, and emotional states facilitate social understanding and might foster smooth and rewarding intergroup interactions and coordination (Knoblich & Sebanz, 2006). Hence, people might not be as responsive to out-group members' needs and goals and be less likely intuitively to grasp their intentions; they might also have a harder time to synchronize their own behavior with that of out-group members and to complement it during cooperation (Sebanz, Bekkering, & Knoblich, 2006). Therefore, investigating the limits and facilitating factors of motor resonance in an intergroup context is important in order to understand the challenges of intergroup interactions. Moreover, our findings provide a possible mechanism for effective interventions to improve cross-group interactions. For example, adopting a perspective-taking mindset has been shown to be conducive to smooth and pleasant intergroup interaction experiences. The findings reported here suggest a potential mechanism: When they are in a perspective-taking mindset, people are more likely to resonate even with out-group members, which potentially could provide them with a richer, more intuitive understanding of the others' actions and intentions, leading to smoother and more positive intergroup interactions. Taken together, eventually the insights on motor resonance in an intergroup context might inform us as to how one can potentially facilitate smoother intergroup encounters, and how to create circumstances that support intergroup coordination and cooperation.

Finally, our findings contribute to the understanding of motor resonance by shedding light on the factors that facilitate and hamper it. We show that not only is motor resonance influenced by social categorization, but it can also be influenced by culturally learned prejudice and by temporarily induced mindsets. After all, motor resonance, like many other basic perception processes before, has turned out not to be as automatic as previously it was thought to be; instead, it seems to be susceptible to the multifaceted and rich world of higher-order cognitive processes that constitute human social cognition.

Strengths and Weaknesses

In the studies described here, we used EEG mu-suppression as an index for motor resonance. We chose to use EEG because it is easily applicable, and a relatively unobtrusive measure of brain activity. Participants are thus left relatively unconstrained, enabling them to perform complex actions and potentially even to interact with others during EEG recording. Moreover, because each EEG session is relatively affordable compared to other methods of neural resonance such as functional magnetic resonance imaging (fMRI), it allows for large enough samples to combine brain imaging with the use of more traditional measures used in social psychology, such as implicit reaction-time measures of prejudice, or self-report scales.

Motor resonance per definition is a phenomenon of the brain; it is the activation of the neural sensory motor system during observation of another's action. Theoretically it is linked to activity of a specific kind of neurons in these brain regions—the mirror neurons (Rizzolatti & Craighero, 2004). Therefore, in order to understand motor resonance, we need to adopt neuroscience techniques to measure brain activity. That said, motor resonance is related to at least two phenomena that can be studied using more traditional social psychological measures—automatic imitation and motor mimicry (Heyes, 2011), but the specific nature of the relationship between these phenomena is still unclear.

Automatic imitation is a type of stimulus–response compatibility effect similar to well-known effects such as the Stroop effect (Stroop, 1935). In automatic imitation paradigms (e.g., Proctor & Vu, 2006; Stürmer, Aschersleben, & Prinz, 2000), participants show response slowing when they have to perform one kind of action (e.g., opening their hand) in response to a stimulus, while a task-irrelevant and incompatible action (e.g., image of a closing hand) is presented; they respond faster when a compatible action (e.g., image of an opening hand) is presented (e.g., Brass, Bekkering, Wohlschlager, & Prinz, 2000; Heyes, Bird, Johnson, & Haggard, 2005; Stürmer et al., 2000). The task-irrelevant stimuli interfere with performing the required action, because participants co-represent the incompatible action depicted in the picture or video. Theoretically this co-representation and the resulting interference are the result of motor resonance and ultimately the mirror neuron system (Kilner, Paulignan, & Blakemore, 2003; Press, Bird, Walsh, & Heyes, 2008; Van Schie et al., 2008). Catmur, Walsh, and Heyes (2009) showed that temporally disrupting the function of relevant motor areas using TMS impaired automatic imitation, suggesting that motor resonance is indeed underlying implicit imitation. Unfortunately, other than this finding, experimental evidence is still scarce and more research is necessary.

In contrast to automatic imitation, motor mimicry can be observed in naturalistic social situations and real interactions. When we interact with others, we unconsciously mimic their postures and mannerisms (Chartrand & Bargh, 1999) and this seems to have many interpersonal benefits. For example, motor mimicry promotes liking (Chartrand & Bargh, 1999; Hove & Risen, 2009), and feelings of closeness (van Baaren, Holland, Kawakami, & van Knippenberg, 2004). Moreover, motor mimicry also seems to be influenced by group biases: Although people mimic others' expressions, gestures, and body postures, this occurs with less frequency for out-group members (Likowski, Mühlberger, Seibt, Pauli, & Weyers, 2008). This similarity in findings suggests that potentially group biases in motor resonance might translate to reduced motor mimicry, depriving intergroup interactions of the many benefits that mimicry can have on interpersonal rapport.

Intuitively it does make sense that motor mimicry depends on motor resonance, and is based on the same psychological process responsible for automatic imitation, but thus far, there is not much research integrating the three. Hence, it is still relatively unclear how motor resonance, automatic imitation, and motor mimicry are related, and specifically, whether motor resonance is the neurological process that gives rise to the two behavioral phenomena. Therefore, motor resonance is a neural process that might not always and not necessarily be expressed in reaction times or actual motor behavior, and in order to research motor resonance, one has to employ brain-imaging methods such as fMRI, TMS, or EEG. Which one of these methods is the best to use depends on the researcher's preferences, the availability of the equipment, and on the specific requirements and restrictions the specific research question poses on experimental design. Using EEG mu-suppression gave us the opportunity to have relatively large sample sizes, making between-subject designs and the use of self-report measures possible. We also appreciated that this method is relatively unobtrusive and tolerant to movement artifacts. Finally, we think that the ability to present continuous and rich stimuli opens up opportunities to create studies that more accurately simulate the experience during real-life interactions.

Ultimately, we need to link motor resonance back to behavior in actual interactions. We know that motor resonance gives rise to an intuitive understanding of another's experiences (Hurley, 2008; Iacoboni, 2009;). Motor resonance has also been associated with many other cognitive processes, such as learning and the experience of empathy (Gallese et al., 2011), but how does this translate to intergroup behavior? If we did not resonate with out-group members, would we have a harder time predicting their next move, and would we have trouble coordinating our actions with theirs? Would we be less likely to learn from their mistakes, or to help them when they are in need? Ultimately, these are the kind of questions in which we, as social psychologists, are most interested. Understanding motor resonance in an intergroup context is the first step on a journey to understand how it is expressed on a behavioral level, and how group biases in motor resonance affect intergroup experiences and interactions.

Suggestions for Further Reading

Gallese, V., Gernsbacher, M. A., Heyes, C., Hickok, G., & Iacoboni, M. (2011). Mirror neuron forum. *Perspectives on Psychological Science, 6*, 369–407.

Provides an in-depth discussion about controversial topics regarding mirror neurons. Leading experts on the subject debate the functions of mirror neurons.

Goldman, A. I. (2006). *Simulating minds: The philosophy, psychology, and neuroscience of mindreading*. Oxford, UK: Oxford University Press.

A stimulating and engaging account of how we understand others. This book provides a good overview of the philosophical debate surrounding simulation theory, and covers empirical supporting findings from developmental psychology, social psychology, and neurosciences.

Gutsell, J. N., & Inzlicht, M. (2010). Empathy constrained: Prejudice predicts reduced mental simulation of actions during observation of out-groups. *Journal of Experimental Social Psychology, 46*, 841–845.

This article describes the findings of Study 1 discussed in this chapter and provides additional information on how we measured mu-suppression.

Pineda, J. A. (2005). The functional significance of mu rhythms: Translating "seeing" and "hearing" into "doing." *Brain Research Reviews, 50*, 57–68.

Provides a discussion of the mu-rhythm and its relation to mirror neuron activity.

Puzzo, I., Cooper, N. R., Cantarella, S., & Russo, R. (2011). Measuring the effects of manipulating stimulus presentation time on sensorimotor alpha and low beta reactivity during hand movement observation. *NeuroImage, 57*, 1358–1363.

A study exploring methodological questions on how best to use EEG mu-suppression as an index of motor resonance.

Rizzolatti, G., & Craighero, L. (2004). The mirror-neuron system. *Annual Review of Neuroscience, 27*, 169–192.

Provides a detailed description of the mirror neuron system, which provides a good overview of the theory and the supporting empirical evidence.

References

Ackerman, J. M., Shapiro, J. R., Neuberg, S. L., Kenrick, D. T., Becker, D. V., Griskevicius, V. . . . Schaller, M. (2006). They all look the same to me (unless they're angry): From out-group homogeneity to out-group heterogeneity. *Psychological Science, 17*, 836–840.

Aglioti, S. M., Cesari, P., Romani, M., & Urgesi, C. (2008). Action anticipation and motor resonance in elite basketball players. *Nature Neuroscience, 11*, 1109–1116.

Aron, A., Aron, E. N., & Smollan, D. (1992). Inclusion of other in the self scale and the structure of interpersonal closeness. *Journal of Personality and Social Psychology, 63*, 596–612.

Batson, C. D., & Moran, T. (1999). Empathy-induced altruism in a prisoner's dilemma. *European Journal of Social Psychology, 29*, 909–924.

Batson, C. D., Sager, K., Garst, E., Kang, M., Rubchinsky, K., & Dawson, K. (1997). Is empathy-induced helping due to self–other merging? *Journal of Personality and Social Psychology, 73*, 495–509.

Bernstein, M. J., Young, S. G., & Hugenberg, K. (2007). The cross-category effect. *Psychological Science, 18*, 706–712.

Bonini, L., Rozzi, S., Serventi, F. U., Simone, L., Ferrari, P. F., & Fogassi, L. (2010). Ventral premotor and inferior parietal cortices make distinct contribution to action organization and intention understanding. *Cerebral Cortex*, *20*, 1372–1385.

Brass, M., Bekkering, H., Wohlschlager, A., & Prinz, W. (2000). Compatibility between observed and executed finger movements: Comparing symbolic, spatial, and imitative cues. *Brain and Cognition*, *44*, 124–143.

Cacioppo, J. T., & Berntson, G. G. (1994). Relationship between attitudes and evaluative space: A critical review, with emphasis on the separability of positive and negative substrates. *Psychological Bulletin*, *115*, 401–423.

Calvo-Merino, B., Glaser, D. E., Grèzes, J., Passingham, R. E., & Haggard, P. (2005). Action observation and acquired motor skills: An fMRI study with expert dancers. *Cerebral Cortex*, *15*, 1243–1249.

Carr, L., Iacoboni, M., Dubeau, M., Mazziotta, J. C., & Lenzi, G. L. (2003). Neural mechanisms of empathy in humans: A relay from neural systems for imitation to limbic areas. *Proceedings of the National Academy of Sciences of the United States of America*, *100*, 5497–5502.

Catmur, C., Walsh, V., & Heyes, C. M. (2009). Associative sequence learning: The role of experience in the development of imitation and the mirror system. *Philosophical Transactions of the Royal Society B: Biological Sciences*, *364*, 2369–2380.

Cattaneo, L., Caruana, F., Jezzini, A., & Rizzolatti, G. (2009). Representation of goal and movements without overt motor behavior in the human motor cortex: A transcranial magnetic stimulation study. *The Journal of Neuroscience*, *29*, 11134–11138.

Chartrand, T. L., & Bargh, J. A. (1999). The chameleon effect: The perception–behavior link and social interaction. *Journal of Personality and Social Psychology*, *76*, 893–910.

Chong, T. T., & Mattingley, J. B. (2009). Automatic and controlled processing within the mirror neuron system. In J. A. Pineda (Ed.), *Mirror neuron systems* (pp. 213–233). Totowa, NJ: Humana Press.

Cialdini, R. B., Brown, S. L., Lewis, B. P., Luce, C., & Neuberg, S. L. (1997). Reinterpreting the empathy–altruism relationship: When one into one equals oneness. *Journal of Personality and Social Psychology*, *73*, 481–494.

Davis, M. H., Conklin, L., Smith, A., & Luce, C. (1996). Effect of perspective taking on the cognitive representation of persons: A merging of self and other. *Journal of Personality and Social Psychology*, *70*, 713–726.

Di Pellegrino, G., Fadiga, L., Fogassi, L., Gallese, V., & Rizzolatti G. (1992). Understanding motor events: A neurophysiological study. *Experimental Brain Research*, *91*, 176–180.

Dovidio, J. F., Vergert, M., Stewart, T. L., Gaertner, S. L., Johnson, J. D., Esses, V. M. et al. (2004). Perspective and prejudice: Antecedents and mediating mechanisms. *Personality and Social Psychology Bulletin*, *30*(12), 1537–1549.

DuCette, J., & Soucar, E. (1974). A further reexamination of the vigilance hypothesis using random shapes as stimuli. *The Journal of Social Psychology*, *92*, 109–113.

Elfenbein, H. A., & Ambady, N. (2002). Is there an in-group advantage in emotion recognition? *Psychological Bulletin*, *128*, 243–249.

Fadiga, L., Fogassi, L., Pavesi, G., & Rizzolatti, G. (1995). Motor facilitation during action observation: A magnetic stimulation study. *Journal of Neurophysiology*, *73*, 2608–2611.

Feldman, M. W., Lewontin, R. C., & King, M. (2003). A genetic melting-pot. *Nature*, *424*, 374.

Fiske, A. P. (1992). The four elementary forms of sociality: Framework for a unified theory of social relations. *Psychological Review*, *99*, 689–723.

Fogassi, L., Ferrari, P. F., Gesierich, B., Rozzi, S., Chersi, F., & Rizzolatti, G. (2005). Parietal lobe: From action organization to intention understanding. *Science, 308*, 662–667.

Frederickson, G. M. (2003). *Racism: A short history*. Princeton, NJ: Princeton University Press.

Galinsky, A. D., & Ku, G. (2004). The effects of perspective-taking on prejudice: The moderating role of self-evaluation. *Personality and Social Psychology Bulletin, 30*, 594–604.

Galinsky, A. D., Maddux, W. W., Gilin, D., & White, J. B. (2008). Why it pays to get inside the head of your opponent: The differential effects of perspective taking and empathy in negotiations. *Psychological Science, 19*, 378–384.

Galinsky, A. D., & Moskowitz, G. B. (2000). Perspective-taking: Decreasing stereotype expression, stereotype accessibility, and in-group favoritism. *Journal of Personality and Social Psychology, 78*, 708–724.

Gallese, V., Fadiga, L., Fogassi, L., & Rizzolatti, G. (1996). Action recognition in the premotor cortex. *Brain, 119*, 593–609.

Gallese, V., Gernsbacher, M. A., Heyes, C., Hickok, G., & Iacoboni, M. (2011). Mirror neuron forum. *Perspectives on Psychological Science, 6*, 369–407.

Gutsell, J. N., & Inzlicht, M. (2010). Empathy constrained: Prejudice predicts reduced mental simulation of actions during observation of outgroups. *Journal of Experimental Social Psychology, 46*, 841–845.

Gutsell, J. N., & Inzlicht, M. (2013). Perspective taking reduces group biases in neural motor resonance. Manuscript under review.

Gutsell, J. N., Tullett, A. M., Inzlicht, M., & Plaks, J. E. (2013). The effects of beliefs in genetic overlap on motor resonance. Manuscript in preparation.

Hamilton, W. D. (1964). The genetical theory of social behaviour. *Journal of Theoretical Biology*, 7, 1–52.

Hari, R., 2006. Action–perception connection and the cortical mu rhythm. *Progress in Brain Research, 159*, 253–260.

Harris, L. T., & Fiske, S. T. (2006). Dehumanizing the lowest of the low: Neuroimaging responses to extreme out-groups. *Psychological Science, 17*, 847–853.

Henry, P. J., & Sears, D. O. (2002). The Symbolic Racism 2000 Scale. *Political Psychology, 2*, 253–283.

Heyes, C. (2011). Automatic imitation. *Psychological Bulletin, 137*, 463–483.

Heyes, C., Bird, G., Johnson, H., & Haggard, P. (2005). Experience modulates automatic imitation. *Cognitive Brain Research, 22*, 233–240.

Hogeveen, J., & Obhi, S. S. (2011). Altogether now: Activating interdependent self-construal induces hypermotor resonance. *Cognitive Neuroscience, 2*, 74–82.

Hove, M. J., & Risen, J. L. (2009). It's all in the timing: Interpersonal synchrony increases affiliation. *Social Cognition, 27*, 949–961.

Hurley, S. (2008). The shared circuits model (SCM): How control, mirroring, and simulation can enable imitation, deliberation, and mindreading. *Behavioral and Brain Sciences, 31*, 1–58.

Iacoboni, M. (2005). Understanding others: Imitation, language, and empathy. In S. Hurley & N. Chater (Eds.), *Perspectives on imitation: From cognitive neuroscience to social science* (pp. 77–99). Cambridge, MA: The MIT Press.

Iacoboni, M. (2009). Imitation, empathy, and mirror neurons. *Annual Review of Psychology, 60*), 653–670. doi:10.1146/annurev.psych.60.110707.163604

James, W. (1890) *Principles of psychology*. 2 vols. NewYork: Holt.

Kilner, J. M., Paulignan, Y., & Blakemore, S. J. (2003). An interference effect of observed biological movement on action. *Current Biology, 13*, 522–525.

Kohler, E., Keysers, C., Umilta, M. A., Fogassi, L., Gallese, V., & Rizzolatti, G. (2002). Hearing sounds, understanding actions: Action representation in mirror neurons. *Science, 297*, 846–848.

Knoblich, G., & Sebanz, N. (2006). The social nature of perception and action. *Current Directions in Psychological Science, 15*, 99–104.

Ku, G., Wang, C. S., & Galinsky, A. D. (2010). Perception through a perspective-taking lens: Differential effects on judgment and behavior. *Journal of Experimental Social Psychology, 46*, 792–798.

Kuhlman, W. N. (1978). Functional topography of the human mu rhythm. *Electroencephalography and Clinical Neurophysiology, 44*, 83–93.

Likowski, K. U., Mühlberger, A., Seibt, B., Pauli, P., & Weyers, P. (2008). Modulation of facial mimicry by attitudes. *Journal of Experimental Social Psychology, 44*, 1065–1072.

Mathur, V. A., Harada, T., Lipke, T., & Chiao, J. Y. (2010). Neural basis of extraordinary empathy and altruistic motivation. *NeuroImage, 51*, 1468–1475.

Mukamel, R., Ekstrom, A. D., Kaplan, J., Iacoboni, M., & Fried, I. (2010). Single-neuron responses in humans during execution and observation of actions. *Current Biology, 20*, 750–756.

Oberman, L. M., Hubbard, E. M., McCleery, J. P., Altschuler, E. L., Ramachandran, V., & Pineda, J. A. (2005). EEG evidence for mirror neuron dysfunction in autism spectrum disorders. *Cognitive Brain Research, 24*, 190–198.

Oberman, L. M., Ramachandran, V. S., & Pineda, J. A. (2008). Modulation of mu suppression in children with autism spectrum disorders in response to familiar or unfamiliar stimuli: The mirror neuron hypothesis. *Neuropsychologia, 46*, 1558–1565.

Öhman, A., Flykt, A., & Esteves, F. (2001). Emotion drives attention: Detecting the snake in the grass. *Journal of Experimental Psychology: General, 130*, 466–478.

Peeters, G., & Czapinski, J. (1990). Positive–negative asymmetry in evaluations: The distinction between affective and informational negativity effects. In W. Stroebe & M. Hewstone (Eds), *European review of social psychology* (Vol. 1, pp. 33–60). Chichester, UK: Wiley.

Pfurtscheller, G., & Aranibar, A. (1979). Evaluation of event-related desynchronization (ERD) preceding and following voluntary self-paced movement. *Electroencephalography and Clinical Neurophysiology, 46*, 138–146.

Pineda, J. A. (2005). The functional significance of mu rhythms: Translating "seeing" and "hearing" into "doing." *Brain Research Reviews, 50*, 57–68.

Pineda, J. A., & Hecht, E. (2009). Mirroring and mu rhythm involvement in social cognition: Are there dissociable subcomponents of theory of mind? *Biological Psychology, 80*, 306–314.

Plaks, J. E., Malahy, L. W., Sedlins, M., & Shoda, Y. (2011). Folk beliefs about human genetic variation predict discrete versus continuous racial categorization and evaluative bias. *Social Psychological and Personality Science, 3*, 31–39.

Press, C., Bird, G., Walsh, E., & Heyes, C. (2008). Automatic imitation of intransitive actions. *Brain and Cognition, 67*, 44–50.

Preston, S. D., & de Waal, F. B. M. (2002). Empathy: Its ultimate and proximate bases. *Behavioral and Brain Sciences, 25*, 1–20.

Prinz, W. (1987). Ideo-motor action. In H. Heuer & A. F. Sanders (Eds.), *Perspectives on perception and action* (pp. 47–76). Hillsdale, NJ: Lawrence Erlbaum Associates, Inc.

Proctor, R. W., & Vu, K.-P. L. (2006). *Stimulus-response compatibility principles: Data, theory and application*. Boca Raton, FL: Taylor & Francis.

Puzzo, I., Cooper, N. R., Cantarella, S., & Russo, R. (2011). Measuring the effects of manipulating stimulus presentation time on sensorimotor alpha and low beta reactivity during hand movement observation. *NeuroImage*, *57*, 1358–1363.

Rizzolatti, G., & Craighero, L. (2004). The mirror-neuron system. *Annual Review of Neuroscience*, *27*, 169–192.

Rizzolatti, G., Fadiga, L., Fogassi, L., & Gallese, V. (1996). Premotor cortex and the recognition of motor actions. *Cognitive Brain Research*, *3*, 131–141.

Rothbart, M., Evans, M., & Fulero, S. (1979). Recall for confirming events: Memory processes and the maintenance of social stereotypes. *Journal of Experimental Social Psychology*, *15*, 343–355.

Sartori, L., Cavallo, A., Bucchioni, G., & Castiello, U. (2011). Corticospinal excitability is specifically modulated by the social dimension of observed actions. *Experimental Brain Research*, *211*, 557–568.

Sebanz, N., Bekkering, H., & Knoblich, G. (2006). Joint action: Bodies and minds moving together. *Trends in Cognitive Sciences*, *10*, 71–76.

Shih, M., Wang, E., Trahan Bucher, A., & Stotzer, R. (2009). Perspective taking: Reducing prejudice towards general outgroups and specific individuals. *Group Processes & Intergroup Relations*, *12*, 565–577.

Sime, J. D. (1983). Affiliative behaviour during escape to building exits. *Journal of Environmental Psychology*, *3*, 21–41.

Skowronski, J. J., & Carlston, D. E. (1989). Negativity and extremity biases in impression formation: A review of explanations. *Psychological Bulletin*, *105*, 131–142.

Sporer, S. L. (2001). The cross-race effect: Beyond recognition of faces in the laboratory. *Psychology, Public Policy, and Law*, 7, 170–200.

Stroop, J. R. (1935). Studies of interference in serial verbal reactions. *Journal of Experimental Psychology*, *18*, 643–662.

Stürmer, B., Aschersleben, G., & Prinz, W. (2000). Correspondence effects with manual gestures and postures: A study of imitation. *Journal of Experimental Psychology: Human Perception and Performance*, *26*, 1746–1759.

Taylor, S. E. (1991). Asymmetrical effects of positive and negative events: The mobilization–minimization hypothesis. *Psychological Bulletin*, *110*, 67–85.

Tiedens, L. Z., & Jimenez, M. C. (2003). Assimilation for affiliation and contrast for control: Complementary self-construals. *Journal of Personality and Social Psychology*, *85*, 1049–1061.

Turner, J. C., Brown, R. J., & Tajfel, H. (1979). Social comparison and group interest in ingroup favouritism. *European Journal of Social Psychology*, *9*, 187–204.

Umilta, M. A., Kohler, E., Gallese, V., Fogassi, L., Fadiga, L., Keysers, C., & Rizzolatti, G. (2001). I know what you are doing. A neurophysiological study. *Neuron*, *31*, 155–165.

van Baaren, R. B., Holland, R. W., Kawakami, K., & van Knippenberg, A. (2004). Mimicry and prosocial behavior. *Psychological Science*, *15*, 71–74.

Van Bavel, J. J., Packer, D. J., & Cunningham, W. A. (2008). The neural substrates of in-group bias: A functional magnetic resonance imaging investigation. *Psychological Science*, *19*, 1131–1139.

Van Schie, H. T., van Waterschoot, B. M., & Bekkering, H. (2008). Understanding action beyond imitation: Reversed compatibility effects of action observation in imitation and joint action. *Journal of Experimental Psychology: Human Perception and Performance*, *34*, 1493–1500.

Weisbuch, M., & Ambady, N. (2008). Affective divergence: Automatic responses to others' emotions depend on group membership. *Journal of Personality and Social Psychology*, *95*, 1063–1079.

Wiesel, E. (n.d.). Quoted in *U.S. News & World Report*, October 27, 1986.

Wilson, M., & Knoblich, G. (2005). The case for motor involvement in perceiving conspecifics. *Psychological Bulletin*, *131*, 460–473.

Xu, X., Zuo, X., Wang, X., & Han, S. (2009). Do you feel my pain? Racial group membership modulates empathic neural responses. *The Journal of Neuroscience*, *29*, 8525–8529.

Ybarra, O., & Stephan, W. G. (1996). Misanthropic person memory. *Journal of Personality and Social Psychology*, *70*, 691–700.

16

INSPIRED BY THE QUESTION, NOT THE MEASURE

Exploiting Neurobiological Responses in the Service of Intergroup Research

Wendy Berry Mendes

My Path

My interest in using psychophysiology to illuminate intergroup processes began long before entering graduate school. During my undergraduate days, I was on a path toward applied statistics, with an interest in applying quantitative models to understanding human behavior, specifically criminal behavior. This interest led me to seek an intern position with the Federal Bureau of Investigation (FBI), crime analysis division. The position required exposure to high-level security information, and as a result required a fairly extensive background check. As part of my pre-employment screening, I had to complete an interview during which my autonomic nervous system (ANS) responses were monitored—more commonly known as a "polygraph test." Of course, polygraph simply means multiple graphs, but people often (and incorrectly) use this term synonymously with "lie-detection test." For me, this experience cemented the idea that our mental states and bodily responses are inextricably linked, and that there are times when our neurobiological responses can expose thoughts and feelings that we would prefer people not know we have. The story of my "polygraph" experience illustrates some critical psychophysiological principles that I will review later in this chapter.

The interview was scheduled in the late afternoon at the main headquarters of the FBI in Los Angeles. When I arrived at the office of the man interviewing me (let us call him John), he was friendly and gracious and suggested we get some tea before the interview. As we made (what I thought was) idle chatter, John relayed a story from his past week, in which he had to dismiss a soon-to-be commissioned agent (let us call him Ted), because Ted had admitted to John that he was not "100% honest all the time." I mimicked John's horrified expression at hearing that someone who desired to work at *the FBI* would not be anything but 100% honest, but my mind also started racing with the high standards that this position required.

As we were walking back to the interview room, John also casually offered that I really had nothing to worry about given that my background check had been pristine, and many of my personal references had remarked on my integrity and how I was the type of person who would "never gossip or say anything negative about others." Again, my mind raced with feelings of pride that my friends would offer such positive statements of my character, but also I wondered, who the hell did they talk to?! No matter, I was feeling relaxed, confident, and supported by my new friend. As for John, I suspect he was thinking something entirely different. Indeed, for him, the trap was set for a *guilty knowledge test.*

The session began with the application of a number of sensors to measure respiration, heart rate, skin conductance, and skin temperature. There was a lot of equipment, but without question, the most important aspect of the situation was the psychological components that preceded the interview. The session started like any psychophysiology experiment would, with me relaxing in a seated position to establish a *baseline* recording of responses. Of course, my responses were not as "relaxed" as they would have been if I were at home reading a book, but in experimental lab studies, we typically care about changes from a relatively neutral state to a task or event. After a few minutes, John began asking fairly serious questions regarding my extensive traveling history, in some cases to Communist countries—the concern was that I was a drug smuggler or, more dramatically, involved in espionage; I was, boringly, not either one. Then, in the midst of this intense questioning, John asked, "In the past 7 years, have you been 100% honest all the time?" Having a short-term memory that extended beyond 1 hour, the answer was easy—John had provided the only answer I could give based on the "Ted" story—I swallowed and offered, "Yes." The questions continued—drug use and activity, criminal activity, and other assorted illegal activities—and then this one: "In the past 7 years, have you ever said something negative about a friend?" Again, I offered the only response that made sense given our earlier conversation about my personal references—"No." After a few more questions, John told me to relax and then announced that the first trial was over.

I turned, expecting a beaming congratulatory smile, but instead John looked worried. Apparently, he explained, the polygraph suggested some deception, and so we would need to "try this again." He told me that two of my responses were problematic: the "100% honest" and the "negative statements about friends." John recommended that I try to think about my answers before responding, to make sure that my mind was not wandering. We then repeated the same line of questions, only this time, I was incredibly aware of my heart racing the moment he started asking "in the past 7 years. . ." After the second trial, John let out a long sigh and then expressed concern that this "doesn't seem like it will work out for you after all." He wanted to be optimistic, though, and suggested that we try it one more time.

The third trial was the worst. I could feel my entire body on edge: cold, sweaty palms, dry mouth, racing heart, all peaking the moment he started asking "in the past 7 years. . ." My body had betrayed me and exposed me for my white lies and gossipy ways; I began to contemplate other career options—business school seemed like a good idea. However, this time when I turned around, John smiled and told me I had passed the polygraph and that all was clear for my hire. So what happened?

I had been an unwitting participant in a *guilty knowledge test*, in which the interviewer obtains autonomic nervous system responses during control and "lie" answers. The more you can increase the respondent's belief in the importance of the "lie" questions, the better this technique works. John needed me to lie so that he could compare the intensity of the physiological responses when I was lying to answers when I was, presumably, telling the truth. If I were skilled at beating the "polygraph test," then the "lie" answers would have shown little reactivity. Given the likelihood that someone in their early 20s has told a white lie or said something negative about a friend, the "lie" questions offer an important comparison to the target questions. By increasing my anxiety over the two "lie" questions, John had responses when I was lying to compare to the target questions.[1]

For me, this experience was foundational. I was fascinated that my body could betray me, and that no matter how much I tried to control my internal responses, I was a slave to them in many respects. Most notably, however, was the realization that it was the *combination* of the psychological context and physiological changes that were important during the polygraph test. This experience planted the seed of interest that emerged a few years later when I went to graduate school.

Neurobiological Measures

There are many different biological changes that can be reliably measured and might be profitably applied to the study of intergroup relations. Indeed, the first obstacle is deciding, among the many options, how and where to focus one's energy and resources. It is true that you cannot do everything, but as it relates to neurobiological methodology, in the Isaiah Berlin dichotomy, I am a fox and not a hedgehog. There are certainly advantages and disadvantages to being a neurobiological researcher who is a hedgehog—narrowly, but expertly, focused on a single domain; compared to a fox—broadly focused, but at the risk of not being deeply proficient across all topics. However, I would argue that our discipline needs both types of researchers, since only with foxes and hedgehogs working together can progress occur both in the development of more precise neurobiological measures (depth), as well as linking across systems to paint a comprehensive picture of the intersection of the mind, brain, body, and behavior (breadth).

The Obstacles

Prior to describing some of the work in my lab, it would only be fair to describe the various obstacles that one might encounter when trying to integrate these measures successfully into one's research. My intentions in describing these obstacles are not to dissuade the interested researcher, but rather to manage the expectations of what lies ahead.

The first obstacle is deciding what to measure. Students and scholars interested in using neurobiology as a method to inform their work are well-advised to start small and focused, and only consider branching out after mastering a single domain. Another option would be to pair up with a colleague with neurobiological expertise from a different area or department; for example, public health and medical schools often have experts in neurobiological domains that are not always obvious collaborators. As with any interdisciplinary endeavor, progress will be slower and more difficult at all stages of research. You will have different methodological and theoretical perspectives, speak a different language from each other, and may not have aligned goals on where to publish. As such, integrating neurobiological responses into your research program should only be done after much consideration and with cautious expectations. It is also worth noting here that neural or biological responses should not be privileged over other outcome variables—they are neither more nor less scientific, nor do they provide a direct window into an individual's mind—but instead they are best considered as part of a multi-componential understanding of how social context influences mental states and behavior.

A second obstacle in integrating neurobiology into your research program is determining what the physiological responses indicate at a psychological level—that is, what can we infer about mental states by examining physiological responses? I would argue that this is the most difficult obstacle for psychologists to overcome. Our discipline would not be satisfied if we simply reported changes in heart rate or increases in sweat gland activity without some interpretation of what the responses indicate at a psychological level. But how certain can we be that, for example, a correlation between a hormone and a personality dimension is not simply epiphenomenal, or that a shift in the parasympathetic response indicates an underlying change in emotions? The unsatisfying answer is, "It depends." And this is where knowledge of psychophysiological theory becomes paramount.

Understanding Psychophysiological Theory

There are several thorough chapters that review psychophysiological inference that should be mandatory reading for researchers interested in these methods (e.g., Cacioppo, Tassinary, & Berntson, 2000, 2007). Here, I summarize and augment what others have written about psychophysiological inference. At the broadest level, we can begin to think of neurobiological measures in terms of their *sensitivity* and *specificity*. These constructs are conceptually orthogonal, but a cursory reading might lead one to conclude that measures that are more sensitive tend not to be as specific, and vice versa, though this would be a faulty conclusion.

Sensitivity is the extent to which the neurobiological response reliably changes as a function of shifts in mental states (emotions, thoughts, intentions, etc.) with the idea that subtle or minor shifts in mental states would affect *highly sensitive* measures, but measures *lower in sensitivity* would require more intense mental states to observe changes. Skin conductance (SC) provides a good example of a measure that is highly sensitive because very subtle and low-level changes in affect, cognition, and intentions can influence SC changes. In contrast, cortisol changes would be considered very low in sensitivity because cortisol increases typically require a very intense psychological experience. For example, in a meta-analysis of over 200 studies, the experimental paradigms that were reliably linked to cortisol increases were ones described as *uncontrollable, socially evaluative, and threatening* (Dickerson & Kemeny, 2004). Passive events, like watching scary movies or the threat of electric shock, did not reliably increase cortisol levels. Casual social interactions, even with out-group or stigmatized partners, are not likely to engender cortisol increases. Instead, researchers often observe cortisol levels that decline over the course of an experiment (reflecting the decline in cortisol as a function of the diurnal cycle), and they are left to interpret what differential declines in cortisol could mean psychologically, often making the vague and clumsy assumption that greater declines indicate "less stress" or "more relaxation." This can muddy psychological inference, given that there is limited evidence that differential declines from resting states are psychologically meaningful.

Specificity refers to the extent to which the neurobiological response is related to a particular mental state, with *low specificity* indicating that a neurobiological response is related to many mental states, and *high specificity* indicating that a neurobiological response is related to fewer or (possibly) a single mental state. Skin conductance, though highly sensitive, is not very specific. Many psychological states, some even polar opposites of each other, can engender a skin conductance change. I often give the example that, if while I measured your skin conductance, a person whom you loathe enters the room, your skin conductance would increase; if, on the other hand, a person whom you love enters the room, your skin conductance would also increase. In other words, the measure is not specific to negative or positive affect—instead it responds to intensity, general arousal, or effort.

Perfect specificity—or one-to-one correspondence between a psychological state and physiological response—is rare, and in most cases, perfect specificity is observed only in constrained contexts. For example, skin conductance responses (SCR), as described above, are sensitive, but non-specific; however, when coupled with a fear-conditioning paradigm, SCRs can be reasonably interpreted as an indication of fear. In a fear-conditioning paradigm, participants view pictures paired with an aversive stimulus, like an electrical shock. After several pairings of the pictures and shocks, the aversive stimulus is removed, and the participants' responses to the conditioned stimuli are examined. Upon presentation of the conditioned stimuli, SCRs are examined, and often the interpretation is that SCRs to the conditioned stimulus indicate "fear." Note the very constrained paradigm of fear conditioning that enables one to then make the inference that SCRs are synonymous with "fear." The situational constraints of the fear-conditioning paradigm limit the number of mental states that can be experienced, which leaves changes in SCRs as most likely indicating fear. The point here is that physiological responses gain meaning when the psychological context is considered.

Stumbling with Psychophysiology

I started my doctoral training with Jim Blascovich who, with Joe Tomaka, had just published an article in *Advances in Experimental Social Psychology* on the theory of challenge and threat, which claimed that cardiovascular responses during stressful tasks could be differentiated to index whether someone was experiencing an approach state, *challenge*, or a defeat state, *threat* (Blascovich & Tomaka, 1996; see also Blascovich, Chapter 12 in this volume). At this point, there had been a handful of papers that made the argument that cardiovascular reactivity during *motivated performance situations*—defined as self-relevant active tasks requiring cognitive responses—could be differentiated, based on whether participants perceived the task as more demanding than their resources to cope (*threatening*) or perceived their resources as greater than the demands of the task (*challenging*).

The threat versus challenge distinction integrated early appraisal work by Lazarus and Folkman (1984), who argued that how individuals perceive the demands of an event, coupled with their resources to cope, determines how successfully they manage the stressor; and Dienstbier's physiological arousal theory (1989), which argued that the two primary stress systems (sympathetic–adrenal–medullary [SAM] and hypothalamic–pituitary–adrenal [HPA] axes) can differentiate *tough* or resilient patterns of reactivity (large SAM activation, with smaller HPA responses) from *weak* patterns (smaller increases in SAM responses coupled with greater activation of HPA). The theory claimed that the mental states of challenge and threat could be differentiated by examining, among other biological responses, cardiovascular reactivity. Specifically, in challenge states, there would be large increases in sympathetic nervous system (SNS) responses: The heart would work more efficiently, thus increasing cardiac output; and the vasculature system would expand (dilate) allowing for more oxygenated blood to travel to the brain and periphery. These changes could be indexed by examining cardiac output (a volume-based measure of blood ejected from the heart each minute) and total peripheral resistance, estimated from the ratio of average blood pressure and cardiac output. In contrast, threat states would be associated with increased SNS responses—but often not as large as those observed in challenge states—less efficient cardiac responses, and increased vascular resistance.

In my first year in graduate school, we began discussing applying this theory to intergroup interactions. Along with Brian Lickel and Sarah Hunter, we began a series of studies attempting to show that intergroup interactions were *threatening*, as indicated by a profile of cardiovascular responses. Finding supportive data turned out to be much more difficult than we initially expected. One of the obstacles was to create a task that was engaging enough to activate the sympathetic nervous system—a requirement for challenge and threat distinctions—but was not so stressful that everyone experienced threat. This is the exact place an experimental psychologist does not want to be—designing an experiment around the constraints of the measures. There were several false starts with this research program. First we tried to have participants simply imagine a partner (either the same race or a different race) and then give a video-recorded speech about working with him/her. The speech activated the SNS, but we did not observe challenge and threat differences—for the most part, participants were challenged, and they did not perceive the task as distressing. We then tried giving participants a picture of their partner (racial in-group or out-group) to make the partner's race salient. Again, we observed no differences between conditions. We then intuited that the interaction might have to be live, so we began to use confederates (see Norton, Dunn, Carney, & Ariely, 2012, for a recent empirical example of this point).

When we switched to live interactions, we finally started seeing the expected differences in reactivity. We used this paradigm in over a dozen studies, in which we demonstrated that partners who were stigmatized in some way (either due to race, socio-economic status, or physical stigmas) engendered threat reactivity in non-stigmatized participants (Blascovich, Mendes, Hunter, Lickel, & Kowai-Bell, 2001; Mendes, Blascovich, Lickel, & Hunter, 2002; Mendes, Major, McCoy, & Blascovich, 2008; see also Blascovich, Chapter 12 in this volume). We also found nuances that shed light on the psychological processes that may lead to *threat* reactivity; for instance, partners who violated expectations because they were counter-stereotypical or atypical (e.g., a wealthy Latino or an Asian American with a thick Southern accent) also engendered threat (Mendes, Blascovich, Hunter, Lickel, & Jost, 2007). This last finding was critical because it demonstrated that uncertainty, and not simply negative affect, could drive *threat* responses.

I offer my naïve intellectual journey partly as a cautionary tale of what not to do. Of course measurement and methodological frameworks are important (see Greenwald, 2012, for a lively discussion), but the theory or question should drive the choice of methods, not vice versa. In my lab, new students and post-doctoral fellows focus on a specific question before we decide what makes sense to include (if anything) in terms of neurobiological measures. I strongly encourage this approach, and I think we feel liberated in the questions we pursue when we are not constrained by one methodological framework.

Integrating Neurobiological Measures to Test Intergroup Theories

Over the past decade in my lab, my collaborators and students have been actively engaged in a program of research examining, at the broadest level, social interactions between two strangers, during which we measure a variety of neurobiological responses to understand prejudice behavior and experiences of discrimination. Broadly, our studies seek to understand intergroup interactions from multiple perspectives:

- the *perceiver*, the non-stigmatized or "advantaged" group member;
- the *target*, the minority, stigmatized, or disadvantaged group member; and more recently
- the *dyad*, referring to reciprocal, complementary, or coordinated action between two strangers.

Using neurobiological measures in these interaction studies has offered illuminating and sometimes contradictory findings when compared to self-report or controlled behavior.

When looking at the broader context of prejudiced behavior, by most historical accounts during the last half of the 20th century and the start of the new millennium, there has been a decrease in expressed prejudice toward racial minority group members in the United States. There are laws that ban discriminatory practices in employment, housing, and healthcare. Many affirmative action policies are intended to correct previous racial injustices by explicitly encouraging, and in some cases requiring, decisions to favor racial minorities, all else being equal. Focusing on African Americans, over the past 50 years, the United States has observed dramatic changes: from Jim Crow laws and school segregation in the 1950s and 1960s, to African Americans serving as heads of industries, Supreme Court justices, presidents of universities, and the highest executive position of power in the United States.

Laws and policies are one way to gauge the progress of minority or stigmatized people, attitudes that people hold are another. One optimistic sign of progress toward a less biased nation can be seen in Americans' attitudes regarding their willingness to vote for an African American for President. In 1958, the majority of voters surveyed by Gallup, 53%, said they would not. In 2002, the percentage of voters who reported that they would not vote for an African American candidate was down to 6%; and in 2006, a *Newsweek* poll reported the percentage below 3%. These numbers, coupled with objective indicators of African Americans occupying the highest offices in the country, offer an optimistic sign of how the United States has progressed over the past 50 years, and have led some journalists and politicians to proclaim that "racism is dead" (Bowden, 2008).

But as social scientists, we have known for decades that ingrained attitudes, preferences, and associations can linger under the surface long after individuals appear to have eschewed these beliefs. Also, we know that when individuals actively try to suppress thoughts, emotions, or beliefs, they often ironically bubble up to the surface, and that this is more likely to occur when the ability to regulate our reactions is compromised (Wegner, 2009). This suggests that explicit desires to be egalitarian may be at odds with automatic responses to stigmatized or out-group members. When this tension exists, there are situations when visceral responses might be difficult or impossible to regulate—resulting in responses that might be counter to one's *expressed* beliefs.

Here, I will describe a series of interrelated studies concerning *meta-monitoring* during intergroup interactions. That is, during interactions with members from different social groups, individuals often engage in self-presentational and monitoring strategies that are distinct from the strategies they use when interacting with in-group members. We attempted to show that these strategies—probably intended to make the interaction smoother—can undermine the authenticity of the interaction and may be part of the root of strained intergroup interactions.

The strategies in which perceivers and targets engage are likely to be different, as are their goals (Bergsieker, Shelton, & Richeson, 2010), but in both cases, the extra monitoring may present an additional strain on the interaction that can be observed using a variety of neurobiological methods. I will describe the different research programs below, starting with the *perceiver*, who I contend often engages in *overcorrection*, an exaggerated positive reaction toward the target. Next, I will discuss the *target*, who may engage in *hypervigilance*, scanning the environment and the behaviors of the interaction partner to attempt to determine his/her motives, intent, and genuine beliefs. Finally, I will describe a study in which, while simultaneously examining the perceiver and target, we attempted to manipulate whether or not the dyad could engage in these meta-monitoring strategies, as a way to examine the counterintuitive hypothesis that, by reducing cognitive resources *prior* to a social interaction, dyads might show more authentic reactions to each other and ultimately develop more mutual trust and shared understanding. Importantly, the neurobiological measures that we obtained were critical in providing insight into how individuals were regulating their affective responses, given their self-reports and behavior presented contradictory findings.

Perceiver Perspective

A good amount of research has accumulated examining neurobiological responses from perceivers who are interacting with or observing out-group (target) others. Early intergroup research dates back to the mid-1950s, when Rankin and Campbell (1955) reported that skin conductance responses were higher when White participants were touched by a Black compared to a White experimenter. But since the late 1990s, research on intergroup interactions using neurobiological methods has increased exponentially. Part of the increase in the number of studies is probably due to advances in functional magnetic resonance imaging (fMRI) and other psychophysiological techniques, coupled with greater precision in theory on intergroup anxiety and threat (e.g., Blascovich & Mendes, 2010; Shelton & Richeson, 2006; Stephan & Stephan, 2000).

If one reads only the physiological or neural literature on intergroup interactions, one might come away with the impression that interactions with out-group members are reliably associated with more stress, threat, and negative affect, such as fear and anxiety. If, however, one examined literature exclusively using self-reports, a different conclusion would be made— that we are moving toward being a more egalitarian and less xenophobic society. In several studies, we simultaneously examined these responses so we could explore the relations between them.

In one study (Mendes et al., 2007), we recruited European American adults ($N = 78$; ages 20–55, $M = 31.5$) to come to the lab for a 90-minute experiment. At least 48 hours prior to coming to the lab, the participants completed some measures on-line that included a Black–White Implicit Association Test (IAT) (Greenwald, McGhee, & Schwartz, 1998), to provide a measure of implicit racial bias. Upon arrival at the lab, we obtained an initial consent for participation in the study that described everything up to the stress task; a health intake interview was conducted; and the participant rested for at least 20 minutes prior to giving an initial saliva sample. We then described the task that they would be asked to complete—a mock job interview to two evaluators—and at this time, we obtained a second consent. It is important not to contaminate baseline responses by informing participants of the stress task at the onset of the study. Upon consent to continue, the experimenter then introduced the two evaluators.

The only manipulation in this study was whether the two evaluators (one male, one female) were two White or two Black evaluators. The evaluators then described the task in more detail and left the room for the participant to prepare the speech. After the preparation period, the evaluators came back into the room, sat across from the participant, and engaged in a mock job interview (modeled after the Trier Social Stress Test; see Kirschbaum, Pirke, & Hellhammer, 1993). The task lasted just over 15 minutes, from the time of the initial explanation until the evaluators left the room. A second saliva sample was obtained 5 minutes after the task, with the intent being to obtain the peak neuroendocrine change due to the evaluation task. Once the study was complete, the saliva samples were assayed for cortisol and dehydroepiandrosterone-sulfate (DHEA-S), an adrenal steroid that can counter-regulate the effects of cortisol. We also had research assistants, blind to evaluators' race, code the videos for affective, behavioral, and observable stress responses.

When we focused on the cortisol reactivity data, we did not observe either a main effect for evaluators' race nor moderation by racial bias. Instead, regardless of the evaluators' race or the participant's level of implicit racial bias, participants showed a large increase in cortisol as a result of the stress task. However, when we considered a more complete profile of neuroendocrine responses by examining DHEA-S and anabolic balance (ratio of DHEA-S and cortisol), we observed the expected role of racial bias. Among White participants evaluated by Black evaluators, racial bias was related to DHEA-S changes, such that higher racial bias was related to lower DHEA-S and anabolic balance.

If one only considered these findings, the conclusion would be clear: among White participants, the higher the racial bias, the worse the neuroendocrine outcome. However, when we examined participants' non-verbal behaviors and emotions as rated by observers unaware of the evaluators' race, a very different conclusion emerged. With observer ratings, we noted main effects for evaluators' race, and racial bias moderated the response. Specifically, White participants displayed *more* positive emotion and smiled more when evaluated by Black interviewers than those evaluated by White interviewers, and greater racial bias was associated with *more* positive emotion and more smiling (Blascovich, Mendes, & Seery, 2002; Mendes et al., 2007; Mendes & Koslov, in press). In sum, when we examined physiological responses, we observed interactions with out-group members engendering greater threat and negative reactions; but when examining explicit self-reports and behavior, we observed more positive responses toward the out-group partners.

These effects have been documented by others, and a variety of labels have been used—overcorrection, color blindness, positive biases—but we wanted to understand the roots of these *correction* effects. One road that we took was to use psychophysiology as both a measure of bias as well as a way to understand emotion regulation capacity, with the idea being that when individuals had the resources to correct for racial biases, they would do so—and often make a miscalibration error of *overcorrecting*; but that when resources were depleted, individuals would not be able to monitor their responses, and instead the *overcorrection* effect would be eliminated.

To explore this idea, we (Mendes & Koslov, in press) designed an experiment in which participants had to complete a stressful task—an evaluative job interview in the presence of two stoic interviewers—while we measured their cardiovascular responses. Immediately after the task, we asked participants to review a stack of resumés in which there were low-, medium-, and high-quality candidates, of which one-third had African American names and the other two-thirds had European American names. We instructed the participants to select the eight best resumés, forcing them to select at least two medium-quality resumés. If one approached this task with either an egalitarian perspective or a completely color-blind approach, they would choose either no African American resumés or one medium-quality African American resumé. If someone were *overcorrecting*, they would pick two or more medium-quality African American resumés.

Our prediction was that individuals would overcorrect if they had the resources to do so, and in this case, we defined "resources" as showing cardiovascular (CV) reactivity consistent with challenge. We used the CV responses obtained from the interview to categorize participants into one of two groups based on their reactivity. About 60% of the participants responded to the speech task with a *challenge* profile, whereas the other participants responded with a *threat* profile. When we examined resumé choice based on challenge and threat classification, we found that individuals higher in racial bias and who were challenged *overcorrected*

—chose two or more medium-quality African American resumés; but if participants were higher in bias and were *threatened*, they did not over-select African American resumés. Low-bias individuals, whether challenged or threatened, showed egalitarian responses. These data demonstrated that among individuals with higher racial bias, when they have the resources to do so, there is overcorrection. Of course, this study was limited due to the individual difference approach we used: We did not randomly assign participants to challenge or threat states, but rather categorized them based on their CV reactivity. We therefore followed up on this study by randomly assigning participants to a resource-depletion condition or a control condition.

The follow-up study comprised a two-experiment ruse in which the first part of the experiment examined White participants' preferences between two celebrities, and the second experiment examined racial bias by measuring neuroendocrine reactivity during an evaluation by two Black interviewers. During the first session, we depleted participants' ability to correct their preferences by having them complete a celebrity-choice task while under cognitive load— keeping track of how many times they heard a piano tone on an audiotape with four different instruments playing the same tone. The control participants were played the same audiotape, but were told that its purpose was to muffle outside sounds. In the second session, participants completed a mock job-interview task in front of two Black evaluators. We collected a saliva sample at baseline and after the interview task and later assayed these for cortisol.

Consistent with the overcorrection hypothesis, participants in the no-depletion (control) condition indicated a preference for Black over White celebrities, whereas participants in the depletion condition did not show this preference. Furthermore, when we examined racial bias, as indicated by cortisol reactivity during the speech, we found that cortisol levels were associated with greater preference for Black celebrities, but only in the no-depletion condition. This effect was reversed in the depletion condition: Large cortisol increases were associated with fewer Black celebrities chosen, though this slope was not significant. We also measured racial bias by asking the evaluators to rate how anxious participants appeared during the interview. Paralleling the cortisol data, the more anxious participants appeared to the Black evaluators, the more these participants had chosen Black celebrities in the no-depletion condition. Interestingly, the correlation between the two measures of "racial bias" (cortisol reactivity and interviewers' ratings of anxiety) were non-significantly correlated at $r = .10$, underscoring the distinctiveness of these measures even though they yielded similar correlations with the outcome measures.

Targets' Perspective

This series of studies led us to question if African Americans perceived Whites' positive treatment as disingenuous. There were several previous studies that examined this idea and concluded that positive feedback from Whites might be discounted by minority members. For example, Crocker, Voelkl, Testa, and Major (1991) found that Black participants' self-esteem dropped after positive evaluations from White partners (see also Major & Kunstman, Chapter 17 in this volume). Similarly, Cohen, Steele, and Ross (1999) provided Black participants with praise on a written essay, which resulted in Black participants *de-identifying* from the academic domain more than Whites receiving the same feedback. The possible limitation with these previous approaches is that minority participants might be reluctant to report negative affect for fear of looking like a complainer (Kaiser & Miller, 2001). We thought physiological measures might yield important insight into this question.

We (Mendes et al., 2008) examined this question, in part, by having Black and White participants receive socially evaluative *positive* feedback from either a Black or White same-sex (confederate) partner. Participants then engaged in a time-pressured, cooperative task that included a joint monetary bonus if the dyad "performed well together." These task constraints were not arbitrary, and indeed when we did not include joint bonuses or tasks with time pressures, we did not observe significant increases in sympathetic nervous system responses.

The cardiovascular reactivity from the task yielded significant interactions for all of our key physiological responses. Specifically, for all race-partner combinations, except one, positive feedback resulted in reactivity consistent with challenge profiles—increased cardiac output and decreased total peripheral resistance. This is consistent with the idea that positive feedback can be perceived as a type of *resource* for participants, triggering more challenge than threat responses. However, for African Americans, positive feedback from a White partner was associated with *threat* reactivity—decreased cardiac output and increased total peripheral resistance.

We also coded the behavior during the cooperative task for *vigilance*—operationalized as how often the participant looked away from the computer task to watch their partner. This measure yielded the same interaction—the only differences were found in the Black participant–White confederate dyads, in which the Black participants monitored their White partners more often than any other dyads. Not surprisingly, given this distracted behavior, performance was lower, with Black participants who received positive feedback performing significantly worse when paired with White partners than when paired with Black partners. Notably, very few self-reports of emotions and stress showed any effects that paralleled what we observed for physiology, behavior, and performance. Whether participants were not aware of these changes, or were aware but did not want to report on them, is unknown; but it would be interesting to try to integrate a guilty knowledge test (see "My Path," above) in an intergroup paradigm to see if that test could shed light on participants' awareness of intergroup anxiety effects that are manifested physiologically.

Intergroup Dyads

Thus far I have described studies that extended early work showing that intergroup interactions typically engender more *threat* reactivity than same-race interactions, and showed that one possibility of why this occurs is due to more *meta-monitoring* during intergroup interactions. In the final study I describe, which is based on a dissertation by one of my graduate students (Koslov, 2010), we explored this idea within a *dyadic* study, in which we recruited same-race or different-race dyads and measured their autonomic nervous system (ANS) reactivity while working together. This paradigm allowed us to examine questions regarding physiological synchrony—the coordinated physiological responses between dyads, typically referred to as *physiological covariation.* Importantly, we manipulated cognitive resource depletion *prior* to the interaction, to examine the counterintuitive idea that depleting resources might increase the genuineness of the interaction by preventing meta-monitoring.

Conducting a dyad study in which one hopes to measure physiological synchrony poses several challenges. First, to obtain temporal precision, one must coordinate physiological collection from the dyad exactly, either with time stamps or integrating the signals from the two participants into the same collection unit. Because ANS changes can occur within seconds of an experienced emotional state, one needs precise recordings of when changes are occurring from each member of the dyad. Additionally, some of the measurements we use, such as impedance cardiography and skin conductance, employ an external current. If measurement devices with external currents are used on both participants and the members of the dyad touch each other, the signal will be lost due to electrical interference.

We used this signal loss to our advantage in the dyad study; one of the tasks we developed was described as an "American sign language task," which required participants to touch each other's hands. By measuring the amount of electrical noise in the physiological signal, we were able to measure precisely how often and for how long, to the millisecond, these strangers touched. While this meant that we did not have all of the ANS responses during the task, we viewed this as a worthwhile trade-off for obtaining a precise measure of touch time.

The papers reporting these results are currently in progress, so I will only briefly outline some of the results reported in Koslov's (2010) dissertation. Consistent with our prediction that cognitive depletion would increase amiability in cross-race dyads, we found that Black and White partners in the control condition touched hands less than those in the depletion condition. In addition, Blacks' race-based rejection sensitivity moderated this effect: The higher they were in race-based rejection sensitivity, the less they touched the hands of White evaluators in the control conditions; this effect was reversed in the depletion condition, where we observed more touching among those higher in race-based rejection sensitivity.

With regard to physiological covariation in general, we found that White participants paired with other White participants showed strong physiological covariation—White participants' pre-ejection period (PEP) responses were highly correlated with their partner's PEP responses. For cross-race dyads, however, we observed non-significant covariation in the control condition; but in the depletion condition, Black and White dyads showed similar covariation to White dyads.

As physiological measurements become more commonplace and affordable, and as more researchers learn to incorporate these measures, dyadic- and group-level processes will be ripe for experimental paradigms. Additionally, I think these types of data are well suited for sophisticated dyadic analyses like those developed by Kenny, West, and their colleagues (Kenny, Kashy, & Cook, 2006; West, in press) that can isolate actor and observer effects. There are many critical intergroup theories that could benefit from these measurements, and I look forward to seeing other labs develop these paradigms.

Conclusions

Applying neurobiological measures to understand intergroup relations is not new, but we are entering an exciting era: Costs for the measures are decreasing, better technology will allow for smaller and more accurate devices, ambulatory measures are affordable, and psychophysiological theory continues to advance. Diving headfirst into using neurobiological measures will not be without frustration, energy, and financial costs, and like any study, there is no guarantee that your results will be consistent with your hypotheses or yield anything of value. Indeed, the unwieldiness of psychophysiological responses might make it more difficult to find effects than with other measures. That stated, psychophysiological measurements may advance social psychology, and intergroup relations in particular, given that individuals are often unwilling or unable to express their true beliefs.

Note

1 It should be noted that the validity of using psychophysiological responses to detect lies remains controversial, and in 2003, a National Academy of Sciences of the United States of America panel concluded that the use of these techniques to identify people who may show deception in the future was without merit. Additionally, the panel noted that there was support for using these measures when paired with a context like the *guilty knowledge task*.

Suggestions for Further Reading

Blascovich, J., Mendes, W. B., Vanman, E., & Dickerson, S. S. (2011). *Social psychophysiology for social and personality psychology*. Affective Science Series. London, UK: Sage.

This book was specifically written as a "how-to" guide for those new to psychophysiology. The scope is limited to ANS measures, EMG, and cortisol, and provides a basic vocabulary to allow researchers to build on.

Cacioppo, J. T., Tassinary, L. G., & Berntson, G. G. (2007). Psychophysiological science: Interdisciplinary approaches to classic questions about the mind. In J. T. Cacioppo, L. G. Tassinary, & G. G. Berntson (Eds.), *Handbook of psychophysiology* (3rd ed., pp. 1–16). New York: Cambridge University Press.

This is the quintessential guide to psychophysiological measurement and covers theory, principles, physiology, and measurement issues. The handbook is in its 3rd edition as of this writing, but is updated approximately every decade.

Major, B., Mendes, W. B., & Dovidio, J. (in press). Intergroup relations and health disparities: A social psychological perspective. *Health Psychology.*

A comprehensive review paper examining how social psychological theories may inform health researchers interested in understanding health disparities.

Mendes, W. B. (2009). Assessing the autonomic nervous system. In E. Harmon-Jones and J. S. Beer (Eds.), *Methods in social neuroscience* (pp. 118–147). New York: Guilford Press.

A chapter dedicated solely to the use and interpretation of the autonomic nervous system responses.

Mendes, W. B., & Jamieson, J. (2012). Embodied stereotype threat: Exploring brain and body mechanisms underlying performance impairments. In M. Inzlicht & T. Schmader (Eds.), *Stereotype threat: Theory, process, and application* (pp. 51–68). New York: Oxford University Press.

A chapter examining neurobiological responses as possible mediators to stereotype threat effects.

References

Bergsieker, H. B., Shelton, J. N., & Richeson, J. A. (2010). To be liked versus respected: Divergent goals in interracial interactions. *Journal of Personality and Social Psychology, 99*, 248–264.

Blascovich, J., & Mendes, W. B. (2010). Social psychophysiology and embodiment. In S. T. Fiske, D. T. Gilbert, & G. Lindzey (Eds.), *The handbook of social psychology* (5th ed.). New York: Wiley.

Blascovich, J., Mendes, W. B., Hunter, S. B., Lickel, B., & Kowai-Bell, N. (2001). Perceiver threat in social interactions with stigmatized others. *Journal of Personality and Social Psychology, 80*, 253–267.

Blascovich, J., Mendes, W. B., & Seery, M. D. (2002). Intergroup threat: A multi-modal approach. In D. M. Mackie & E. R. Smith (Eds.), *From prejudice to intergroup emotions: Differentiated reactions to social groups* (pp. 89–109). New York: Psychology Press.

Blascovich, J., & Tomaka, J. (1996). The biopsychosocial model of arousal regulation. *Advances in Experimental Social Psychology, 28*, 1–51.

Bowden, R. S. (2008, November 16). Racism is dead. *American Thinker.* Retrieved from http://www.americanthinker.com/2008/11/racism_is_dead.html

Cacioppo, J. T., Tassinary, L. G., & Berntson, G. G. (2000). Psychophysiological science. In J. T. Cacioppo, L. G. Tassinary & G. G. Berntson (Eds), *Handbook of psychophysiology* (2nd ed., pp. 1–23). New York: Cambridge University Press.

Cacioppo, J. T., Tassinary, L. G., & Berntson, G. G. (2007). Psychophysiological science: Interdisciplinary approaches to classic questions about the mind. In J. T. Cacioppo, L. G. Tassinary & G. G. Berntson (Eds), *Handbook of psychophysiology* (3rd ed., pp. 1–16). New York: Cambridge University Press.

Cohen, G. L., Steele, C. M., & Ross, L. D. (1999). The mentor's dilemma: Providing critical feedback across the racial divide. *Personality and Social Psychology Bulletin, 25*, 1302–1318.

Crocker, J., Voelkl, K., Testa, M., & Major, B. (1991). Social stigma: The affective consequences of attributional ambiguity. *Journal of Personality and Social Psychology*, *60*, 218–228.

Dickerson, S. S., & Kemeny, M. E. (2004). Acute stressors and cortisol responses: A theoretical integration and synthesis of laboratory research. *Psychological Bulletin*, *130*, 355–391.

Dienstbier, R. A. (1989). Arousal and physiological toughness: Implications for mental and physical health. *Psychological Review*, *96*, 84–100.

Greenwald, A. G. (2012). There is nothing so theoretical as a good method. *Perspectives on Psychological Science*, 7, 99–108.

Greenwald, A. G., McGhee, D. E., & Schwartz, J. L. K. (1998). Measuring individual differences in implicit cognition: The Implicit Association Test. *Journal of Personality and Social Psychology*, *74*, 1464–1480.

Kaiser, C. R., & Miller, C. T. (2001). Stop complaining! The social costs of making attributions to discrimination. *Personality and Social Psychology Bulletin*, *27*, 254–263.

Kenny, D. A., Kashy, D. A., & Cook, W. L. (2006). *Dyadic data analysis*. New York: Guilford Press.

Kirschbaum, C., Pirke, K. M., & Hellhammer, D. H. (1993). The "Trier Social Stress Test": A tool for investigating psychobiological stress responses in a laboratory setting. *Neuropsychobiology*, *28*, 76–81.

Koslov, K. R. (2010). *Pounding hearts and brittle smiles: The effect of resource depletion on outgroup positivity*. Dissertation, Harvard University, Cambridge, MA.

Lazarus, R. S. & Folkman, S. (1984). *Stress, appraisal, and coping*. New York: Springer Publishing Company.

Mendes, W. B., Blascovich, J., Hunter, S., Lickel, B., & Jost, J. T. (2007). Threatened by the unexpected: Physiological responses during social interactions with expectancy-violating partners. *Journal of Personality and Social Psychology*, *92*, 698–716.

Mendes, W. B., Blascovich, J., Lickel, B., & Hunter, S. (2002). Challenge and threat during social interactions with White and Black men. *Personality and Social Psychology Bulletin*, *28*, 939–952.

Mendes, W. B., & Koslov, K. (in press). Brittle smiles: Positive biases toward stigmatized and outgroup targets. *Journal of Experimental Psychology: General*. doi: 10.1037/a0029663

Mendes, W. B., Major, B., McCoy, S., & Blascovich, J. (2008). How attributional ambiguity shapes physiological and emotional responses to social rejection and acceptance. *Journal of Personality and Social Psychology*, *94*, 278–291.

Norton, M. I., Dunn, E. W., Carney, D. R., & Ariely, D. (2012). The persuasive "power" of stigma? *Organizational Behavior and Human Decision Processes*, *117*, 261–268.

Rankin, R. E., & Campbell, D. T. (1955). Galvanic skin responses to Negro and white experimenters. *The Journal of Abnormal and Social Psychology*, *51*, 30–33.

Shelton, J. N., & Richeson, J. A. (2006). Interracial interactions: A relational approach. In M. P. Zanna (Ed.), *Advances in experimental social psychology* (Vol. 38, pp. 121–181). New York: Academic Press.

Stephan, W. G., & Stephan, C. W. (2000). An integrated threat theory of prejudice. In S. Oskamp (Ed.), *Reducing prejudice and discrimination* (pp. 23–34). Mahwah, NJ: Lawrence Erlbaum.

Wegner, D. M. (2009). How to think, say, or do precisely the worst thing for any occasion. *Science*, *325*, 48–50.

West, T. V. (in press). Repeated-measure methods with dyads. In J. A. Simpson & L. Campbell (Eds.), *The Oxford handbook of close relationships*. New York: Oxford University Press.

17

SUSPICION IN INTERRACIAL INTERACTIONS

Using Measures of Cardiovascular Reactivity to Index Threat

Brenda Major and Jonathan W. Kunstman

Introduction

Can positive feedback and acceptance be experienced as a threat? The mere suggestion sounds contrary to common sense. Yet we contend that ethnic minority recipients sometimes are threatened by positive feedback from Whites. Specifically, we suggest that because of social pressures on Whites to respond without prejudice, positive treatment from Whites can be attributionally ambiguous for racial minorities—they may be uncertain whether the treatment genuinely reflects liking or respect, or is motivated by a desire to avoid appearing prejudiced. This uncertainty, we hypothesize, can lead minority recipients to experience positive feedback from Whites as threatening, with potentially negative effects for their psychological and physical health. We describe three experiments that test this hypothesis using measures of cardiovascular reactivity (CVR) to index threat. The work described here builds on more than 30 years of research by the first author addressing how members of stigmatized groups perceive, experience, and cope with social devaluation.

The story behind this research starts in 1989, when Jennifer Crocker and Brenda Major (1989) teamed up to write a paper challenging the widely held view that membership in a stigmatized group necessarily leads to lower personal self-esteem. In an attempt to explain why members of stigmatized groups often have "paradoxically" high self-esteem, they proposed that intergroup interactions are often *attributionally ambiguous* for members of stigmatized groups. Specifically, they speculated that members of stigmatized groups are often aware that they are vulnerable to being victims of prejudice and discrimination. As a consequence of this awareness, when they experience rejection or poor outcomes in intergroup contexts, they may be unsure whether they deserved those outcomes or whether they resulted from prejudice and discrimination. Drawing on attributional models of emotion (e.g., Weiner, 1982), Crocker and Major further hypothesized that although attributional ambiguity may have many negative effects, it also offers a self-protective benefit to stigmatized individuals. In particular, when rejected or facing poor outcomes, being able to attribute negative treatment to something external to the self, such as another person's prejudice, rather than to internal, stable factors of oneself, such as one's lack of ability or personality, might buffer the personal self-esteem of members of stigmatized groups.

In their first experimental test of this hypothesis, Crocker, Major, and their students randomly assigned White and Black participants to receive positive or negative interpersonal feedback from a White peer who either did, or did not, know their race (Crocker, Voelkl, Testa, & Major, 1991). Consistent with the researchers' self-esteem buffering hypothesis, Black participants who were rejected by a White peer who did not know their race experienced a drop in self-esteem, whereas those rejected by a White peer who did know their race experienced no change in self-esteem. More surprising were responses of Black students who were accepted by a White peer. As might be expected, these students experienced a *boost* in self-esteem when evaluated favorably by a peer whom they thought did not know their race. However, if they thought the peer who evaluated them favorably *did* know their race, they experienced a large *drop* in self-esteem relative to baseline levels.

To explain this latter finding, Crocker and Major speculated that Black students may have been suspicious of the evaluator's motives for giving them accepting feedback. In particular, they may have been unsure whether the feedback was sincere or motivated by something else, such as their partner's desire to appear unbiased or unprejudiced. Ambiguity about their partner's motives may have created a sense of uncertainty about whether the positive feedback was genuine. This uncertainty about whether they were truly liked or accepted may have posed a *threat to the self*. Uncertainty—particularly as it relates to the self—is linked with feelings of unease, anxiety, and distress and is highly aversive (Baumeister, 1985; McGregor, Haji, Nash, & Teper, 2008; van den Bos, 2009). Thus, uncertainty about the others' motives for giving positive feedback may have undermined Black students' self-esteem.

Can Positive Feedback Be Attributionally Ambiguous?

The idea that positive feedback could be threatening or damaging to self-esteem initially seemed implausible to us. Anyone will tell you that people feel good after receiving positive feedback, and respond more favorably to being accepted than rejected by another person. Social psychological research generally corroborates this commonsense assumption (e.g., Leary et al., 2003). Almost all of this research, however, has examined responses to social evaluation either within same-race interactions or without regard to intergroup context. Thus, it remained to be seen whether social acceptance and rejection are construed differently in interracial as compared to intraracial interactions. Although the study by Crocker et al. (1991) suggested they might be construed differently, Whites' reactions to *intraracial* feedback, but Blacks' reactions to *interracial feedback* were examined in that study. The imbalanced design of this study obviated direct comparisons between majority group and minority group members' reactions to interracial and intraracial feedback.

There are reasons to believe that minorities may regard positive feedback from Whites with suspicion and uncertainty. In contemporary society, majority group members face strong social pressures to respond without prejudice toward members of racial and ethnic minorities. As a result, many Whites are concerned that in interracial settings, they will do or say the wrong thing and be perceived as prejudiced (Plant, 2004; Vorauer, Main, & O'Connell, 1998). In response to these social concerns, White people sometimes amplify their positive responses toward members of racial and ethnic minorities to convey a non-prejudiced image (Mendes & Koslov, in press). For example, Whites rate an essay more positively when it is ostensibly written by a Black student compared to when the same essay is believed to be written by a White student (Harber, 2004), and sometimes "bend over backward" to show minorities and members of other stigmatized groups that they are not prejudiced (e.g., Gaertner & Dovidio, 1981; Mendes & Koslov, in press; Norton, Sommers, Apfelbaum, Pura, & Ariely, 2006).

One consequence of this heightened positivity is that the meaning of Whites' positive treatment of ethnic minorities, as well as the motives behind it, may be ambiguous. The dilemma of discernment that results is likely to be especially pernicious for minority recipients of these positive responses. Minority group members may be unable to tell whether accepting, welcoming responses from Whites are genuine or should be viewed with skepticism. When they meet a new White person, minority group members may question, is this person being nice because s/he likes me, or is this person simply trying to avoid appearing prejudiced? The resulting ambiguity surrounding Whites' motives for positive treatment can create uncertainty about the social environment and the self. This self-uncertainty may be threatening, leading to negative effects on self-esteem.

Results of several recent experiments are consistent with this theorizing. Hoyt, Aguilar, Kaiser, Blascovich, and Lee (2007) conducted an experiment in which Latina participants interacted with two ostensibly White partners in a virtual environment. After completing a leadership task, participants received positive feedback from their partners. To vary the salience of race and resultant attributional ambiguity for feedback, Hoyt et al. (2007) manipulated how Latina participants thought they appeared to their partners. In half the conditions, participants' own avatars were White, like those of their partners, thereby eliminating ethnicity as a possible motivator for positive feedback. In the other conditions, participants' avatars were Latina, thus raising the possibility that ethnicity and Whites' concerns with appearing biased may have influenced their White partners' feedback.

Compared to when they believed their avatar was White, Latinas who believed their partners saw them (their avatar) as Latina were more likely to attribute their partners' positive feedback to their ethnicity, and also reported lower self-esteem. Importantly, attributions to ethnicity mediated the effect of condition on impaired self-esteem.

In another study, Mendoza-Denton, Goldman-Flythe, Pietrzak, Downey, and Aceves (2010) found that receiving positive feedback on a writing exercise lowered performance self-esteem and increased concerns with belonging among Black students high in race-based rejection sensitivity (RS-race) when they believed their race was known to the White evaluator. In contrast, when Black participants received positive feedback from a White evaluator and believed their race was unknown, RS-race was unrelated to belonging and well-being. Attributional ambiguity about the genuineness of feedback also can undermine minorities' knowledge of their own strengths and limitations (Aronson & Inzlicht, 2004). Findings of these studies are consistent with our idea that when issues of race and concerns with appearing biased are believed to influence positive feedback, its meaning is ambiguous. As a result, overtures that would otherwise signal acceptance and respect may become sources of threat.

Cardiovascular Reactivity Measures of Threat

To test our idea, we needed a way to measure directly whether something is "threatening" to the self. Although the concept of threat plays a central role in many social psychology theories—for example, dissonance theory (Festinger, 1957); stereotype threat theory (Steele & Aronson, 1995)—most attempts to measure threat directly have been unsuccessful. Measuring threat via self-report instruments is notoriously difficult (e.g., Matheson & Cole, 2004; Steele & Aronson, 1995). To further complicate matters, we also sought to measure threat in response to positive feedback, where basic social norms of reciprocity may discourage overt negativity. Hence, even if minority group members are suspicious of positive treatment from White people, situational pressures may restrain their negative explicit responses. To address this measurement problem, we theorized that implicit indices of threat may be advantageous because they provide a subtle measure that is not susceptible to normative pressures or other self-presentational concerns.

Our search for a direct, implicit, and unobtrusive measure of threat led us to Blascovich and colleagues' biopsychosocial model of challenge and threat (BPM-CT) (Blascovich & Mendes, 2000; Blascovich & Tomaka, 1996; see also Blascovich, Chapter 12 in this volume). According to the BPM-CT, threat is a maladaptive psychological state that occurs when a person unconsciously or consciously appraises the demands of a situation as exceeding his or her coping resources. Threat is an avoidance or inhibitional state that is linked to vigilance, emotions such as anxiety and fear, and often poor cognitive performance (Herrald

& Tomaka, 2002; Mendes et al., 2008). In contrast, challenge is an adaptive psychological state that occurs when resources are perceived as meeting or exceeding demands (see Lazarus & Folkman, 1984). Challenge is an approach state linked to high-arousal emotions, both positive (e.g., joy) and negative (e.g., anger). According to the theory, a variety of factors can increase the perceived demands of a situation, and thereby its likelihood of being appraised as threatening. One of those factors is uncertainty (Blascovich & Mendes, 2000).

Importantly, the BPM-CT predicts that psychological states of challenge and threat can be indexed via specific patterns of cardiovascular reactivity (CVR) that are present when an individual is engaged in motivated performance situations such as a self-relevant task. Engagement is typically indexed by increased heart rate and ventricular contractility relative to the baseline. When an individual appraises the situation as a challenge, it activates the sympathetic–adrenal–medullary (SAM) axis, leading to increased cardiac output (CO; liters of blood pumped out of the heart per minute) and decreased total peripheral resistance (TPR; total constriction of the arterioles). In contrast, when a person appraises the situation as a threat, the HPA (hypothalamic–pituitary–adrenal) cortical axis is also activated, releasing cortisol. Threat is characterized by little or no increases in CO and by vasoconstriction, resulting in an increase in TPR. Cardiac output and TPR are measured independently and are often reported that way. However, they can be combined statistically to form an overall Threat and Challenge Index (TCI) that is useful in analyses. The CVR measures of threat and challenge proposed in the BPM-CT have been shown reliably to predict "threat" and "challenge" behavioral responses in a range of intergroup contexts and among members of majority and minority groups (Blascovich, Mendes, Hunter, Lickel, & Kowai-Bell, 2001; Mendes, Blascovich, Hunter, Lickel, & Jost, 2007; Mendes, Blascovich, Lickel, & Hunter, 2002; Townsend, Major, Sawyer, & Mendes, 2010; see also Blascovich, Chapter 12 in this volume).

Cardiovascular (CV) indices seemed ideally suited for measuring minority group members' threat response to positive feedback from Whites. As noted above, not only are measures of CVR implicit—and therefore immune to the influence of self-presentation—they are also beneficial because they index threat online, in response to a specific social interaction. By comparison, if we employed other implicit measures like lexical-decision tasks (e.g., Fazio, Jackson, Dunton, & Williams, 1995) or the Implicit Association Test (IAT) (Greenwald, McGhee, & Schwartz, 1998), we would be measuring threat *after* minority participants interacted with White peers. Threat would also need to be assessed in a general fashion (e.g., semantic associations between threat and Whites generally). Cardiovascular indices, therefore, were advantageous not only because they provided an implicit measure of threat, but also because they provided a precise measure of minorities' specific responses to positive feedback from a White person in a particular social context.

Studies

Overview

We conducted three experiments to examine whether, when, and why ethnic minorities might experience threat in response to positive, accepting feedback from a White peer. All of these experiments used the CVR measures pioneered by Blascovich and Tomaka (1996) to assess threat. Our first experiment compared the physiological responses of Blacks and Whites while they interacted with a same-race or other-race partner who had just given them accepting or rejecting feedback (Mendes, Major, McCoy, & Blascovich, 2008). In our second and third studies (Major, Sawyer, Kunstman, Townsend, & Mendes, 2013), we examined how Latinas' perceptions of Whites' motives for non-prejudiced behavior influenced their physiological reactions while interacting with a White peer who had given them positive feedback. In one study we examined whether Latinas' meta-perceptions of Whites' motives predicted their responses while interacting with a White, versus a Latina, partner who gave them accepting feedback. In a final study we examined whether meta-perceptions of Whites' motives predicted Latinas' responses while interacting with a White partner who gave them accepting or rejecting feedback.

Although the central focus of our work was on minorities' physiological responses, it is worth noting that in each of these studies, we also collected behavioral measures (Study 1) and self-report measures (Studies 2 and 3) to provide further insight into minorities' responses to positive feedback from Whites. These additional measures are useful because they inform the psychological experiences and behavioral responses that correspond to the state of "threat." For example, in Study 2, we measured minorities' perceptions of their partners; and in Study 3, we measured participants' self-esteem. By linking the physiological experience of "threat" with participants' thoughts about themselves and their partners, these measures place the threat pattern of CV response in a specific social-cognitive context. As such, these data collectively bridge the oft distinct literatures on self-esteem (e.g., Crocker et al., 1991; Mendoza-Denton et al., 2010) and psycho-physiology.

Positive Feedback in Intraracial versus Interracial Contexts

Major and her colleagues began by directly comparing minority and majority group members' responses to positive feedback from a same-race or other-race peer (Mendes et al., 2008). We assigned Black and White students to receive positive, accepting, interpersonal feedback from a previously unknown same-race partner or other-race partner based on a brief interaction. All participants in this experiment believed that their partner knew their race. After learning that their partner liked them and wanted to be their friend, participants completed a "word-finding" task with their partner (similar to the game of Boggle). We measured performance on the task as well as CV responses during the task. Cardiovascular reactivity measures were calculated by subtracting CV responses during the task from CV responses measured during a 5-minute baseline at the beginning of the experiment.

Our results showed that receiving positive, accepting feedback from a *same-race* partner affected Black and White participants in exactly the same way. Both Blacks and Whites reported feeling positive emotions and performed well on the task after learning that a same-race partner liked them. While performing the task, they also displayed a pattern of CVR consistent with a challenge motivational state. That is, they showed an increase in CO accompanied by a decrease in TPR.

In contrast, Black and White students responded quite differently to positive feedback from a partner of the other race. White students responded just as positively to being liked by a Black partner as they responded to being liked by a White partner—they reported positive emotions, performed well on the task, and showed a challenge pattern of CVR. Thus, Whites seemed to value interracial acceptance as much as intraracial acceptance. In contrast, Black students showed a very different pattern. They performed significantly worse on the task when the accepting partner was White as compared to Black. They also were more vigilant behaviorally when the accepting partner was White as compared to Black. We assessed vigilance by coding participants' non-verbal behavior (looking at the partner) while they performed the task. Of particular relevance to this chapter, following positive feedback from a White partner, Black students also showed a *threat* pattern of CVR. That is, they showed a less healthy set of CV responses—a decrease in CO and an increase in TPR—while interacting with a White peer who had evaluated them favorably.

This physiological response contrasted sharply with the challenge pattern of CVR that Black students showed while interacting with a Black peer who had evaluated them favorably. Black students who were evaluated favorably by their White partner also performed the worst on the task. The threat pattern displayed by Black students who had received accepting, friendly feedback from a White peer is consistent with the drop in self-esteem observed by Crocker et al. (1991) and by Hoyt et al. (2007), and shows the effect on an implicit rather than an explicit measure. It also provided direct evidence that under certain circumstances, positive treatment from Whites is threatening to racial and ethnic minority group members. Our next studies sought to investigate the source of this threat more directly.

Measuring Meta-perceptions of Whites' Motives for Positive Feedback

According to our theorizing, minority group members experience positive interpersonal feedback from (previously unknown) White peers as threatening because they doubt its sincerity. Specifically, we believe that ethnic and racial minorities often suspect that positive feedback is motivated more by Whites' *external concerns with not appearing prejudiced* than by their *internal desires not to be prejudiced*. The more minorities perceive Whites as externally rather than internally motivated, the less certain they are of the authenticity of positive feedback in interracial interactions, and the more likely they are to experience positive feedback from Whites as threatening.

We needed to know, however, the content of minority group members' perceptions of Whites' motives for non-prejudiced behavior (meta-perceptions). To this end, we created a measure of minorities' perceptions of Whites' motivations for responding without prejudice (Major, Sawyer, & Kuntsman, 2013), modeled on the Internal and External Motivation to Respond without Prejudice Scale (IMS/EMS; Plant & Devine, 1998). Our scale, the Perceived Internal and External Motivations for Responding without Prejudice Scale (PIMS/PEMS), includes five statements measuring agreement with the belief that Whites' non-prejudiced behavior is motivated by internal, egalitarian values (PIMS; e.g., "When White people act in a nonprejudiced way toward members of racial/ethnic minority groups, it is because it is personally important to them not to be prejudiced"); and five statements assessing agreement with the belief that Whites' non-prejudiced behavior is motivated by external concerns with appearing prejudiced (PEMS; e.g., "When White people act in a nonprejudiced way toward members of racial/ethnic minority groups, it is because they feel pressure from others to act nonprejudiced"). We administered this scale to large numbers of ethnic minority college students (primarily Latino/a American) in the western United States.

We found significant variability in students' perceptions of Whites' motives for engaging in non-prejudiced behavior. Some students scored high on the PIMS and low on the PEMS, whereas others showed the reverse pattern. Others scored high on both scales or low on both scales. Overall, scores on the PIMS and PEMS were unrelated. On average, participants scored higher on the PIMS than the PEMS, suggesting that Latino/a college students generally perceive Whites' non-prejudiced behavior as motivated more by a desire to be egalitarian than by a desire to comply with external pressures. Even so, a significant number of students saw Whites' non-prejudiced behavior as primarily motivated by external rather than internal factors.

To index suspicion, we subtracted scores on the PIMS from scores on the PEMS to yield a difference score reflecting the extent to which minorities suspect that Whites' positive feedback is inauthentic; i.e., perceive it as motivated more by external than internal factors. We called this the Suspicion of Motives Index (SOMI). Examination of the relationship between SOMI and other related measures revealed theoretically meaningful but modest correlations in the predicted directions, thus illustrating the discriminant and convergent validity of the SOMI. For example, ethnic minority students who scored higher on the SOMI also tended to score higher on measures of ethnic stigma consciousness

(Pinel, 1999), sensitivity to race-based rejection (Mendoza-Denton, Downey, Purdie, Davis, & Pietrzak, 2002), and distrust of Whites (Landrine & Klonoff, 1996). They also tended to make more external attributions for positive events in general. Scores on the SOMI were unrelated to measures of ethnic identification (Luhtanen & Crocker, 1992), self-esteem, or psychological well-being.

Our next question was whether minorities who were suspicious would be more threatened by positive feedback from Whites in interracial interactions than those who were not suspicious. To answer this question, we conducted two studies to determine whether scores on the SOMI predicted Latinas' physiological and psychological reactions to receiving positive interpersonal feedback from a previously unknown White peer.

Suspicion of Motives and Physiological Responses to Positive Feedback

Our next study tested the hypothesis that SOMI would predict a CVR pattern characteristic of threat among Latinas who were the recipients of positive feedback from a White partner, but not from a Latina partner. College women who had self-identified as Latina were brought to the lab where they met either a White or a Latina partner (a confederate blind to hypotheses). After a 5-minute baseline physiological data collection period, we asked them to give a 3-minute speech to their ostensible partner about "why they would be a good friend." Afterwards, participants received positive feedback from their partner in which partners indicated that they would like to become friends with the participant. Participants then completed a cognitive memory task (a digit-span task) in the presence of their partner, while the partner presumably scored their answers. In accordance with the BPM-CT, we measured CVR during the task to ensure the requisite psychological engagement necessary to elicit states of threat or challenge (Blascovich & Tomaka, 1996).

We averaged the task minutes to compute key measures of CO and TPR (see Blascovich & Mendes, 2000). We then used these aggregate CV responses to compute a single Threat–Challenge Index (TCI) that combines TPR and CO responses. This allowed us to assess participants' patterns of CV responses in a single regression analysis. We computed the TCI by assigning a negative value to CO scores, and then standardized and combined CO and TPR scores such that higher scores indicated a physiological response pattern characteristic of threat, whereas lower scores indicated a response pattern of challenge (see Blascovich, Seery, Mugridge, Norris, & Weisbuch, 2004).

We predicted that the more suspicious minority group members were of the motives behind positive feedback from Whites (i.e., the higher their SOMI scores), the more likely they would be to evince a pattern of CVR characteristic of threat (i.e., higher TPR relative to CO) during the task. Consistent with our hypotheses, the higher participants scored on the SOMI, the more threatened they were while interacting with the White peer who had evaluated them favorably, as indexed by their CVR during the task. In contrast, scores on the SOMI did not predict CVR among Latinas interacting with a Latina peer who had evaluated them favorably.

We also asked participants to rate their partner on a set of trait adjectives after completing the task. We predicted that the more uncertain minority group members were of the motives behind positive feedback from Whites, the more ambivalent they would feel toward their White (but not their Latina) partner. As a result, they might perceive her as honest *and* dishonest, as genuine *and* disingenuous, etc. To assess this, we calculated a measure of attitudinal ambivalence toward the partner using procedures outlined by Priester and Petty (1996). Consistent with predictions, SOMI scores were positively and significantly related to attitudinal ambivalence when the partner was White, but were unrelated to ambivalence when the partner was Latina.

In sum, in keeping with attributional ambiguity theory and the effects first documented in Crocker et al. (1991), minorities' suspicions surrounding Whites' motives directly predicted psychological and physiological reactions to positive treatment from Whites. Among those who believed Whites are more externally motivated to avoid appearing prejudiced than internally motivated not to be prejudiced, positive feedback from a White (but not a Latina) partner increased uncertainty about how to perceive a White interaction partner and led to avoidant and maladaptive patterns of CVR. Thus, for suspicious minorities, positive treatment from Whites seems to create uncertainty and threat.

Our final study employed a similar set of procedures with several key changes. First, in this study, all participants (Latinas) interacted with a White peer. Second, after a brief interaction, half were assigned to receive positive, accepting feedback from their partner and the other half were assigned to receive negative, rejecting feedback. All then performed the same cognitive task as in Study 2 in the presence of their partner while we recorded their CV responses. Subsequently, we measured participants' self-esteem and compared it to a pre-study baseline to assess changes.

We predicted that scores on the SOMI would interact with feedback valence to predict CVR and self-esteem. Consistent with predictions, among those who received positive feedback, greater suspicion was positively and significantly associated with a threat pattern of CVR during the task. In contrast, in the negative feedback condition, the relationship between the SOMI and CVR was not significant. Thus, among Latina participants, SOMI predicted threat after positive feedback—but not after negative feedback—from a White peer. We also observed a significant interaction between scores on the SOMI and feedback valence on self-esteem. Among those who received positive feedback, higher SOMI scores were associated with significantly greater *decreases* in social self-esteem (relative to the baseline). In contrast, among participants who received negative feedback, the relationship between SOMI and self-esteem was non-significant. Thus, the more suspicious Latina participants were of Whites' motives, the lower their self-esteem after receiving positive feedback, but not after receiving negative feedback.

Interestingly, irrespective of their level of suspicion, Latina participants' self-esteem was relatively high after receiving negative feedback from a White partner. Recall that in this study all participants believed their partner knew their ethnicity. Thus, these self-esteem findings are similar to those observed among Blacks in the earlier study by Crocker et al. (1991) who believed their evaluator knew their race. The findings suggest that in the negative feedback condition, Latinas' self-esteem may have been buffered by thinking that their rejection could have been due to their partner's prejudice, rather than to something internal and stable about themselves. It is also noteworthy that self-esteem among suspicious (high SOMI scores) Latinas who received positive feedback was much lower than self-esteem among Latinas in the other three conditions. The low self-esteem among these suspicious (and threatened) participants is consistent with the low self-esteem observed in the Crocker et al. (1991) study among Black participants who received positive feedback from a White partner whom they thought knew their race. Collectively, these results provide empirical support for our theorizing that when race and Whites' concerns with appearing prejudiced are perceived to bias positive feedback, acceptance from Whites can threaten minority group members' self-esteem.

Although the indices of CVR were the central dependent measures in our studies, it is worth noting that the complementary psychological measures of uncertainty and self-esteem are important for interpreting our results. Cardiovascular patterns of "threat" could be linked to numerous psychological states including anxiety, fear, distress, and sadness. Without the informative power of our self-report measures, any of these emotional patterns may have seemed theoretically applicable for suspicious minority group members. In our studies, the parallel effects observed for ambivalence toward the partner and self-esteem help to clarify what suspicious participants were experiencing when they received positive feedback from White interaction partners. This underscores the importance of collecting theoretically relevant psychological measures to interpret physiological results (see also Chalabaev, Sarrazin, & Brisswalter, Chapter 8 in this volume; Scheepers, Chapter 13 in this volume).

As opposed to merely disregarding Whites and dismissing them as fake, those high in SOMI saw their White partners as simultaneously trustworthy and dishonest, genuine and disingenuous. Essentially, high SOMI ethnic minorities who received positive feedback from a White peer were uncertain how to perceive—and react to—their partners. When considered in conjunction with the threat patterns of CVR and decline in self-esteem, these results imply that for suspicious minority group members, positive overtures from Whites are difficult to interpret and lead to doubt about the meaning of the feedback, which undermines positive self-regard. The co-occurrence of these effects is important because they locate the psychophysiological responses in a specific social-cognitive context and paint a nuanced picture of minorities' "threat" experience. Without these additional self-report responses, it would be unclear precisely how minorities felt about themselves and others while experiencing a state of threat.

Added Value of Psychophysiological Measures

In our view, measures of CVR provide a number of methodological benefits to those who wish to study intergroup interactions. First, changes in CVR represent automatic responses that are outside participants' conscious control. Hence, CV measures are not subject to potential demand and self-presentational concerns that can sometimes bias self-reports. In our work, eliminating the effect of self-presentation on participants' responses was essential for several reasons. In intergroup interactions, people may feel pressure from politically correct norms to respond positively toward members of different groups. Measures of CVR bypass these self-presentational concerns and provide unbiased accounts of participants' true responses.

In addition, people feel pressure to respond favorably to those who treat them well. Thus, expressing hostility in response to positive treatment or derogating a partner who has given one "accepting" feedback represents a violation of basic social norms. As a result, in novel social situations, even though suspicious minorities may feel threatened by positive treatment from Whites, norms of reciprocity may discourage them from sharing these negative feelings or evaluating their partner harshly. And indeed, participants in our studies did not evaluate their accepting White partner negatively. Measures of CVR thus were advantageous for our work because they provided an index of threat that circumvented self-presentational concerns with reciprocity and general intergroup interactions.

A second reason why measures of CVR were beneficial to us is because they provide online assessments of participants' inner states. Rather than rely on retrospective reports that could be biased by recall, measures of CVR are recorded in the moment and can be linked to specific tasks and partner responses. For example, in the current work, we were able directly to measure changes in participants' CVR while they were interacting with a White peer who had evaluated them positively. These direct changes in CVR as a function of feedback lend support to causal inferences about the paradoxically negative effects that positive feedback may have on suspicious minority group members.

Moreover, the online nature of CV measures is also beneficial for researchers interested in testing indirect (mediational) relationships between variables in an experimental session. One could imagine collecting additional measures of cognitive performance, non-verbal behavior, or other downstream psychological processes to test how early changes in CVR influence subsequent responses. By collecting CV data in real time prior to subsequent outcome variables, these types of measures satisfy conditions of temporal precedence suggested for strong tests of mediation (e.g., Baron & Kenny, 1986; MacKinnon, Fairchild, & Fritz, 2007).

Third, CV indices are also advantageous because of their sensitivity. Although effect sizes in the challenge/threat literature are regularly medium to large in magnitude (e.g., Mendes et al., 2008), these measures are also capable of detecting subtle physiological changes. For example, we regularly detected small to medium-sized effects when investigating suspicious minorities' CV responses to positive treatment from Whites, while observing no changes on self-report variables. Without the precision afforded by CV measures, exploring this phenomenon would have been far more difficult.

A fourth strength of CV measures is that they can reliably measure motivational states that are difficult to quantify in social settings. When quantifying motivation, researchers often rely on measurements of task persistence. Although persistence works well in achievement domains (e.g., academic performance), it is less applicable to social situations like interracial interactions. In these intergroup contexts, researchers need to employ other measures to operationalize motivation. Here CV responses of challenge and threat are particularly helpful. Not only have these measures been extensively validated, they are also unobtrusive, which makes them ideally suited for measuring online motivational changes in social situations (see Blascovich & Mendes, 2000; Blascovich & Tomaka, 1996; see also Blascovich, Chapter 12 in this volume). By measuring motivation in this way, researchers gain a nuanced picture of how social engagement changes within intergroup interactions.

A fifth advantage of using CV measures in intergroup relations research is that they have implications for long-term physical health and well-being. As a growing body of research attests, there are profound connections between experiences of discrimination and health deficits (see Krieger, 1999; for reviews, see Pascoe & Smart Richman, 2009; Williams & Mohammed, 2009). The vascular constriction associated with the threat pattern of CVR represents a less efficient circulatory response than responses associated with the "challenge" state. Repeated exposure to stressors that evoke threat is believed to increase risks for various CV problems, including hypertension among African Americans (e.g., Anderson, McNeilly, & Myers, 1993; Roberts & Rowland, 1981; Saab & Schneiderman, 1993). Measures of CVR are beneficial because they can inform theories linking interracial relations and perceived discrimination with long-term health problems.

In summary, we believe there are numerous benefits associated with measuring CVR for researchers interested in intergroup relations. Cardiovascular responses circumvent response biases related to self-presentation, social norms, and biased recall. Measures of CVR are automatic, continuous, and online. Indices of CVR are also sensitive and capable of detecting subtle changes in physiological responding. Changes in CO and TPR associated with states of threat and challenge are also validated and unobtrusive, making them ideal for studying motivation in social intergroup interactions. Moreover, these patterns of CVR have been linked with long-term health outcomes.

Despite these advantages, we also discovered some challenges associated with doing research using CV measures. First, there is a big learning curve. Doing studies using psychophysiological measurements is more time-consuming, labor-intensive, and difficult than doing studies that rely only on self-report measures. It takes time to learn how to collect data reliably—and to train research assistants to do so—and time to learn the biology behind the measures. Second, we discovered that it takes patience to do this type of research. Once the data are collected, it is time-consuming and labor-intensive to get the data into a form suitable for analyses. Scoring the resultant waveforms from BP, ECG and impedance cardiography (or teaching research assistants to do so reliably) is a considerable investment. This time lag is difficult for those of us who cannot wait to see our results! Furthermore, compared to studies using traditional self-report, the considerable resources necessary to set up a lab, train assistants, and collect, score, and compile data can make these intensive psychophysiological studies relatively "high risk." Studies using measures of CVR provide valuable and unique insight into the links between mind and body, but they take time and effort.

Third, we discovered that we had to change how we designed our studies. Simply measuring self-reported attributions and self-esteem in response to positive or negative feedback, as we did in our initial studies (e.g., Crocker et al., 1991), was not feasible. Challenge and threat patterns of physiological response emerge as individuals mobilize personal resources to contend with situational demands. Hence, active situations are necessary to spur differential patterns. Thus, we had to create experiments in which we put participants in a motivated performance situation—in a self-relevant task—while we assessed their CVR. This is why participants performed a cognitive task (e.g., digit-span, Boggle game) after they received feedback from their partner. As a result, the "cover stories" and procedures involved in doing this type of research are often complex.

Fourth, we found that measuring BP and electro- and impedance cardiography can impede the measurement of other types of behavior. For example, the attachment of electrodes, bands, and measurement cuffs impedes participants' movements, making assessment of certain motor behaviors difficult. In addition, movement can produce changes in CVR that make data difficult to score and interpret. We also found that attaching electrodes, bands, and measurement cuffs made some participants feel uncomfortable and self-conscious, particularly since attaching some of the sensors requires participants to partially disrobe or expose skin. One implication of this is that we found it necessary to rely on female experimenters, since both men and women reported being more comfortable having a female, rather than a male, attach bands and electrodes to their bodies.

Finally, we frequently observed a lack of correspondence between self-report and physiological measures of the same construct. For example, participants whose physiology indicated that they were threatened or stressed in response to an experimental manipulation often reported that they were not at all stressed or threatened. Although this lack of correspondence between self-report and physiological measures violates the expectations of editors and reviewers, it is a commonly observed phenomenon that parallels the discordance often observed between implicit (automatic) and explicit (controlled) measures of the same psychological construct. This discordance, although disconcerting, may provide insight into people's reactions to interracial interactions. Someone who is experiencing threat at a more unconscious or automatic level, for example, may show non-verbal or physiological indicators of avoidance and anxiety, whereas someone who is experiencing threat at a more conscious or controlled level may engage in deliberate actions to cope with interracial situations. People's automatic responses may also reveal threat that they are unwilling or unable to report on controllable self-report measures. Understanding this discordance is an important agenda for researchers seeking to use CV measures to test psychological theory.

Conclusions

The research described here builds on a course of research that began over 20 years ago with the provocative findings of Crocker et al. (1991). Why do racial and ethnic minority group members sometimes respond negatively to positive treatment from majority group members? Although the answer to this question probably has many facets, one factor appears to be minorities' beliefs about Whites' motives for non-prejudiced behavior. We hypothesized that when they are suspected to be motivated by a desire not to appear prejudiced, rather than by genuine liking or respect, positive overtures from Whites are sources of uncertainty and threat. We found that using CVR to assess automatic, real-time psycho-physiological responses to feedback, combined with directly measuring ethnic minorities' meta-perceptions of Whites' motives for non-prejudiced behavior, provided a useful way to test our hypothesis. Collectively, these studies help us to understand why and when positive feedback from majority groups can paradoxically have negative effects on minorities' self-perceptions.

Suggestions for Further Reading

Blascovich, J., & Mendes, W. B. (2000). Challenge and threat appraisals: The role of affective cues. In J. Forgas (Ed.), *Feeling and thinking: The role of affect in social cognition* (pp. 59–82). Cambridge, UK: Cambridge University Press.

This chapter outlines the biopsychosocial model of challenge and threat and provides additional information about CV indices in social psychological research.

Crocker, J., Voelkl, K., Testa, M., & Major, B. (1991). Social stigma: The affective consequences of attributional ambiguity. *Journal of Personality and Social Psychology, 60*, 218–228.

This paper provided the first experimental evidence that minority group members sometimes show negative affective responses to positive feedback from White people.

Major, B., Quinton, W. J., & McCoy, S. K. (2002). Antecedents and consequences of attributions to discrimination: Theoretical and empirical advances. *Advances in Experimental Social Psychology, 34*, 251–330.

This paper provides an overview of findings and theoretical advancements related to research on attributions to discrimination and attributional ambiguity.

Mendes, W. B., Blascovich, J., Hunter, S. B., Lickel, B., & Jost, J. T. (2007). Threatened by the unexpected: Physiological responses during social interactions with expectancy-violating partners. *Journal of Personality and Social Psychology, 92*, 698–716.

This work provides additional information related to the use of CV responses to index challenge and threat in social interactions.

Mendes, W. B., Major, B., McCoy, S., & Blascovich, J. (2008). How attributional ambiguity shapes physiological and emotional responses to social rejection and acceptance. *Journal of Personality and Social Psychology, 94*, 278–291.
This work integrates the challenge and threat model of motivation with attributional ambiguity theory to reveal Black and White participants' distinct responses to social acceptance and rejection.

Plant, E. A., & Devine, P. G. (2001). Responses to other-imposed pro-Black pressure: Acceptance or backlash? *Journal of Experimental Social Psychology, 37*, 486–501.
This paper illustrates how differences in Whites' motivations to respond without prejudice moderate responses to anti-prejudice social norms and can lead to backlash against egalitarian social pressures.

References

Anderson, N. B., McNeilly, M., & Myers, H. F. (1993). A biopsychosocial model of race differences in vascular reactivity. In J. J. Blascovich & E. S. Katkin (Eds.), *Cardiovascular reactivity to psychological stress and disease* (pp. 83–108). Washington, DC: American Psychological Association.

Aronson, J., & Inzlicht, M. (2004). Ups and downs of attributional ambiguity. *Psychological Science, 15*, 829–836.

Baron, R. M., & Kenny, D. A. (1986). The moderator-mediator variable distinction in social psychological research: Conceptual, strategic, and statistical considerations. *Journal of Personality and Social Psychology, 51*, 1173–1182.

Baumeister, R. F. (1985). The championship choke. *Psychology Today, 19*, 48–52.

Blascovich, J., & Mendes, W. B. (2000). Challenge and threat appraisals: The role of affective cues. In J. Forgas (Ed.), *Feeling and thinking: The role of affect in social cognition* (pp. 59–82). Cambridge, UK: Cambridge University Press.

Blascovich, J., Mendes, W. B., Hunter, S. B., Lickel, B., & Kowai-Bell, N. (2001). Perceiver threat in social interactions with stigmatized others. *Journal of Personality and Social Psychology, 80*, 253–267.

Blascovich, J., Seery, M. D., Mugridge, C. A., Norris, R. K., & Weisbuch, M. (2004). Predicting athletic performance from cardiovascular indexes of challenge and threat. *Journal of Experimental Social Psychology, 40*, 683–688.

Blascovich, J., & Tomaka, J. (1996). The biopsychosocial model of arousal regulation. In M. P. Zanna (Ed.), *Advances in experimental social psychology* (Vol. 28, pp. 1–15). New York: Academic Press.

Crocker, J., & Major, B. (1989). Social stigma and self-esteem: The self-protective properties of stigma. *Psychological Review, 96*, 608–630.

Crocker, J., Voelkl, K., Testa, M., & Major, B. (1991). Social stigma: The affective consequences of attributional ambiguity. *Journal of Personality and Social Psychology, 60*, 218–228.

Fazio, R. H., Jackson, J. R., Dunton, B. C., & Williams, C. J. (1995). Variability in automatic activation as an unobtrusive measure of racial attitudes: A bona fide pipeline? *Journal of Personality and Social Psychology, 69*, 1013–1027.

Festinger, L. (1957). *A theory of cognitive dissonance.* Stanford, CA: Stanford University Press.

Gaertner, S. L., & Dovidio, J. F. (1981). Racism among the well-intentioned. In

E. G. Clausen & J. Bermingham (Eds.), *Pluralism, racism, and public policy: The search for equality* (pp. 208–222). Boston, MA: G. K. Hall and Co.

Greenwald, A. G., McGhee, D. E., & Schwartz, J. K. L. (1998). Measuring individual differences in implicit cognition: The Implicit Association Test. *Journal of Personality and Social Psychology, 74,* 1464–1480.

Harber, K. D. (2004). The positive feedback bias as a response to out-group unfriendliness. *Journal of Applied Social Psychology, 34,* 2272–2297.

Herrald, M. M., & Tomaka, J. (2002). Patterns of emotion-specific appraisal, coping, and cardiovascular reactivity during an ongoing emotional episode. *Journal of Personality and Social Psychology, 83,* 434–450.

Hoyt, C. L., Aguilar, L., Kaiser, C. R., Blascovich, J., & Lee, K. (2007). The self-protective and undermining effects of attributional ambiguity. *Journal of Experimental Social Psychology, 43,* 884–893.

Krieger, N. (1999). Embodying inequality: A review of concepts, measures, and methods for studying health consequences of discrimination. *International Journal of Health Services, 29,* 295–352.

Landrine, H., & Klonoff, E. A. (1996). *African American acculturation: Deconstructing race and reviving culture.* Thousand Oaks, CA: Sage Publications.

Lazarus, R. S., & Folkman, S. (1984). *Sress, appraisal and coping.* New York: Springer.

Leary, M. R., Gallagher, B., Fors, E., Buttermore, N., Baldwin, E., Kennedy, K., & Mills, A. (2003). The invalidity of disclaimers about the effects of social feedback on self-esteem. *Personality and Social Psychology Bulletin, 29,* 623–636.

Luhtanen, R., & Crocker, J. (1992). A collective self-esteem scale: Self-evaluation of one's social identity. *Personality and Social Psychology Bulletin, 18,* 302–318.

McGregor, I., Haji, R., Nash, K. A., & Teper, R. (2008). Religious zeal and the uncertain self. *Basic and Applied Social Psychology, 30,* 183–188.

MacKinnon, D. P., Fairchild, A. J., & Fritz, M. S. (2007). Mediation analysis. *Annual Review of Psychology, 58,* 593–614.

Major, B., Sawyer, P. J., Kuntsman, J. W. (2013). Minority perceptions of Whites' motives for responding without prejudice: The perceived internal and external motivation to avoid prejudice scales. *Personality and Social Psychology Bulletin, 39,* 404–414.

Major, B., Sawyer, P. J., Kunstman, J. W., Townsend, S. S. M., & Mendes, W. B. (2013). Why are you being so nice to me? How suspicion of Whites' motives shapes minorities' responses to positive feedback. Manuscript submitted for publication.

Matheson, K., & Cole, B. (2004). Coping with a threatened group identity: Psychological and neuroendocrine responses. *Journal of Experimental Social Psychology, 40,* 777–786.

Mendes, W. B., Blascovich, J., Hunter, S. B., Lickel, B., & Jost, J. T. (2007). Threatened by the unexpected: Physiological responses during social interactions with expectancy-violating partners. *Journal of Personality and Social Psychology, 92,* 698–716.

Mendes, W. B., Blascovich, J., Lickel, B., & Hunter, S. (2002). Challenge and threat during social interactions with White and Black men. *Personality and Social Psychology Bulletin, 28,* 939–952.

Mendes, W. B., & Koslov, K. (in press). Brittle smiles: Positive biases toward stigmatized and outgroup targets. *Journal of Experimental Psychology.* doi: 10.1037/a0029663

Mendes, W. B., Major, B., McCoy, S., & Blascovich, J. (2008). How attributional ambiguity shapes physiological and emotional responses to social rejection and acceptance. *Journal of Personality and Social Psychology, 94,* 278–291.

Mendoza-Denton, R., Downey, G., Purdie, V. J., Davis, A., & Pietrzak, J. (2002).

Sensitivity to status-based rejection: Implications for African American students' college experience. *Journal of Personality and Social Psychology, 83*, 896–918.

Mendoza-Denton, R., Goldman-Flythe, M., Pietrzak, J., Downey, G., & Aceves, M. J. (2010). Group-value ambiguity: Understanding the effects of academic feedback on minority students' self-esteem. *Social Psychological and Personality Science, 1*, 127–135.

Norton, M. I., Sommers, S. R., Apfelbaum, E. P., Pura, N., & Ariely, D. (2006). Color blindness and interracial interactions: Playing the political correctness game. *Psychological Science, 17*, 949–953.

Pascoe, E. A., & Smart Richman, L. (2009). Perceived discrimination and health: A meta-analytic review. *Psychological Bulletin, 135*, 531–554.

Pinel, E. C. (1999). Stigma consciousness: The psychological legacy of social stereotypes. *Journal of Personality and Social Psychology, 76*, 114–128.

Plant, E. A. (2004). Responses to interracial interactions over time. *Personality and Social Psychology Bulletin, 30*, 1458–1471.

Plant, E. A., & Devine, P. G. (1998). Internal and external motivation to respond without prejudice. *Journal of Personality and Social Psychology, 75*, 811–832.

Priester, J. M., & Petty, R. E. (1996). The gradual threshold model of ambivalence: Relating the positive and negative bases of attitudes to subjective ambivalence. *Journal of Personality and Social Psychology, 71*, 431–449.

Roberts, J., & Rowland, M. (1981). Vital and health statistics (Series 11, Number 221). Hypertension in adults 25–74 years of age: United States, 1971–1975 (DHEW Publication Number PHS 81–1671). Washington, DC: U.S. Government Printing Office.

Saab, P. G., & Schneiderman, N. (1993). Biobehavioral stressors, laboratory investigation, and the risk of hypertension. In J. J. Blascovich & E. S. Katkin (Eds.), *Cardiovascular reactivity to psychological stress and disease* (pp. 49–82). Washington, DC: American Psychological Association.

Steele, C. M., & Aronson, J. (1995). Stereotype threat and the intellectual test performance of African Americans. *Journal of Personality and Social Psychology, 69*, 797–811.

Townsend, S. S. M., Major, B., Sawyer, P. J., & Mendes, W. B. (2010). Can the absence of prejudice be more threatening than its presence? It depends on one's worldview. *Journal of Personality and Social Psychology, 99*, 933–947.

van den Bos, K. (2009). Making sense of life: The existential self trying to deal with personal uncertainty. *Psychological Inquiry, 20*, 197–217.

Vorauer, J. D., Main, K. J., & O'Connell, G. B. (1998). How do individuals expect to be viewed by members of lower status groups? Content and implications of meta-stereotypes. *Journal of Personality and Social Psychology, 75*, 917–937.

Weiner, B. (1982). The emotional consequences of causal attributions. In M. Clark & S. T. Fiske (Eds.), *Affect and cognition: The Seventeenth Annual Carnegie Symposium on Cognition* (pp. 185–209). Hillsdale, NJ: Erlbaum.

Williams, D. R., & Mohammed, S. A. (2009). Discrimination and racial disparities in health: Evidence and needed research. *Journal of Behavioral Medicine, 32*, 20–47.

18

FROM BEHAVIOR TO BRAIN AND BACK AGAIN

Case Studies on the Use of fMRI to Investigate Intergroup Threat and Trust

Katie N. Rotella and Jennifer A. Richeson

Over the past century, the outward expression of ethnic/racial prejudice toward most socio-cultural groups has evolved in many societies. Specifically, the once completely normal overt expression of negative attitudes toward members of other ethnic groups became largely unacceptable, socially inappropriate, and thus quite rare (again, toward most ethnic/racial groups). The biases of the past, however, did not completely disappear from the minds or behavior of many individuals. Instead, their expression morphed in ways that have required social psychologists to invent and/or embrace new methodologies to study them.

This chapter presents three "case studies" to demonstrate how members of our lab have used one such method—functional magnetic resonance imaging (fMRI)—to deepen our understanding of several contemporary dynamics of intergroup relations. In the first two cases, we describe how fMRI methods were used to explore potential mechanisms underlying previously observed behavioral effects. Specifically, in the first case, we describe the use of fMRI to consider a possible mechanism underlying the relation between interracial contact and impairments in executive control. In the second case, we describe a relatively new line of research wherein recent fMRI data have been useful in revealing a neural signature for an intergroup trust bias found among members of socially stigmatized groups. In both of these cases, furthermore, the fMRI data have been helpful in directing further behavioral studies of the respective phenomenon. In the third case, we used social neuroscience methods to clarify, if not confirm, our interpretation of a behavioral finding. Specifically, fMRI was insightful in revealing that eye-gaze direction does indeed seem to moderate the extent to which racial out-group members are perceived as threatening—as suggested by a series of behavioral studies. We offer brief descriptions of these three cases in order to reveal the potential utility of incorporating neuroscience methods into the social psychological study of prejudice, intergroup bias, and/or social stigma.

Functional Magnetic Resonance Imaging

In each of the cases described herein, we made use of functional magnetic resonance imaging (fMRI): This measures activity in the brain by tracking a proxy, namely the blood oxygen-level dependent (BOLD) signal. In order to function, neurons require glucose and oxygen, which are delivered to them in the blood carried by the brain's network of capillaries. Blood hemoglobin contains iron molecules whose magnetic properties differ depending on whether the blood is oxygenated or not. The assumption is that neuronal firing (presumably for the purpose of information processing) depletes the cells of oxygen, requiring the delivery of oxygenated blood to the particular neurons in order for them to continue their work. Consequently, when specific neurons (and thus brain regions) are recruited for information processing, they should experience a rush of oxygenated blood. Functional magnetic resonance imaging (fMRI) detects these changes in the flow of oxygenated compared with de-oxygenated blood (Huettel, Song, & McCarthy, 2008; Thulbern, Waterton, Matthews, & Radda, 1982), which is thought to reflect neural activity. Psychologists can, therefore, exploit this ability of fMRI in order to infer which brain regions have been recruited in the service of the various component processes thought to be involved in different psychological phenomena. Intergroup relations researchers, in other words, can use fMRI to explore the neural substrates associated with viewing, making judgments about, or even engaging in exchanges with out-group members.

Case 1: The Effect of Interracial Contact on Executive Function

Extant social psychological research has found that interracial interactions are often distressing for members of dominant social groups (e.g., Devine & Vasquez, 1998; Stephan, W. & Stephan, C., 1985). This distress results in at least some Whites preferring interracial interactions that require only limited levels of intimacy, such as sharing a table in a library, or that involve highly scripted situations such as the interaction between a diner and a waiter (Avery, Richeson, Hebl, & Ambady, 2009; Towles-Schwen & Fazio, 2003). Presumably, the more structured the interaction, the less cognitive effort required to navigate it, and consequently, the lower the possibility that something may go awry.

One implication of this idea is that interracial contact is indeed more cognitively taxing than same-race contact; indeed, this is exactly what was tested and found in an early behavioral study in our lab (Richeson & Shelton, 2003). Specifically, Richeson and Shelton (2003) tested this possibility by having White participants engage in a brief interview-like interaction with either a White or Black confederate of the experimenters before completing the Stroop color-naming task (Stroop, 1935). The Stroop task requires participants to identify a series of color words presented in either a congruous or an incongruous color (e.g., "red" written in blue type). Identifying the word requires the inhibition of automatized responses, which itself requires the use of executive attentional control (Engle,

2002; MacLeod, 1991). Interestingly, we found that White participants performed worse on the Stroop task after interacting with a Black confederate, compared with a White confederate, and further, this was especially true for participants with higher levels of racial bias (Richeson & Shelton, 2003).

This result was consistent with Baumeister and colleagues' then emerging theory of self-control (Baumeister, Bratslavsky, Muraven, & Tice, 1998; Muraven & Baumeister, 2000). According to the theory, attempts at self-regulation draw upon a common, albeit limited, resource. Any act of self-regulation partially depletes this resource, making subsequent attempts at self-control more difficult, at least in the short term (Muraven & Baumeister, 2000). For example, participants who had to resist the temptation to eat chocolate-chip cookies and instead eat radishes later spent less time working on an unrelated, difficult task (Baumeister et al., 1998).

Given that the Stroop color-naming task is known to require self-control, one plausible explanation for the results of the Richeson and Shelton (2003) study is that individuals' self-control (i.e., executive control) resources were depleted more by interracial contact than by same-race contact. This possibility was bolstered by previous research suggesting that executive control resources may be necessary for individuals to avoid expressing stereotypes and prejudiced behaviors (von Hippel, Silver, & Lynch, 2000). That said, because Richeson and Shelton (2003) observed greater Stroop difficulty after interracial contact among individuals with higher levels of racial bias, it was also possible that individuals' difficulty on the subsequent Stroop task was due to increased anxiety associated with the interracial encounter. Indeed, research has found that heightened anxiety can also result in poor performance on executive-control tasks (Cohen, 1980; Hartley & Adams, 1974) and individuals with higher levels of automatic racial bias (i.e., as measured by reaction-time tasks like the Implicit Association Test [IAT]) may find interracial interactions more anxiety-provoking (Blascovich, Mendes, Hunter, Lickel, & Kowai-Bell, 2001). Hence, it was not clear through which mechanism interracial contact may yield impaired Stroop performance.

Testing the Resource Depletion Model by Exploring the Neural Basis of Cognitive Control

Richeson and Shelton (2003) revealed that interracial interaction seems to impair executive functioning and, drawing on Baumeister and colleagues' work, resource depletion seemed a likely mechanism for this effect. Unfortunately, traditional psychology measures could not test this plausible mediation model (a) because it was not known what exactly constituted the limited resource; and (b) because resource depletion itself was measured by proxy—namely, by testing individuals for impairments on tasks that are known (or thought) to require executive function resources. Cognitive neuroscientists, however, had identified a circuit of brain structures that are involved in executive control and, importantly, are activated while individuals perform the Stroop color-naming task. These brain regions include both ventral and lateral regions of the prefrontal cortex (PFC) (Banich et al., 2000; Kimberg & Farah, 1993) and the anterior cingulate cortex (ACC) (Carter et al., 1998; Gehring & Knight, 2000). In our investigation, we were particularly interested in the dorsolateral prefrontal cortex (DLPFC) given that it has been shown to be involved in response inhibition during behavioral tasks like the Stroop and go/no-go tasks (Casey et al., 1997; MacDonald, Cohen, Stenger, & Carter, 2000; Smith & Jonides, 1995). While PFC regions are thought to execute control, the ACC is thought to monitor for the need for control, as when a discrepancy arises between an individual's pre-potent or dominant response in a situation and the response intended and/or required by the task at hand (Carter et al., 1998; Gehring & Knight, 2000).

Drawing upon this body of research from cognitive neuroscience, we thought it plausible to use fMRI to produce evidence bearing on the veracity of the resource depletion mechanism. Specifically, if the recruitment of executive control resources during interracial interactions (presumably to regulate thoughts, emotions, or even behavior) depletes the resource, thereby impairing individuals' performance on subsequent tasks that require such resources, then we might observe the recruitment of the aforementioned brain regions (i.e., lateral PFC, ACC) when individuals are simply exposed to photographs of other-race compared with same-race individuals.

Richeson et al. (2003) recruited White undergraduate students to participate in a two-session study, ostensibly on serial cognition. Participants were greeted by a White experimenter and completed the IAT (Greenwald, McGhee, & Schwartz, 1998), an unobtrusive measure of racial bias. Participants then entered a new room where they were greeted by and had a brief interaction with a Black experimenter. After about 5 minutes, the White experimenter retrieved the participants who then completed the Stroop color-naming task (Stroop, 1935). After that, they were sent home. After a few days, participants were contacted and were asked to participate in an ostensibly unrelated fMRI study on the perception of spatial location. No sooner than 2 weeks after the interaction component of the study, participants came into the imaging center where they engaged in the fMRI task. That is, they were shown images of White and Black male faces, and neural activity (or, as outlined earlier, a correlate of neural activity) in response to the faces was assessed.

Before we move on to the results, we should note that this study could only be conducted across two sessions (one behavioral, one fMRI) because the fMRI apparatus (at the time) prevented scanning during an actual face-to-face interaction (i.e., participants in the scanner must be lying down on their backs and remain quite still during scanning). Hence, the neural activity revealed in the scanner does not reflect the actual activity from these participants' interactions during their same-race or interracial interactions. Nevertheless, our methodology allowed us to examine whether the cognitive processes engaged during actual interracial, compared with same-race, interactions mirror those engaged when White individuals process the faces of Black, compared with White, individuals.

The results of the study were consistent with predictions. First, replicating Richeson and Shelton (2003), participants with higher racial bias (as assessed by the IAT) were significantly more impaired on the Stroop task compared with those with lower bias. Further, fMRI results showed that BOLD signal intensities (i.e., neural activity) were stronger in response to Black faces than to White faces in both bilateral regions of DLPFC and ACC (see Figure 18.1). These patterns of activation suggest that viewing Black faces was associated with greater engagement of brain areas known to be associated with executive control than viewing White faces. What was even more intriguing, however, was that neural activity in regions of the DLPFC and right ACC while viewing Black faces was positively correlated with individuals' racial bias scores. In addition, activity in a region of the right DLPFC while viewing Black, but not White, faces was also positively correlated with individuals' level of Stroop impairment following the earlier interracial interaction. Sobel tests indicated that this activity in the right DLPFC (to Black faces) statistically mediated the relation between racial bias and Stroop interference after interracial contact.

Taken together, the results of our fMRI study supported the idea that White individuals, especially those with higher levels of automatic racial bias, engage in executive control processes during encounters with Black individuals which, in turn, leave them less able to perform optimally on subsequent tasks that require self-control, such as the Stroop task. In other words, the use of fMRI techniques

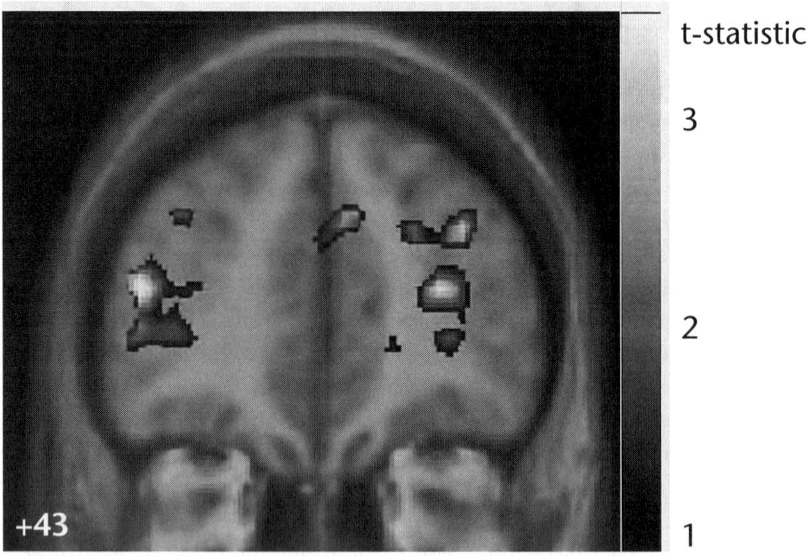

FIGURE 18.1 Statistical Activation Map Showing Regions of Right and Left Dorsolateral Prefrontal Cortex (DLPFC), as well as Right Anterior Cingulate Cortex (ACC), in which Activation was Greater to Black Faces than to White Faces

Richeson et al. (2003)

offered compelling and direct evidence for the hypothesis that the interracial dyadic interactions result in the depletion of executive attentional resources for many individuals.

The Utility of fMRI

The major advantage of employing social neuroscience methods in this line of research was the ability to garner direct evidence for (or potentially against) a theoretical model that could not be generated with social cognitive measures alone. Specifically, by using fMRI, we were able to assess the extent to which brain regions known to be involved in cognitive control were recruited during the processing of faces of Black and White individuals and, furthermore, whether such activity could statistically account for the correlation between racial bias and Stroop interference after interracial contact that was found in our behavioral studies. In other words, this research was instructive in pointing toward self-control depletion as the mechanism underlying this intriguing behavioral result.

It is important to acknowledge, however, that this fMRI study did not provide all of the answers to our research questions. First, the brain activity observed was in response to viewing still photographs of Black and White men and, thus, was only a proxy for the cognitive processes that may actually occur during live interracial encounters. Second, the results of our fMRI study were necessarily correlational; racial bias scores predicted brain activity in response to faces of Black men that, in turn, predicted performance on the Stroop task. This study alone, therefore, could not provide causal evidence for the resource depletion mechanism. Hence, armored with this correlational fMRI evidence, we turned to behavioral experiments to provide even stronger evidence for the role of self-control in engendering the Stroop performance difference after interracial, compared with same-race, contact. Specifically, in a series of experiments, we manipulated the self-control demands required of an interracial interaction and observed whether there was also a subsequent increase or decrease in Stroop impairment (Richeson & Trawalter, 2005). The results of these behavioral experiments in tandem with the fMRI work collectively provided a strong case for the resource depletion mechanism.

Case 1 Summary

In sum, in this line of research, we turned to fMRI relatively early on in order to provide direction regarding the hypothesized mechanism underlying a behavioral result. This approach was plausible only because extant research by cognitive neuroscientists had already identified brain structures that are involved in the very component processes implicated by that proposed mechanism. The results of our fMRI study, therefore, provided a stronger foundation from which to conduct behavioral experiments that could offer causal evidence for (or against) the resource depletion mechanism.

Although our fMRI study was indeed helpful in elucidating the mechanisms underlying our behavioral effect, one obvious drawback of the study is the fact that brain activity could not be measured during the interracial interaction itself. Current MRI machines require participants to lie as still as possible (alone) in a prone position inside of a small bore—a situation not well-suited for face-to-face interaction given that, among other constraints, participants cannot move freely, and thus engage in the types of non-verbal behaviors typical of interactions. While future technology may be able to address this concern—for instance, it may allow for head and bodily movement during scanning—Richeson et al. (2003) demonstrated that even with these limitations, fMRI can be useful to increase understanding of social psychological phenomena including dynamics of interracial interactions.

Case 2: Perceived Group Victimhood and Intergroup Trust

Most social neuroscience research on intergroup relations has focused on the neural activity associated with members of dominant social groups, presumably the perpetrators of stereotyping and prejudice. Indeed, within the broader area of intergroup relations, most research has adopted this perspective. Nevertheless, it is equally important to consider the processes of mind and brain that shape the experiences of members of socially stigmatized groups (see e.g., Derks, Inzlicht, & Kang, 2008)—that is, the targets of stereotyping and prejudice. This second case study describes a relatively new line of research in our lab in which we have begun to examine some of the neural correlates of in-group, compared with out-group, trust from the perspective of individuals who feel that their socio-cultural group has been (or continues to be) victimized in society.

Because trust involves making oneself vulnerable to another party, it is an inherently risky undertaking, and therefore trust is only bestowed upon a select few. Research based on a variety of self-report, implicit, and behavioral measures reveals that trust in fellow in-group members appears to be stronger than trust in out-group members (Alesina & La Ferrara, 2002; Voci, 2006; Yuki, Maddux, Brewer, & Takemura, 2005), and may even have a biological foundation (De Dreu et al., 2010; see also De Dreu, Chapter 4 in this volume).

In particular, people who perceive their in-group to be discriminated against may be most likely to trust in-group members over out-group members (Noor, Brown, & Prentice, 2008). Such perceived group victimhood (PGV) can be incorporated into group members' sense of self, and even have an impact on those born long after a conflict has ended (Brown, Wohl, & Exline, 2008; Noor et al., 2008). Research has found that people who view their group as victimized and/or at risk often engage in behaviors to strengthen the in-group, such as promoting bonds with the group, donating money to in-group causes, or promoting the group's culture to members (Jetten, Branscombe, Schmitt, & Spears, 2001; Wohl, Branscombe, & Reysen, 2010). Our research sought to investigate whether PGV leads to one particular form of in-group strengthening—increased in-group trust and loyalty, which may ultimately produce a particularly resilient intergroup trust bias.

We hypothesized that participants who have been primed to be high in PGV or who are dispositionally high in PGV would reveal greater in-group favoritism in trust compared to participants who have neither been primed nor are dispositionally high in PGV. In one of our studies (Rotella, Richeson, Chiao, & Bean, 2013), Jewish students were asked to participate in what was ostensibly an investigation into interpersonal decision making. When participants arrived for the study, they were asked first to review some data collected in a previous academic term and rate how accurate the data appeared to be. These supposed data were presented in histograms purporting to represent responses to the question, "What most comes to mind when thinking about the Jewish community?" The high-victimhood prime contained two references to past Jewish victimization, specifically the Holocaust and European pogroms, embedded among four neutral statements, such as "Orthodox Jews keep kosher." The control prime consisted of the neutral group-relevant statements plus one negative, non-discriminatory, statement. This statement, highlighting the increased risk of having a child with the fatal Tay-Sachs disease, was included in order to ensure that both conditions included negative, group-relevant information.

Participants were then introduced to "partners" with whom they would play a decision-making game via demographic questionnaires. Using stereotypical names such as "Adam Cohen" and "Christopher Schroeder" as well as a question asking about each partner's religious affiliation, the demographic questionnaires revealed to participants that they would participate in a decision-making game with an in-group and an out-group member. Trust was then measured using a modification of the investment game (Berg, Dickhaut, & McCabe, 1995). In this task, each trial consists of participants seeing the name and photograph of one of the other players (their partner for the trial) and deciding whether or not to invest a US$1 endowment in that partner. If the money is invested, the amount is tripled such that the partner receives US$3 and then must decide either to keep the sum (behave untrustworthily—i.e., defect) or return US$1.50 to the participant (behave trustworthily—i.e., cooperate). Participants played 24 randomly ordered trials with each partner. The partners' behaviors were programmed such that they all behaved in an equally trustworthy manner; they cooperated on half of the trials in which participants trusted them with the endowment. Following the investment game, participants were asked to estimate the frequency with which they believed each partner had reciprocated, and completed a measure of group identification (Luhtanen & Crocker, 1992).

Results revealed an interaction between partner group membership (in-group versus out-group) and participant PGV condition. For participants in the control prime condition, there was no difference between the trust accorded to (i.e., number of investments made with) in-group and out-group partners. However, the high-PGV prime led to a distinct in-group-favoring trust bias. Moreover, this bias was expressed not in lower trust in the out-group partner, compared with control participants, but rather in significantly greater trust in the in-group partner. Supplementary analyses revealed, furthermore, that group identification did not differ by PGV condition and thus, this is not a convincing alternative explanation for the differential in-group trust displayed by the high-PGV participants. Moreover, participants also accurately reported that each partner reciprocated to the same degree when trusted by participants, suggesting that both primed and control participants were explicitly aware that the in-group and out-group partners behaved equally trustworthily (as dictated by the computer program governing their responses to participants' trust decisions).

We wondered why, then, does high PGV result in differential trust of in-group, compared with out-group, partners despite explicit awareness of their equal trustworthiness? The pattern of results in this and similar behavioral studies suggests that participants who are high in PGV may blindly disregard evidence that in-group, but not out-group, members may actually be untrustworthy or, at least, no more trustworthy than out-group members. To begin to examine this possibility, we turned to neuroscience methods.

The Neural Basis of Trust Development

One brain area that has been implicated in the processes through which individuals learn to trust another is the caudate nucleus (CN), a C-shaped structure that is part of the striatum. The caudate responds to short-term reward feedback and the magnitude and direction of activity in this region parallel learning and behavioral change. Specifically, the CN responds differently to feedback reflecting whether one's behaviors or decisions bring favorable outcomes (i.e., rewards) or unfavorable outcomes (Delgado, Locke, Stenger, & Fiez, 2003; Haruno et al., 2004). As such, the CN has been implicated as a critical structure in the processing of trial-and-error feedback, which makes it particularly well suited to facilitate learning whether to trust another person. Trust building typically occurs by observing another's behavior over time to determine whether the person is more likely to reciprocate one's trust or, rather, to defect. Research examining brain activation during this process has found that activity in the CN is modulated by an interaction partner's behavior (i.e., his/her reciprocations or defections; King-Casas et al., 2005). Interestingly, however, as people become more confident that trust is either warranted or not through this trial-and-error learning mechanism, CN activity in response to the interaction partner's actual trust-relevant behavior diminishes. Indeed, in a multi-round trust game, CN activity decreased as participants came to learn about their partner's trustworthiness (King-Casas et al., 2005). As people come to know one another, in other words, they begin to rely less on observations of the person's behavior, favoring instead the reputation that the person has already earned.

One implication of this work is that a person who enters into an interaction with a pre-conceived notion regarding another person's trustworthiness, even that of a relative stranger, may reveal little to no evidence of the type of activity in the CN that is associated with trial-and-error feedback learning. In other words, an individual's CN may be insensitive to how trustworthily an interaction partner is actually behaving and, quite intriguingly, so too may be the individual's decisions regarding whether or not to trust the partner. A study by Delgado, Frank, and Phelps (2005) revealed just this type of outcome. Participants read information about three partners suggesting that one was highly moral (and therefore trustworthy), one highly immoral (and therefore untrustworthy), and one was of neutral/unknown moral status. Participants then completed a multi-round trust game, playing with each partner several times while in an fMRI scanner. Importantly, as in our behavioral study, each partner actually behaved equally trustworthily, reciprocating participants' trust on average 50% of the time. As predicted, the CN was differentially active in response to positive feedback (i.e., partner reciprocates) and negative feedback (i.e., partner defects) from the neutral partner, but not in response to feedback from either partner about whom the participant held previous beliefs regarding trustworthiness.[1] In other words, participants' brains seemed to ignore the actual behavior of both the "immoral" and "moral" partners. Additional analyses revealed that they also made their own trust decisions without regard to this information, trusting the "moral" partner more than the "neutral" and "immoral" partners, despite the partners' equally trustworthy play.

We wondered whether individuals who are high in PGV might similarly be blind to the actually trust-relevant behavior of in-group members. That is, does the caudate nucleus (CN) of high-PGV individuals differentiate between positive and negative feedback from in-group members and out-group members? If the CNs of high-PGV individuals do not seem to be sensitive to the actual behavior of in-group partners, then this may help explain why high-PGV individuals also seem to show unwarranted enhanced trust in in-group members, compared with individuals who are not high in PGV. Recall that the intergroup trust bias we observed in our behavioral work among high-PGV individuals was due to their greater in-group favoritism, compared with their low-PGV counterparts, rather than being due to derogation of (i.e., lower trust in) out-group partners. In other words, it is possible that PGV leads individuals to turn off the mechanisms in the brain that are responsible for tracking the feedback provided by fellow in-group interaction partners regarding their actual trustworthiness, presumably because high-PGV individuals have already decided to trust fellow in-group members no matter what, and thus do not need to rely on the CN mechanism to make their in-group trust decisions.

The Effect of Perceived Victimhood on the Neural Mechanisms of Trust

To test this hypothesis, we recently collected data wherein Jewish participants were first primed to be high in PGV and then completed the trust game in an fMRI scanner with either another Jewish partner, a Christian partner, or a partner of unknown religious background (Rotella, Richeson, & Chiao, 2013). Similar to the behavioral studies described previously, this information about partner religious affiliation was communicated through demographic questionnaires that the partners had allegedly completed prior to engaging in the trust game. Based on the findings reported in Delgado et al. (2005), we were particularly interested in caudate activity after trials in which the participant trusted each type of partner, specifically during the period in which participants learned whether the partner reciprocated (i.e., was trustworthy) or defected— namely, the outcome phase. Preliminary results revealed evidence of in-group favoritism in participants' trust behavior (recall that all participants in this study were primed to be high in PGV). Mirroring these findings, analyses of the imaging data revealed differential CN activity in response to positive (i.e., reciprocating) versus negative (i.e., defecting) behavior of both the out-group and "religion-unknown" partners, but *not* in response to the positive and negative behavior of the in-group partner (see Figure 18.2). These data suggest,

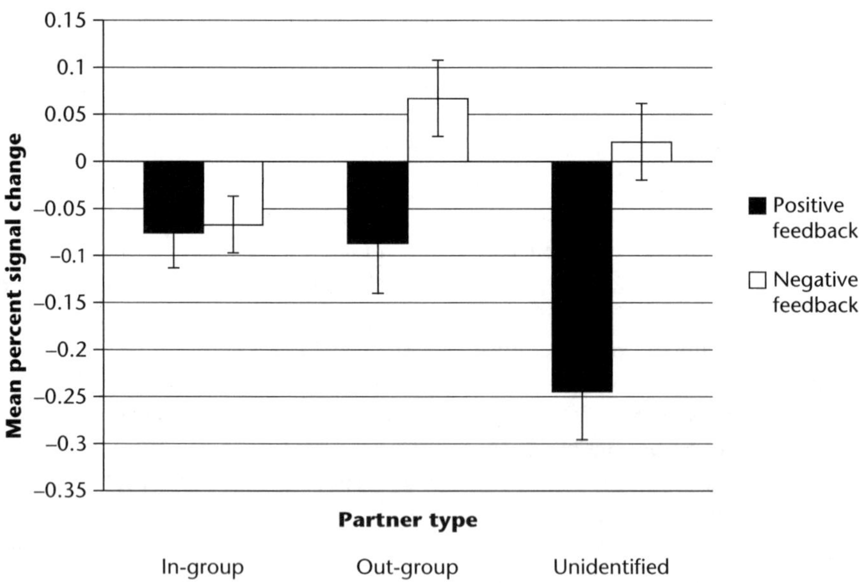

FIGURE 18.2 Percent BOLD-Signal Change in the Caudate Nucleus following Positive (Reciprocate) Versus Negative (Defect) Feedback from In-group, Out-group, and Unidentified Partners

Note: x = –13, y = 8, z = 1.

in other words, that these high-PGV participants failed to use trial-and-error learning accurately when interacting with in-group members, preventing them from properly processing information suggesting that the in-group member was actually undeserving of their enhanced trust.

The Utility of fMRI

In this case, social neuroscience helped elucidate one mechanism through which perceived group victimhood (PGV) may lead to heightened in-group trust. This relatively surprising finding is also leading to additional research (using both behavioral and neuroscience methods) examining the extent to which individuals who are high in PGV orient their attention toward other in-group members rather than out-group members. It is possible that under relatively benign conditions, PGV promotes engagement with in-group members (De Dreu et al., 2010), perhaps in order to improve the group's standing and/or strength (Wohl et al., 2010). Although the focus of our initial fMRI work in this line of research was limited to the CN, due to previous research on the neural correlates of trust development (Delgado et al., 2005), revealing the neural correlates underlying any number of processes that give rise to in-group favoritism in any number of domains will, in turn, help to demonstrate how incredibly powerful social psychological mental states such as PGV may be in shaping even seemingly basic brain processes.

Case 2 Summary

Similar to our research on interracial contact and depletion, fMRI allowed us to investigate not just which brain areas are active during the trust-game task, but also to explore the route of the behavioral trust differences found. In other words, these two cases reveal one way in which neuroscience methods can be used to enrich our understanding of the social psychology of intergroup relations.

Case 3: The Role of Eye Gaze in Race-related Threat Reactions

Our first two case studies revealed the potential for neuroscience methods to benefit specific lines of social psychological research at relatively early stages of development by offering direct evidence regarding the involvement of specific brain regions (and thus presumably, cognitive processes) in the phenomenon under investigation. In other lines of research, however, we have turned to neuroscience methods quite late in the process in order clarify, if not to confirm, a mechanism that is already supported fairly well by behavioral data.

This is the approach we took in one line of research. Specifically, we and others had found relatively consistent evidence that (young) Black men are often readily, if not automatically, appraised as threatening and dangerous (Correll, Park, Judd, & Wittenbrink, 2002; Cottrell & Neuberg, 2005; Eberhardt, Goff, Purdie, &

Davies, 2004; Payne, 2006). Indeed, a number of studies using both explicit, self-report measures as well as more indirect, social-cognitive measures offered evidence in favor of such processing of Black men. In our lab, we sought to broaden this examination by considering whether such threat responses may even be observed in the relatively low-level cognitive processes that govern visual attention. Specifically, previous research had found that humans selectively attend to stimuli of particular socio-emotional import, especially those suggestive of impending threat or danger (e.g., Mogg et al., 2000; Öhman, Flykt, & Esteves, 2001).

In our work, we utilized one method of inferring such patterns of visual attention—namely, the dot-probe detection paradigm (MacLeod, Mathews, & Tata, 1986)—to assess the extent to which Whites automatically attend to the faces of Black, compared with White, males (Richeson & Trawalter, 2008; Trawalter, Todd, Baird, & Richeson, 2008). Specifically, in the dot-probe task, participants must detect the location of a probe that is initially hidden from view behind one of two stimuli that are simultaneously presented on a computer screen, but subsequently revealed when the two stimuli disappear. A short response latency to detect the probe suggests that participants' attention had already been oriented, albeit sometimes unconsciously, to the stimulus that previously obscured it. By contrast, a relatively long response latency suggests that participants' attention had already been oriented to the stimulus that had not obscured the probe. Of course, the relevant stimuli in our task were facial photographs of young Black and White men that were presented for only a brief duration (i.e., only 33 milliseconds [ms]). Consequently, faster responses to detect the dot when it was obscured by a Black face, compared with a White one, reflects a pro-Black attentional bias. Consistent with previous research, we predicted and found that White participants revealed a reliable pro-Black attentional bias, suggesting that (young) Black male faces were construed as more threatening than White male faces (Trawalter et al., 2008).

The Moderating Role of Eye Gaze

Building on this initial study, we sought to examine whether cues that are thought to be more biologically and evolutionarily relevant than race could attenuate this effect. Specifically, we considered whether eye-gaze direction may provide a stronger signal regarding a stimulus's threat potential than race and thus attenuate the pro-Black attentional bias. Indeed, detection of eye-gaze direction is thought to be a biologically prepared ability that provides important information regarding the emotions, intentions, and mental states of others (Baron-Cohen, 1995; Emery, 2000). Almost without exception, the most important social targets are those with whom direct eye contact has been established. Direct eye gaze, furthermore, serves as a signal that targets are motivated to approach, communicating a range of intentions, including romantic interest (Mason, Tatkow, & Macrae, 2005), but also hostility and impending peril (Argyle & Cook, 1976). Direct eye gaze, furthermore, may be especially significant to perceivers when communicated by a relevant out-group member. Accordingly, research has found that direct eye gaze facilitates social categorization better than averted gaze (Macrae, Hood, Milne, Rowe, & Mason, 2002). In other words, we wondered whether social cues that are associated with threat, such as race and eye-gaze direction might combine to signal a target's threat potential and, accordingly, whether eye gaze may modulate the patterns of race-based selective attention we found previously.

To investigate this possibility, we again made use of the dot-probe detection paradigm with Black and White stimulus faces that either held direct eye gaze or averted eye gaze (Trawalter et al., 2008). Given the higher threat value associated with direct eye gaze, especially for out-group targets, we predicted that White participants would reveal a significant pro-Black attentional bias in trials with direct eye-gaze faces, replicating our prior study. In contrast, because averted eye gaze is associated with submission rather than threat, we predicted that White participants would not selectively attend to Black male faces with averted gaze. Results were consistent with these predictions.

Although we believe these results offer compelling evidence for the rapid appraisal of young Black men as threatening, and further, the moderation of such an appraisal by a more relevant cue (i.e., eye-gaze direction), the interpretation of the dot-probe detection task results rests on a series of inferences. As noted previously, the logic of the task requires one to infer that relatively faster probe-detection times are due, at least in part, to a shift in visual attention toward the stimulus that previously obscured the probe. In other words, like many social-cognitive tasks, the dot-probe task does not provide direct evidence that young Black male faces are indeed rapidly appraised as more threatening than those of young White males, and further, that eye-gaze direction may modulate this appraisal. Consequently, we considered both extant neuroscience data bearing on this question and collected our own new data to bolster our confidence that what we were observing in the dot-probe task does indeed reflect heightened vigilance for Black, compared with White, male faces that is attenuated by eye-gaze direction.

Neural Correlates of Threat

To that end, we looked for evidence that regions of the amygdala, a set of nuclei in the medial temporal lobes that are important in fear conditioning, emotional processing, vigilance, and evaluative processing (Adolphs, Tranel, Damasio, H., & Damasio, A., 1994; Amodio, Chapter 9 in this volume; Davis, Hitchcock, & Rosen, 1987; LeDoux, 1996), may also be differentially sensitive to the faces of Black and White men and perhaps also to eye-gaze direction. The impressive body of work on the role of the amygdala in socio-emotional processing also characterizes this structure as a threat detector, reactive to low-level, subtle cues of danger in the environment that may operate at an automatic or unconscious level.

Indeed, research has shown that the amygdala is differentially activated by racial in-group and out-group members (Amodio & Lieberman, 2009; Phelps et al., 2000; Wheeler & Fiske, 2005). In addition, research has found that the observed eye-gaze direction of others (i.e., direct or averted) also modulates amygdala activity (Kawashima et al., 1998; Ochsner, 2004). Building on this work, we (Richeson, Todd, Trawalter, & Baird, 2008) conducted a study wherein White participants viewed photographs of Black and White males (who were depicted with either direct or averted eye gaze) while being scanned by fMRI. Results revealed that amygdala activation was greater in response to Black, relative to White, faces only when the face displayed direct eye gaze. In other words, Black faces depicting averted eye gaze did not evoke greater amygdala activity than White faces depicting averted eye gaze; indeed the pattern was just the opposite, albeit not significantly so. This pattern, furthermore, was nearly identical to the pattern of results found in our dot-probe detection study. Specifically, whereas the faces of Black men bearing direct eye gaze were perceived as more threatening than those of White men (as indexed by both greater amygdala activity and a pro-Black attentional bias), the faces of Black men bearing averted eye gaze were not perceived as any more threatening than those of White men.

Case 3 Summary

Overall, the results of this study served to confirm the findings that were inferred from Richeson and colleagues' earlier work employing the dot-probe detection task. That is, based on our previous behavioral research finding evidence for patterns of selective attention that are thought to reflect threat processing, one may not be surprised that the amygdala revealed similar patterns of activity in response to Black and White men with direct, compared with averted, eye gaze. In other words, these fMRI data primarily served to bolster our confidence in our interpretation of the behavioral data, despite that interpretation being grounded in extant research. That said, these data also stood on their own as one of very few studies (at the time) revealing both race-based differentiation of amygdala activity and the moderation of such differential neural activity by a competing stimulus cue—i.e., eye-gaze direction. In so doing, these results expand understanding of race-related neural activity beyond the simplistic argument that Black targets yield greater amygdala activity than do White targets. In other words, these data served to highlight how finely tuned humans are to the social world and how sensitive we are to relevant stimulus cues.

Conclusion: What Can Neuroscience Add to the Study of Intergroup Relations?

These three varied lines of research highlight just a few of the ways that neuroscience methods can benefit the study of intergroup relations. Specifically, intergroup relations researchers can employ neuroscience methods with the potential to:

1. investigate mechanisms underlying psychological phenomena;
2. test hypotheses otherwise unable to be examined; and
3. build support for conclusions drawn from behavioral studies.

Approach 1

Most notably, neuroscience methods can be used to explore or confirm potential mechanisms underlying behavioral effects by implicating specific brain structures and/or circuits that have been studied extensively by cognitive (and, sometimes, behavioral) neuroscientists (see also Amodio, Chapter 9 in this volume). As demonstrated by our work on interracial contact and depletion, the findings of initial social neuroscience studies can trigger subsequent behavioral experiments that serve to test a specific causal model.

Approach 2

Social neuroscience studies, however, can also generate novel hypotheses that can be tested further using both more traditional social-cognitive and behavioral measures as well as neuroscience methods, as is the case in our ongoing research on perceived group victimhood (PGV).

Approach 3

Last, as exemplified by our work on the processing of race and eye gaze, social neuroscience studies can come at the end of a line of research, primarily serving to underscore and confirm the inferences of the previous behavioral studies. Even in these cases, however, we would argue that social neuroscience studies are not useless. Indeed, it is possible that the results will not confirm the inferences made based on behavioral studies; in our case, we might have not observed differential amygdala activity by race and eye gaze; and thus, we would have had to reconsider the interpretation of our previous work. Moreover, the demonstration in this work that eye-gaze direction moderates the amygdala response to race moves both social psychologists' and cognitive neuroscientists' general understanding(s) of race-related processing forward. Indeed, prior to our (and similar) research finding race-based differences in prefrontal activity (Cunningham et al., 2004; Richeson et al., 2003), many cognitive neuroscientists would not have believed such differential processing to be very likely. In other words, social neuroscience studies not only provide answers to discrete questions, but they often also help generate new questions for both social psychology and cognitive neuroscience!

Of course, neuroscience methods still have their drawbacks—they are costly, time-consuming, and can limit the way some studies are designed (for example, face-to-face intergroup interactions are currently impossible in today's MRI). However, the technology has improved rapidly in recent years, becoming more powerful and more accurate, and psychologists are becoming increasingly creative in their uses of a variety of neuroscientific methods to study psychological questions. While more traditional measures will not be replaced by neuroscience methods, we believe that the addition of tools like fMRI to intergroup relations research has the potential to transform our understanding of stereotyping, prejudice, discrimination, and their effects on both targets and perpetrators.

Suggestions for Further Reading

Adams, R. B. Jr., Franklin, R. G., Rule, N. O., Freeman, J. B., Yoshikawa, S., Kveraga, K. . . . Ambady, N. (2010). Culture, gaze, and the neural processing of fear expressions: An fMRI investigation. *Social Cognitive and Affective Neuroscience, 5*, 340–348.

This paper investigates the moderating role of eye-gaze direction, coupled with in-group versus out-group status, on amygdala activity in response to facial fear displays. European American (in the United States) and Japanese participant samples were examined, and results revealed greater amygdala activity in response to direct, versus averted, eye gaze on the faces of in-group members, but less amygdala activity in response to direct versus averted eye gaze on the faces of out-group members. This cross-cultural study broadens understanding of how various cues—in this case facial emotion and eye-gaze direction—combine with group membership to shape neural activity.

Amodio, D. M., & Lieberman, M. D. (2009). Pictures in our heads: Contributions of fMRI to the study of prejudice and stereotyping. In T. Nelson (Ed.), *Handbook of prejudice, stereotyping, and discrimination* (pp. 347–366). New York: Erlbaum Press.

This paper is a succinct, yet thoughtful, review of the application of various neuroscientific methods to the study of prejudice and stereotyping, particularly EEG and fMRI.

Delgado, M. R., Frank, R. H., & Phelps, E. A. (2005). Perceptions of moral character modulate the systems of reward during the trust game. *Nature Neuroscience, 8*, 1611–1618.

This paper investigates the neural basis of interpersonal trust using an iterated investment game. This is the first paper (to our knowledge) to reveal that individuals' preconceived notions regarding the trustworthiness of a particular individual biases neural processing of that individual's behaviors—at least, processing relevant to the development of trust. Specifically, results revealed a reduction in activity of the caudate nucleus in response to the "trial-and-error-like" feedback from the individual about whom participants had already formed an opinion. This work has important implications for intergroup relations, given that group memberships and their associated stereotypes often serve the function of providing individuals with preconceived beliefs about others.

Norton, M. I., Mason, M. F., Vandello, J. A., Biga, A., & Dyer, R. (in press). An fMRI investigation of racial paralysis. *Social Cognitive and Affective Neuroscience.*

This paper investigates one form of reaction to the threat of intergroup interactions—what the authors call "racial paralysis," or the tendency to avoid decisions involving racial out-group members. The paper reveals that when individuals make racial judgments, several brain regions associated with deliberative processing, inhibition, and executive function are recruited, including the anterior cingulate cortex (ACC) as well as both medial and lateral regions of the prefrontal cortex (e.g., VMPFC, VLPFC, and DLPFC).

Note

1 There was, however, a marginally significant difference in CN responses to positive versus negative feedback from the "immoral" partner, suggesting that expectations of trustworthiness may have more impact than expectations of untrustworthiness.

References

Adolphs, R., Tranel, D., Damasio, H., & Damasio, A. R. (1994). Impaired recognition of emotion in facial expressions following bilateral damage to the human amygdala. *Nature, 372*, 669–672.

Alesina, A., & La Ferrara, E. (2002). Who trusts others? *Journal of Public Economics, 85*, 207–234.

Amodio, D. M., & Lieberman, M. D. (2009). Pictures in our heads: Contributions of fMRI to the study of prejudice and stereotyping. In T. Nelson (Ed.), *Handbook of prejudice, stereotyping, and discrimination* (pp. 347–366). New York: Erlbaum Press.

Argyle, M., & Cook, M. (1976). *Gaze and mutual gaze.* Cambridge, UK: Cambridge University Press.

Avery, D. R., Richeson, J. A., Hebl, M., & Ambady, N. (2009). It doesn't have to be uncomfortable: The role of behavioral scripts in interracial interactions. *Journal of Applied Psychology, 94*, 1382–1393.

Banich, M. T., Milham, M. P., Atchley, R. A, Cohen, N. J., Webb, A., Wszalek, T. . . . Magin, R. (2000). fMRI studies of Stroop tasks reveal unique roles of anterior and posterior brain systems in attentional selection. *Journal of Cognitive Neuroscience, 12*, 988–1000.

Baron-Cohen, S. (1995). *Mindblindness.* Cambridge, MA: The MIT Press/Bradford.

Baumeister, R. F., Bratslavsky, E., Muraven, M., & Tice, D. M. (1998). Ego-depletion: Is the active self a limited resource? *Journal of Personality and Social Psychology, 74*, 1252–1265.

Berg, J., Dickhaut, J., & McCabe, K. (1995). Trust, reciprocity, and social history. *Games and Economic Behavior, 10*, 122–142.

Blascovich, J., Mendes, W. B., Hunter, S. B., Lickel, B., & Kowai-Bell, N. (2001). Perceiver threat in social interactions with stigmatized others. *Journal of Personality and Social Psychology, 80*, 253–267.

Brown, R. P., Wohl, M. J. A., & Exline, J. J. (2008). Taking up offenses: Secondhand forgiveness and group identification. *Personality and Social Psychology Bulletin, 34*, 1406–1419.

Carter, C. S., Braver, T. S., Barch, D. M., Botvinick, M., Noll, D., & Cohen J. D. (1998). Anterior cingulate cortex, error detection, and the online monitoring of performance. *Science, 280*, 747–749.

Casey, B. J., Trainor, R. J., Orendi, J. L., Schubert, A. B., Nystrom, L. E., Giedd, J. N. . . . Rapoport, J. L. (1997). A developmental functional MRI study of prefrontal activation during performance of a go-no-go task. *Journal of Cognitive Neuroscience, 9*, 835– 847.

Cohen, S. (1980). After effects of stress on human performance and social behavior: A review of research and theory. *Psychological Bulletin, 88*, 82–108.

Correll, J., Park, B., Judd, C. M., & Wittenbrink, B. (2002). The police officer's dilemma: Using ethnicity to disambiguate potentially threatening individuals. *Journal of Personality and Social Psychology, 83*, 1314–1329.

Cottrell, C. A., & Neuberg, S. L. (2005). Different emotional reactions to different groups: A sociofunctional threat-based approach to "prejudice." *Journal of Personality and Social Psychology, 88*, 770–789.

Cunningham, W. A., Johnson, M. K., Raye, C. L., Gatenby, J. C., Gore, J. C., & Banaji, M. R. (2004). Separable neural components in the processing of Black and White Faces. *Psychological Science, 15*, 806–813.

Davis, M., Hitchcock, J. M., & Rosen, J. B. (1987). Anxiety and the amygdala: Pharmacological and anatomical analysis of the fear-potentiated startle paradigm. In G. Bower (Ed.), *The psychology of learning and motivation* (Vol. 21, pp. 263–305). San Diego, CA: Academic Press.

De Dreu, C. K. W., Greer, L. L., Handgraaf, M. J. J., Shalvi, S., Van Kleef, G. A., Baas, M. . . . Feith, S. W. W. (2010). The neuropeptide oxytocin regulates parochial altruism in intergroup conflict among humans. *Science, 328*, 1408–1411.

Delgado, M. R., Frank, R. H., & Phelps, E. A. (2005). Perceptions of moral character modulate the systems of reward during the trust game. *Nature Neuroscience, 8*, 1611–1618.

Delgado, M. R., Locke, H. M., Stenger, V. A., & Fiez, J. A. (2003). Dorsal striatum responses to reward and punishment: Effects of valence and magnitude manipulations. *Cognitive Affective Behavioral Neuroscience, 3*, 27–38.

Derks, B., Inzlicht, M., & Kang, S. (2008). The neuroscience of stigma and stereotype threat. *Group Processes and Intergroup Relations, 11*, 163–181.

Devine, P. G., & Vasquez, K. A. (1998). The rocky road to positive intergroup relations. In J. L. Eberhardt & S. T. Fiske (Eds.), *Confronting racism: The problem and the response* (pp. 234–262). Thousand Oaks, CA: Sage.

Eberhardt, J. L., Goff, P. A., Purdie, V. J., & Davies, P. G. (2004). Seeing Black: Race, representation, and visual perception. *Journal of Personality and Social Psychology, 87*, 876–893.

Emery, N. J. (2000). The eyes have it: The neuroethology, function and evolution of social gaze. *Neuroscience and Biobehavioral Reviews, 24*, 581–604.

Engle, R. W. (2002). Working memory capacity as executive attention. *Current Directions in Psychological Science, 11*, 19–23.

Gehring, W. J., & Knight, R. T. (2000). Prefrontal–cingulate interactions in action monitoring. *Nature Neuroscience, 3*, 516–520.

Greenwald, A. G., McGhee, D. E., & Schwartz, J. L. K. (1998). Measuring individual differences in implicit cognition: The Implicit Association Test. *Journal of Personality and Social Psychology, 74*, 1464–1480.

Hartley, L. R., & Adams, R. G. (1974). Effect of noise on the Stroop test. *Journal of Experimental Psychology, 102*, 62–66.

Haruno, M., Kuroda, T., Doya, K., Toyama, K., Kimura, M., Samejima, K. . . . Kawato, M. (2004). A neural correlate of reward-based behavior learning in caudate nucleus: A functional magnetic resonance imaging study of a stochastic decision task. *Journal of Neuroscience, 24*, 1660–1665.

Huettel, S. A., Song, A. W., & McCarthy, G. (2008). *Functional magnetic resonance imaging* (2nd ed.). Sunderland, MA: Sinauer Associates Inc.

Jetten, J., Branscombe, N. R., Schmitt, M. T., & Spears, R. (2001). Rebels with a cause: Group identification as a response to perceived discrimination from the mainstream. *Personality and Social Psychology Bulletin, 27*, 1204–1213.

Kawashima, R., Sugiura, M., Kato, T., Nakamura, A., Hatano, K., Ito, K. . . .Nakamura, K. (1998). The human amygdala plays an important role in gaze monitoring: A PET study. *Brain, 122*, 779–783.

Kimberg, D. Y., & Farah, M. J. (1993). A unified account of cognitive impairments following frontal lobe damage: The role of working memory in complex organized behavior. *Journal of Experimental Psychology: General, 122*, 411–428.

King-Casas, B., Tomlin, D., Anen, C., Camerer, C. F., Quartz, S. R., & Montague, P. R. (2005). Getting to know you: Reputation and trust in a two-person economic exchange. *Science, 308*, 78–83.

LeDoux, J. (1996). Emotional networks and motor control: A fearful view. *Progress in Brain Research, 107*, 437–446.

Luhtanen, R., & Crocker, J. (1992). A collective self-esteem scale: Self-evaluation of one's social identity. *Personality and Social Psychology Bulletin, 18*, 302–318.

MacDonald, A. W., Cohen, J. D., Stenger, V. A., & Carter, C. S. (2000). Dissociating the role of dorsolateral prefrontal and anterior cingulate cortex in cognitive control. *Science, 288*, 1835–1838.

MacLeod, C. M., Mathews, A., & Tata, P. (1986). Attentional bias in emotional disorders. *Journal of Abnormal Psychology, 95*, 15–20.

MacLeod, C. M. (1991). Half a century of research on the Stroop effect: An integrative review. *Psychological Bulletin, 109*, 163–203.

Macrae, C. N., Hood, B. M., Milne, A. B., Rowe, A. C., & Mason, M. F. (2002). Are you looking at me? Eye gaze and person perception. *Psychological Science, 13*, 460–484.

Mason, M. F., Tatkow, E., & Macrae, C. M. (2005). The look of love: Gaze shifts and person perception. *Psychological Science, 16*, 236–239.

Mogg, K., Bradley, B. P., Dixon, C., Fisher, S., Twelftree, H., & McWilliams, A. (2000). Trait anxiety, defensiveness and selective processing of threat: An investigation using two measures of attentional bias. *Personality and Individual Differences, 28*, 1063–1077.

Muraven, M., & Baumeister, R. F. (2000). Self-regulation and depletion of limited resources: Does self-control resemble a muscle? *Psychological Bulletin, 126*, 247–259.

Noor, M., Brown, R. J., & Prentice, G. (2008). Precursors and mediators of reconciliation in Northern Ireland: A new model. *British Journal of Social Psychology, 47*, 481–495.

Ochsner, K. N. (2004). Current directions in social cognitive neuroscience. *Current Opinion Neurobiology*, 14, 254–258.

Öhman, A., Flykt, A., & Esteves, F. (2001). Emotion drives attention: Detecting the snake in the grass. *Journal of Experimental Psychology: General, 130*, 466–478.

Payne, B. K. (2006). Weapon bias: Split second decisions and unintended stereotyping. *Current Directions in Psychological Science, 15*, 287–291.

Phelps, E. A., O'Connor, K. J., Cunningham, W. A., Funayama, E. S., Gatenby, J. C., Gore, J. C., & Banaji, M. R. (2000). Performance on indirect measures of race evaluation predicts amygdala activation. *Journal of Cognitive Neuroscience, 12*, 729–738.

Richeson, J. A., Baird, A. A., Gordon, H. L., Heatherton, T. F, Wyland, C. L., Trawalter, S., & Shelton, J. N. (2003). An fMRI examination of the impact of interracial contact on executive function. *Nature Neuroscience, 6*, 1323–1328.

Richeson, J. A., & Shelton, J. N. (2003). When prejudice does not pay: Effects of interracial contact on executive function. *Psychological Science, 14*, 287–290.

Richeson, J. A., Todd, A. R., Trawalter, S., & Baird, A. A. (2008). Eye-gaze direction modulates race-related amygdala activity. *Group Processes and Intergroup Relations, 11*, 233–246.

Richeson, J. A., & Trawalter, S. (2005). Why do interracial interactions impair executive function? A resource depletion account. *Journal of Personality and Social Psychology, 88*, 934–947.

Richeson, J. A., & Trawalter, S. (2008). The threat of appearing prejudiced and race-based attentional biases. *Psychological Science, 19*, 98–102.

Rotella, K. N., Richeson, J. A., & Chiao, J. Y. (2013). The neural correlates of intergroup trust among highly victimized group members. Manuscript in preparation.

Rotella, K. N., Richeson, J. A., Chiao, J. Y., & Bean, M. G. (2013). Blinding trust: The effect of perceived group victimization on intergroup trust. *Personality and Social Psychology Bulletin, 39*, 115–127.

Smith, E. E., & Jonides, J. (1995). Working memory in humans: Neuropsychological evidence. In M. Gazzaniga (Ed.), *The cognitive neurosciences* (pp. 1009–1020). Cambridge, MA: The MIT Press.

Stephan, W. G., & Stephan, C. (1985). Intergroup anxiety. *Journal of Social Issues, 41*, 157–176.

Stroop, J. R. (1935). Studies of interference in serial verbal reactions. *Journal of Experimental Psychology, 18*, 643–662.

Thulbern, K. R., Waterton, J. C., Matthews, P. M., & Radda, G. K. (1982). Oxygenation dependence of the transverse relaxation time of water protons in whole blood at high field. *Biochimica et Biophysica Acta, 714*, 265–270.

Towles-Schwen, T., & Fazio, R. H. (2003). Choosing social situations: The relation between automatically-activated racial attitudes and anticipated comfort interacting with African Americans. *Personality and Social Psychology Bulletin, 29*, 170–182.

Trawalter, S., Todd, A. R., Baird, A. A., & Richeson, J. A. (2008). Attending to threat: Race-based patterns of selective attention. *Journal of Experimental Social Psychology, 44*, 1322–1327.

Voci, A. (2006). The link between identification and in-group favoritism: Effect of threat to social identity and trust-related emotions. *British Journal of Social Psychology, 45*, 265–284.

von Hippel, W., Silver, L. A., & Lynch, M. E. (2000). Stereotyping against your will: The role of inhibitory ability in stereotyping and prejudice among the elderly. *Personality and Social Psychology Bulletin, 26*, 523–532.

Wheeler, M. E., & Fiske, S. T. (2005). Controlling racial prejudice: Social-cognitive goals affect amygdala and stereotype activation. *Psychological Science, 16*, 56–63.

Wohl, M. J. A., Branscombe, N. R., & Reysen, S. (2010). Perceiving your group's future to be in jeopardy: Extinction threat induces collective angst and the desire to strengthen the ingroup. *Personality and Social Psychology Bulletin, 36*, 898–910.

Yuki, M., Maddux, W. W., Brewer, M. B., & Takemura, K. (2005). Cross-cultural differences in relationship- and group-based trust. *Personality and Social Psychology Bulletin, 31*, 48–62.

INDEX